Airplane Design

Part VIII: Airplane Cost Estimation: Design, Development, Manufacturing and Operating

Dr. Jan Roskam
Ackers Distinguished Professor of Aerospace Engineering
The University of Kansas, Lawrence

2002

DARcorporation
Design • Analysis • Research

120 East 9th Street, Suite 2 • Lawrence, Kansas 66044, U.S.A.

PUBLISHED BY
Design, Analysis and Research Corporation (*DARcorporation*)
120 East Ninth Street, Suite 2
Lawrence, Kansas 66044
U.S.A.
Phone: (785) 832-0434
Fax: (785) 832-0524
e-mail: info@darcorp.com
http://www.darcorp.com

Library of Congress Catalog Card Number: 97-68580

ISBN 1-884885-55-1

In all countries, sold and distributed by
Design, Analysis and Research Corporation
120 East Ninth Street, Suite 2
Lawrence, Kansas 66044
U.S.A.

Printed in the United States of America
First Printing, 1990
Second Printing, 2002

TABLE OF CONTENTS

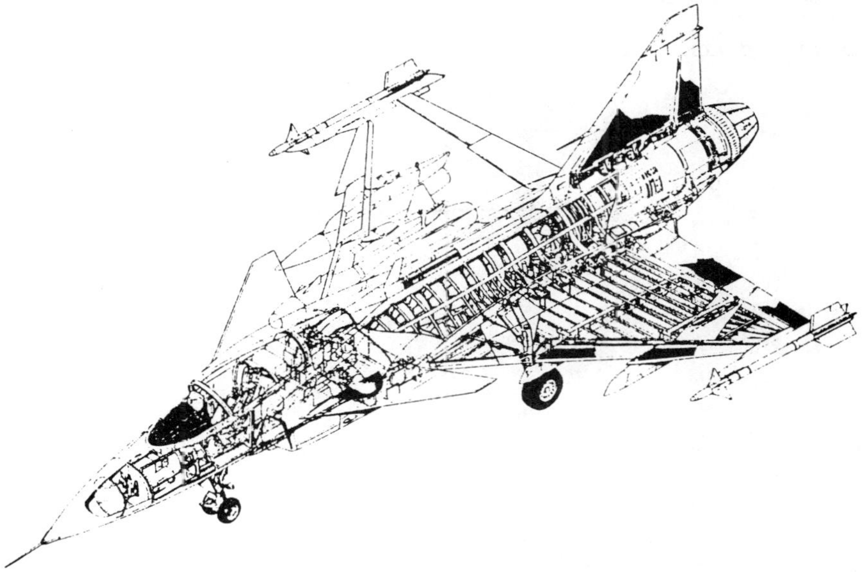

COURTESY: SAAB

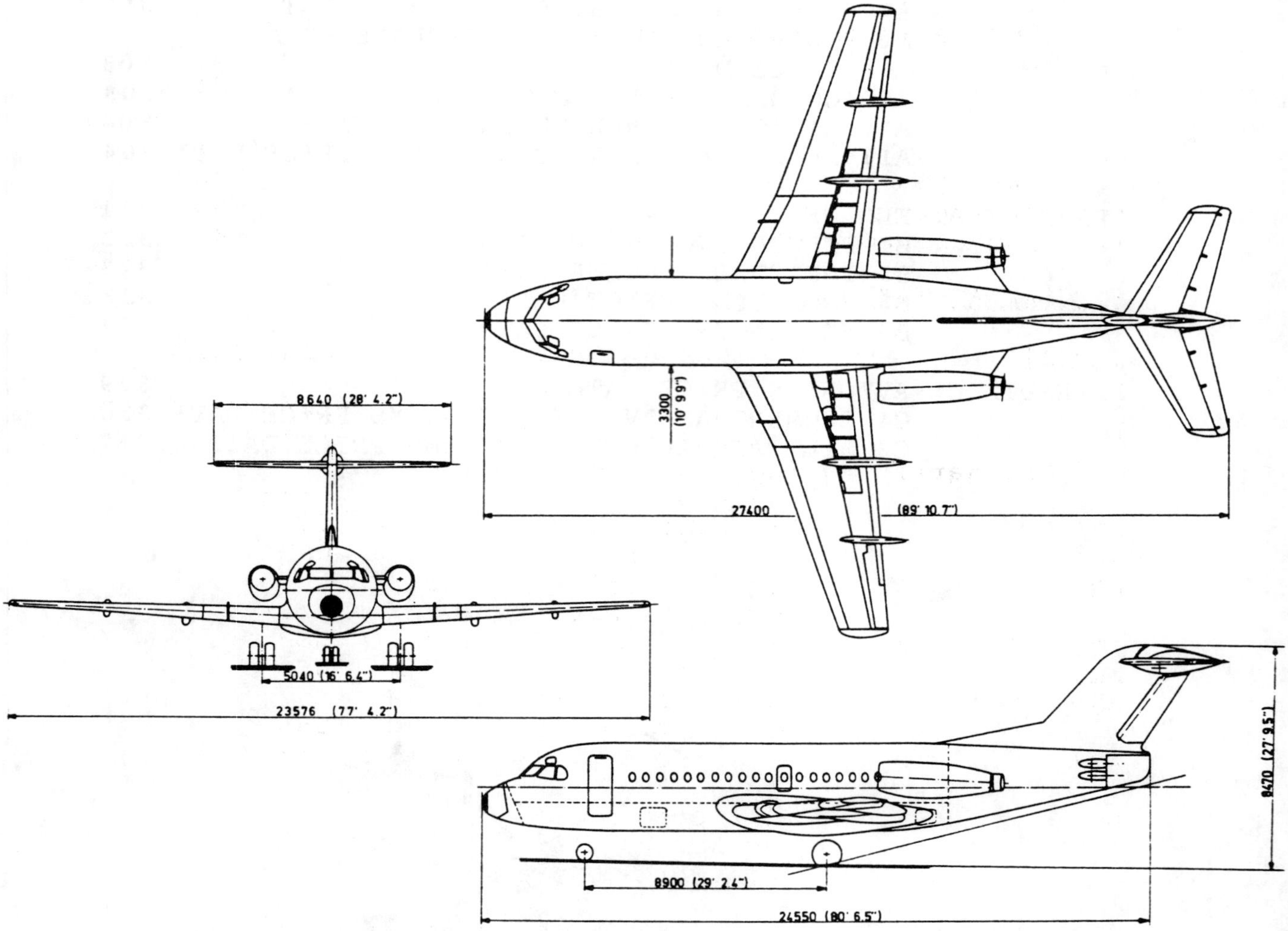

COURTESY: FOKKER

TABLE OF SYMBOLS

The Table of Symbols is organized as follows:

1. GENERAL SYMBOLS

Symbol	Definition	Dimension
a	speed of sound	ft/sec or kts
A	Aspect ratio (wing)	-----
A (alternate)	Regression coefficient in Equation (8.1)	-----
AEP	Airplane Estimated Price	USD
AFP	Airframe price	USD
AH	Number of flight hours for a crew member, per year	hrs/yr
AMP	Airplane Market Price	USD
ANW	Airplane net worth	USD
ASP	Avionics systems price per airplane	USD
ATF	Airplane type factor	-----
b	wing span	ft
B	Regression coefficient in Equation (8.1)	-----
c_f	skin friction coefficient	-----
c_j	Engine specific fuel consumption (jet)	1/hr
c_p	Engine specific fuel consumption (piston or turboprop)	1/hphr
C	See Eqn.(8.3)	-----
C_{aed}	Airframe engineering and design cost	USD
C_{amb}	Applied maintenance burden	USD/nm
C_{apc}	Airplane production cost	USD
C_{apctrl}	Cost of airplane control	USD/nm
C_{aps}	Cost of airplane service	USD/nm

C_{aplf}	Airplane landing fee per landing	USD/landing
C_{apnf}	Navigation fee charged per flight	USD/flight
C_{auc}	Acquisition unit cost	USD/airplane
$C_{avionics}$	Avionics systems cost	USD
C_{cat}	Cost for cabin attendants	USD/nm
C_{crew}	Crew cost	USD/nm
C_{crewpr}	Crew cost for airplane program	USD
C_{dap}	Cost of airframe depreciation	USD/nm
C_{dapsp}	Cost of airplane spare parts depreciation	USD/nm
C_{dav}	Cost of avionics systems depreciation	USD/nm
C_{deng}	Cost of engine depreciation	USD/nm
C_{dengsp}	Cost of engine spare parts depreciation	USD/nm
C_{dgef}	Cost of depreciation of ground facilities and equipment	USD/nm
C_{dprp}	Cost of propeller depreciation	USD/nm
C_{dst}	Development support and test cost	USD
C_{e}	Engine unit cost	USD
$C_{(e + a)}$	Cost of engine and avionics	USD
C_{fin}	Financing cost	USD
C_{frt}	Cost of handling freight	USD/nm
C_{fta}	Cost for flight test airplanes	USD
C_{fto}	Flight test operations cost	USD
C_{ins}	Cost of insurance	USD/nm
C_{int}	Cost of interiors	USD
$C_{lab/ap}$	Labor cost of airframe and systems	USD/nm
$C_{lab/eng}$	Labor cost of engine maintenance	USD/nm

C_{lf}	Direct operating cost due to landing fees	USD/nm
C_{lcuc}	Life cycle unit cost	USD/airplane
C_{man}	Manufacturing labor cost	USD
C_{mat}	Materials cost	USD
$C_{mat/ap}$	Cost of maintenance materials for airplane and systems	USD/nm
$C_{mat/apblhr}$	Cost of airframe and systems maintenance materials	USD/nm
$C_{mat/eng}$	Cost of maintenance materials for engines	USD/nm
$C_{mat/engblhr}$	Cost of engine maintenance materials per block hour	USD/hr
C_{mgef}	Cost of maintaining ground equipment and facilities	USD/nm
C_{mls}	Cost for meal service	USD/nm
$C_{mpersdir}$	Progam cost for direct maintenance personnel	USD
C_{muc}	Manufacturing unit cost	USD/airplane
C_{nf}	Direct operating cost due to navigation fees	USD/nm
$C_{n_{\delta_R}}$	Yawing moment derivative due to rudder deflection	1/rad
$C_{ops_{dir}}$	Direct operating cost for entire airplane program	USD
$C_{ops/hr}$	Operating cost per hour	USD/hr
$C_{ops_{ind}}$	Indirect operating cost for entire airplane program	USD
C_p	Propeller unit cost	USD
$C_{pax/gen}$	General passenger cost	USD/nm
$C_{pax/ins}$	Cost of passenger insurance	USD/nm
C_{pol}	Fuel and oil cost	USD/nm
C_{pro}	Profit (cost)	USD
C_{puc}	Program unit cost	USD/airplane
C_{qc}	Quality control cost	USD
C_{rt}	Direct operating cost due to registry taxes	USD/nm

C_{tool}	Tooling cost	USD
C_{ACQ}	Total cost of acquisition	USD
C_{CONMAT}	Program cost of consumable materials	USD
C_D	Drag coefficient	-----
C_{D_o}	Zero-lift drag coefficient	----
C_{DEPOT}	Program cost associated with depots	USD
C_{DISP}	Total cost of the disposal phase	USD
CEF	Cost Escalation Factor	-----
CEF-ratio	See below Eqn.(2.5) for a definition	-----
C_L	Lift coefficient	-----
$C_{L_{max}}$	Maximum lift coefficient	-----
C_{MAN}	Total cost of the manufacturing phase	USD
C_{MISC}	Program miscellaneous cost	USD
C_{OPS}	Total cost of the operations phase	USD
$C_{PERSDIR}$	Program cost of direct personnel	USD
$C_{PERSIND}$	Program cost of indirect personnel	USD
C_{POL}	Program cost of fuel, oil and lubricants	USD
C_{PRO}	Manufacturer's profit in the manufacturing phase	USD
C_{PROT}	Prototype cost	USD
C_{RDTE}	Total cost of the RDTE phases	USD
C_{SPARES}	Program cost of spares	USD
D	See Eqn.(8.4)	lbs
DOC	Direct Operating Cost	USD/nm
DOC_{depr}	Direct operating cost of depreciation	USD/nm
DOC_{fin}	Direct operating cost of financing	USD/nm
DOC_{flt}	Direct operating cost of flying	USD/nm

DOC_{lnr}	Direct operating cost of landing fees, navigation fees and registry taxes	USD/nm
DOC_{maint}	Direct operating cost of maintenance	USD/nm
DP_{ap}	Airplane depreciation period	yrs
DP_{apsp}	Airplane spare parts depreciation period	yrs
DP_{av}	Avionics systems depreciation period	-----
DP_{engsp}	Engine spare parts depreciation period	yrs
DP_{prp}	Propeller depreciation period	yrs
e	Oswald's efficiency factor	----
E	Endurance	hrs
$EP = C_e$	Engine price	USD
ESPPF	Engine spare parts price factor	-----
$f_{amb/lab}$	overhead distribution factor for labor cost	-----
$f_{amb/mat}$	overhead distribution factor for material cost	-----
f_{depot}	depot cost fraction of C_{OPS}	-----
$f_{ins_{hull}}$	annual hull insurance rate	USD/USD/yr
f_{ioc}	indirect operating cost fraction of DOC	-----
f_{lf}	(landing fee) factor which depends on airplane size	-----
f_{misc}	miscellaneous cost fraction of C_{OPS}	-----
$f_{persdir}$	direct personnel cost fraction of C_{OPS}	-----
$f_{persind}$	indirect personnel cost fraction of C_{OPS}	-----
f_{pol}	fuel, oil and lubricant cost fraction of C_{OPS}	-----
f_{rt}	(registry tax) factor which depends on airplane size	-----

f_{spares}	indirect cost of spares as a fraction of C_{OPS}	-----
F_{apsp}	Airplane spare parts factor	---
F_{cad}	Judgement factor of relative company experience with CAD	
F_{dap}	Airframe depreciation factor	--
F_{dapsp}	Airplane spare parts depreciation factor	-----
F_{dav}	Avionics systems depreciation factor	-----
F_{deng}	Engine depreciation factor	----
F_{dengsp}	Engine spare parts depreciation factor	-----
F_{diff}	Judgement factor of relative program difficulty	-----
F_{dprp}	Propeller depreciation factor	-----
F_{engsp}	Engine spare parts factor	-----
F_{fin}	Finance cost fraction	-----
F_{ftoh}	Overhead factor associated with flight test activities	---
F_{int}	Interior cost factor which depends on number of pax	USD/pax
F_{mat}	Judgement factor to account for differing materials cost	-----
F_{obs}	Judgement factor to account for observables characteristics	
F_{pro}	Profit fraction	-----
F_{tsf}	Cost adjustment factor for extra test and simulation facilities	-----
FD	Fuel density	lbs/gallon
F_{OL}	Factor which accounts for cost of oil and lubricants	-----
FP	Fuel price	USD/gallon
H_{em}	Attained number of hours between engine overhaul	hrs
IOC	Indirect Operating Cost	USD/nm
IOC_{ascf}	Indirect operating cost for airplane and traffic servicing, control and freight	USD/nm

IOC_{gaa}	Indirect operating cost for general administrative expenses	USD/nm
IOC_{pax}	Indirect operating cost for passenger services	USD/nm
IOC_{pse}	Indirect operating cost for promotion, sales and entertainment	USD/nm
IOC_{sta}	Indirect operating cost for maintaining ground equipment and facilities	USD/nm
K_j	Vacation (etc) pay factor	-----
$K_{H_{em}}$	Attained period between overhaul factor	-----
LCC	Life Cycle Cost	USD (United States Dollars)
L/D	Lift-to-drag ratio	-----
L_R	Annual loss rate in number of airplanes lost per 10^5 hrs	1/hrs
M_1	Steady state Mach number	-----
M_{ff}	Overall mission fuel fraction	-----
M_{res}	Reserve fuel fraction	-----
M_{tfo}	Trapped fuel and oil fraction	-----
MHR_{aed}	Engineering manhours	hrs
$MHR_{aed_{program}}$	Engineering manhours for entire program	hrs
MHR_{flthr}	Total maintenance manhours per flight hour	hrs/hr
MHR_{man}	Manufacturing manhours	hrs
$MHR_{map_{bl}}$	Airframe and systems maintenance manhours per block hour	hrs/hr
$MHR_{map_{flt}}$	Airframe and systems maintenance manhours per flight hour	hrs/hr
$MHR_{man_{program}}$	Manufacturing manhours for entire program	hrs

$MHR_{map_{bl}}$	Maintenance manhours for airframe and systems	hrs/blhr
$MHR_{meng_{bl}}$	Maintenance manhours for engines per block hour	hrs/hr
MHR_{tool}	Tooling manhours	hrs
$MHR_{tool_{program}}$	Tooling manhours for entire program	hrs
n	learning curve exponent	-----
n_{c_j}	number of crew members of type j	-----
N_{acq}	Number of airplanes acquired	--
N_{crew}	Number in crew	-----
N_e	Number of engines per airplane	-----
N_{loss}	Number of airplanes lost through accidents over the airplane service life	-----
N_m	Number of airplanes built to production standards	-----
N_{market}	Estimated number of airplanes in a given market	-----
$N_{mission}$	Number of missions flown per year	-----
N_p	Number of propellers per airplane	-----
N_{pax}	Number of passengers	-----
$N_{program}$	Total number of airplanes built in a program	-----
N_{prot}	Number of prototypes	-----
N_{rdte}	Number of airplanes built in rdte phases	-----
N_r	Number of airplanes built per month	-----
N_{res}	Number of airplanes kept in reserve	-----
N_{serv}	Number of airplanes in active service	-----
N_{st}	Number of static test airframes built	-----
N_{yr}	Number of years an airplane is operated	-----

OD	Oil density	lbs/gallon
OHR_{crew}	Overhead rate factor associated with crew pay	-----
OLP	Oil price	USD/gallon
P	P percent learning curve	-----
Pay_{crew}	Annual pay of crew	USD/yr
P_{pot}	Profit potential	USD
$PP = C_p$	Propeller price	USD
$\bar{q}$	dynamic pressure	lbs/ft^2
R	Range	nm
R_{bl}	Block distance	nm
$R_{bl_{ann}}$	Annual block miles flown	nm/yr
R_{cl}	Climb distance	nm
R_{conmat}	Average cost of consumable materials per maintenance manhour	USD/hr
R_{cr}	Crew ratio	-----
R_{de}	Descent distance	nm
R_e	Engineering manhour rate	USD/hr
$R_{l_{ap}}$	Airplane maintenance labor rate	USD/hr
$R_{l_{eng}}$	Engine maintenance labor rate	USD/hr
R_m	Manufacturing manhour rate	USD/hr
R_{man}	Maneuvering distance	nm
$R_{m_{ml}}$	Military maintenance labor rate (incl. overhead)	USD/hr
R_t	Tooling manhour rate	USD/hr
REV	Revenue per nm	USD/nm
ROI	Return on investment	-----
S	Wing (reference) area	ft^2
S_{wet}	Wetted area	ft^2
SAL_j	Annual salary of crew member of type j	USD/yr

t_{bl}	block time	hrs
t_{cl}	climb time	hrs
t_{cr}	cruise time	hrs
t_{de}	descent time	hrs
t_{flt}	average flight time	hrs
t_{gm}	ground maneuver time	hrs
t_{pft}	Number of flight test hours flown by manufacturer	hrs
t_{man}	maneuvering time	hrs
t_{mis}	average mission time	hrs
tx_{inv}	investment tax rate	-----
tx_{rev}	income (revenue) tax rate	-----
T	Thrust	lbs
TEF	Travel expense factor	-----
TOC	Total operating cost	USD/nm
T_{TO}	Total takeoff thrust	lbs
T/W	Thrust to weight ratio	-----
$U_{ann_{bl}}$	Annual utilization in blockhours	hrs
$U_{ann_{flt}}$	Annual utilization in flight hours	hrs
V	Speed	keas
V_{bl}	Block speed	kts
V_{cr}	Cruise speed	kts
V_{de}	Descent speed (forward)	kts
V_{flt}	Average flight speed	kts
V_{man}	Speed while maneuvering	kts
V_{max}	Maximum design speed	kts
V_{mc}	Minimum control speed	kts

V_C	Design cruising speed	kts
V_s	Stall speed	kts
W	Weight	lbs
W_{ampr}	Aeronautical Manufacturers Planning Report Weight	lbs
W_{crew}	Crew weight	lbs
W_{eng}	Weight per engine	lbs
W_{feq}	Fixed equipment weight	lbs
$W_{ol_{bl}}$	Block oil used	lbs
$W_{pwrplt} = W_{pwr}$	Powerplant weight	lbs
W_{struct}	Airplane structural weight	----
W_A	Airframe weight	lbs
W_E	Airplane empty weight	lbs
$W_{F_{bl}}$	Block fuel used	lbs
$W_{F_{used}}$	Mission fuel used	lbs
W_{PL}	Payload weight	lbs
W/S	Wing loading	lbs/ft^2
W_{TO}	Takeoff Weight	lbs
W_{US}	Useful load	lbs
y_T	Engine-out moment arm	ft

<u>2. Greek Symbols</u>

δ_R	rudder deflection	rad
η_p	propeller efficiency	-----
ρ	air density	$slugs/ft^3$

<u>3. Subscripts</u>

aed	airframe engineering and design
api	airconditioning and pressurization

apu	auxiliary power unit
arm	armament and fire control
avionics	avionics
AMPR, ampr	Aeronautical Manufacturers Planning Report (Also called DCPR)
cl	climb
clean	clean airplane: gear and flaps up (retracted)
cr	cruise
C	Design Cruise
CAD	Computer Aided Drafting
dst	development support and testing
DCPR	Defense Contractors Planning Report (Also called AMPR)
DISP	Disposal
e	engine
els	electrical system
ess	engine starting system
E	Empty
fin	finance
fta	flight test airplanes
fto	flight test operations
F	Fuel
H	Maximum Level
iae	instruments, avionics and electronics
j	number which indicates type of crew member, j=1, 2, 3
L	Landing
m	manufacturing phase
mat	materials
max	maximum design
OPS	Operations
pro	profit
program	for entire program
PL	Payload
qc	quality control

r	rdte phase
rdte, RDTE	Research, Development, Test and Evaluation
tfo	trapped fuel and oil
tool	tooling
tsf	test and simulation facilities
TO	Takeoff

4. Acronyms and Abbreviations

AAA	Advanced Aircraft Analysis
ANG	Air National Guard
Ch	Chapter
CL	Climb
CR	Cruise
DESC	Descent
ECM	Electronic Counter Measures
FBO	Fixed Base Operator
IFF	Identification Friend or Foe
keas	knots equivalent airspeed
LCC	Life Cycle Cost
LTR	Loiter
Pt	Part
SLS	Sea level standard
TO	Takeoff
USAF	United States Air Force
USMC	United States Marine Corps
USN	United States Navy

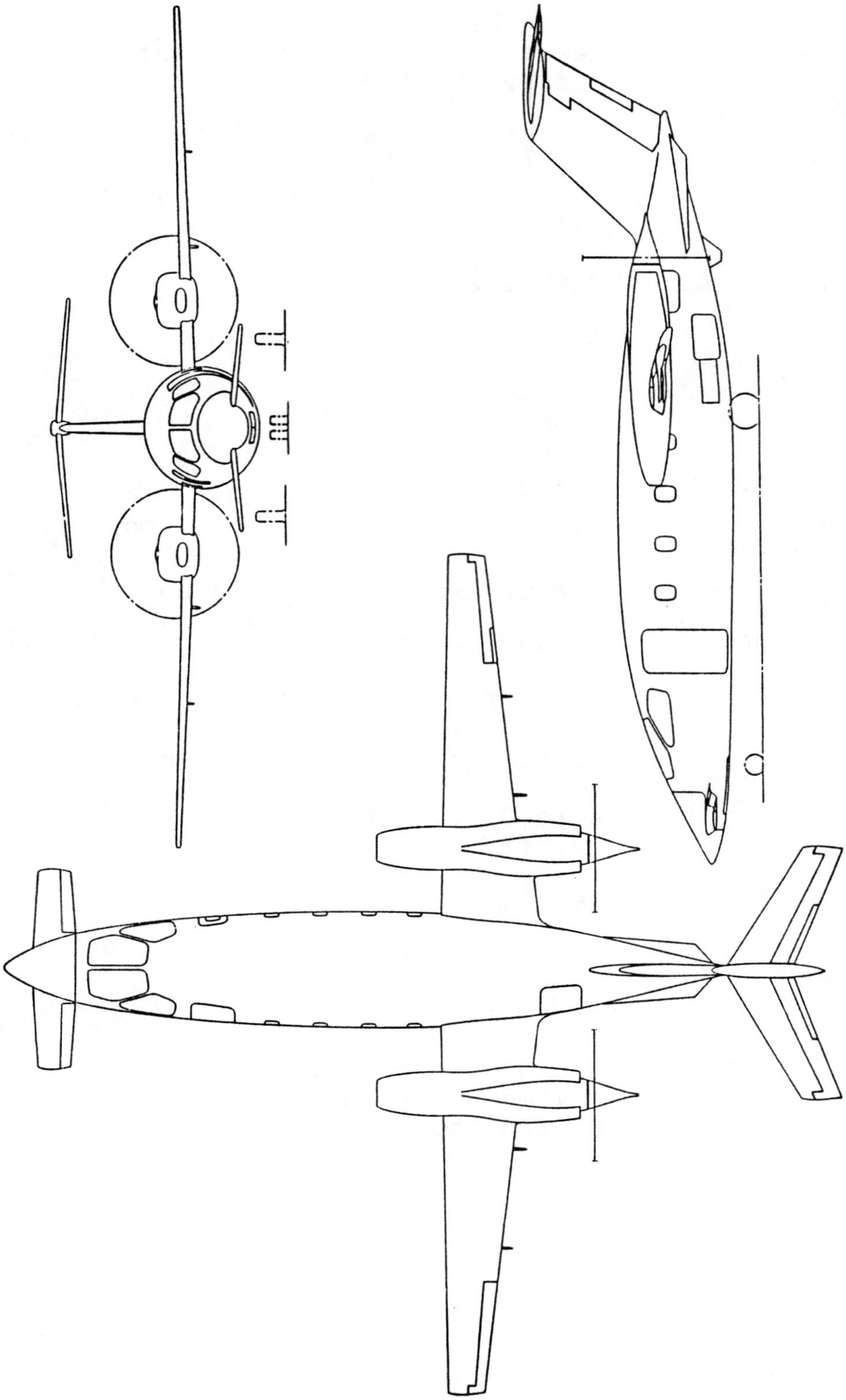

COURTESY: PIAGGIO

ACKNOWLEDGEMENT

Writing a book on airplane design is impossible without the supply of a large amount of data. The author is grateful to the following companies for supplying the necessary data to incorporate in this book:

Beech Aircraft Corporation
Boeing Commercial Airplane Company
Learjet Corporation
Lockheed Aircraft Corporation

A significant amount of airplane cost, price and revenue information has been accumulated by the author from the following magazines:

Interavia (Swiss monthly)
Flight International (British weekly)
Business and Commercial Aviation (USA, monthly)
Aviation Week and Space Technology (USA, weekly)

The author is particularly grateful to Mr. John W. Olcott, Publisher/Editor of Commercial and Business Aviation, for his permission to copy material from the 1989 Planning and Purchasing Handbook in Appendix C of this book.

The author also acknowledges the work done by two of his students: Mr. Larry Bellmard and Mrs. Donna Gerren in checking calculations and proofreading the material.

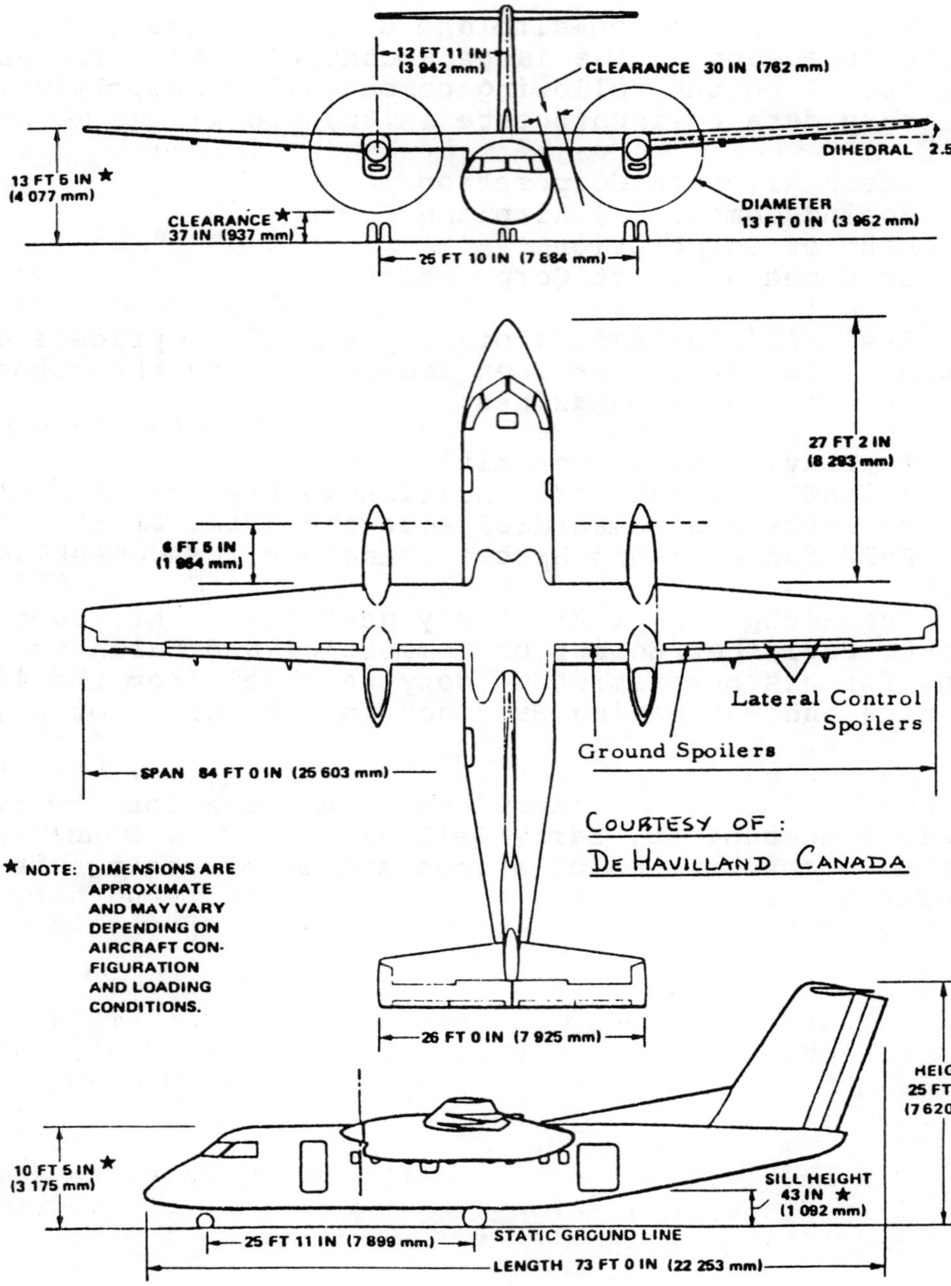
12 FT 11 IN
(3 942 mm)
CLEARANCE 30 IN (762 mm)
DIHEDRAL 2.5°
13 FT 5 IN ★
(4 077 mm)
CLEARANCE ★
37 IN (937 mm)
DIAMETER
13 FT 0 IN (3 962 mm)
25 FT 10 IN (7 684 mm)
27 FT 2 IN
(8 293 mm)
6 FT 5 IN
(1 964 mm)
Lateral Control
Spoilers
Ground Spoilers
SPAN 84 FT 0 IN (25 603 mm)
COURTESY OF:
DE HAVILLAND CANADA
★ NOTE: DIMENSIONS ARE APPROXIMATE AND MAY VARY DEPENDING ON AIRCRAFT CONFIGURATION AND LOADING CONDITIONS.
26 FT 0 IN (7 925 mm)
HEIGHT ★
25 FT 0 IN
(7 620 mm)
10 FT 5 IN ★
(3 175 mm)
SILL HEIGHT
43 IN ★
(1 092 mm)
25 FT 11 IN (7 899 mm)
STATIC GROUND LINE
LENGTH 73 FT 0 IN (22 253 mm)

1. INTRODUCTION

The purpose of this series of textbooks on Airplane Design is to familiarize aerospace engineering students with the design methodology and design decision making involved in the process of designing airplanes. The series of books is organized as follows:

Part I:	PRELIMINARY SIZING OF AIRPLANES
Part II:	PRELIMINARY CONFIGURATION DESIGN AND INTEGRATION OF THE PROPULSION SYSTEM
PART III:	LAYOUT DESIGN OF COCKPIT, FUSELAGE, WING AND EMPENNAGE: CUTAWAYS AND INBOARD PROFILES
PART IV:	LAYOUT DESIGN OF LANDING GEAR AND SYSTEMS
PART V:	COMPONENT WEIGHT ESTIMATION
PART VI:	PRELIMINARY CALCULATION OF AERODYNAMIC, THRUST AND POWER CHARACTERISTICS
PART VII:	DETERMINATION OF STABILITY, CONTROL AND PERFORMANCE CHARACTERISTICS: FAR AND MILITARY REQUIREMENTS
PART VIII:	AIRPLANE COST ESTIMATION: DESIGN, DEVELOPMENT, MANUFACTURING AND OPERATING

In this part (Part VIII) frequent reference is made to other parts in this Airplane Design series: Refs 1-7.

The purpose of Part VII is to present preliminary cost estimating methods for newly designed airplanes. The methods are presented in such a manner that they can be applied to commercial and to military airplanes of all types.

Chapter 2 presents a discussion of cost definitions and concepts. The idea of Life Cycle Cost is defined and its relation to the design decision making process is outlined.

The major cost sources of LCC are treated in some detail in Chapters 3 through 6 in the form of systematic cost estimating procedures and examples for:

* Research, development, technology and evaluation cost in Chapter 3

* Manufacturing and Acquisition cost in Chapter 4

* Operating cost for commercial airplanes in Chapter 5

* Operating cost for military airplanes in Chapter 6

An example of LCC determination for a military airplane is given in Chapter 7.

It is important that students of airplane design understand the role of optimization. Chapter 8 contains some fundamental discussions on this subject. The topic of 'design-to-cost' is also introduced in this chapter. Finally, a list of 88 design guidelines for 'low cost' is included in Chapter 8.

The decision on whether or not to proceed with full scale development of a new design involves a variety of factors. The purpose of Chapter 9 is to give an overview of key factors in airplane program decision making. This includes a list of 19 lessons learned on past programs.

Actual price data for airplanes and airplane components are hard enough to find for practicing engineers. For students this is often a very time consuming problem. To help with the problem of getting 'on track' with price data several appendices are included with specific information on prices:

* Appendix A includes airplane price data

* Appendix B includes engine and propeller price data

* Appendix C includes avionics systems price data

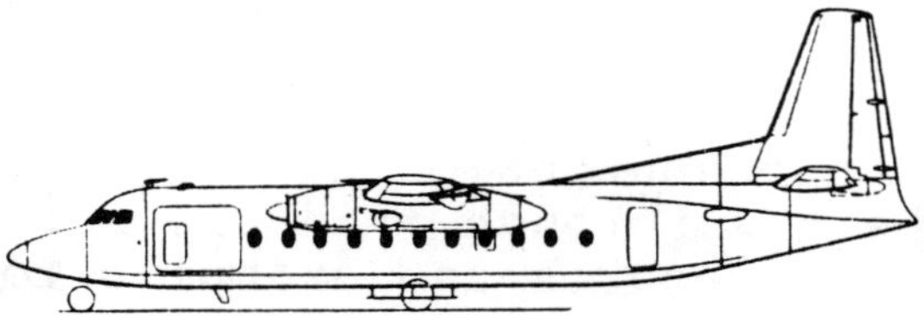

2. COST: DEFINITIONS AND CONCEPTS

The purpose of this chapter is to introduce aerospace engineering students to several important definitions and concepts, all dealing with airplane cost. The material is organized as follows:

2.1 Cost, Price and Profit
2.2 Airplane Program, Life Cycle and Life Cycle Cost
2.3 Examples of Aeronautical Enterprises
2.4 Some Observations on Profit and Cost

2.1 COST, PRICE AND PROFIT

First, it is necessary to define the meaning of the terms: COST, PRICE and PROFIT.

Definition 1: The COST of an airplane is the total amount of expenditure of resources, usually measured in dollars, needed to manufacture that airplane.

Definition 2: The PRICE of an airplane is the amount of dollars paid for the airplane by customers.

Definition 3: Profit = Price - Cost (2.1)

How cost, price and profit are viewed depends on which position is occupied in the economic process. Five examples of different viewpoints are:

Example 1: Position of the airplane manufacturer.

Assume that the cost to produce a certain airplane is USD 100,000. If the selling price is USD 110,000 a profit is made of USD 10,000. In the presence of competition it will be of prime interest to the manufacturer to offer his airplane at the lowest possible price (and still make a profit) while assuring that his airplane will be profitable or affordable to operate by his customer. This means that design, development and manufacturing costs must be kept as low as possible.

Example 2: Position of the commercial airplane operator.

To the commercial airplane operator, the price paid for an airplane is a cost (acquisition cost). Another cost is that incurred to operate the airplane.

The operator hopes to sell enough revenue ton miles (RTM) or revenue passenger miles (RPM) to make the operation of the airplane profitable. To make this possible, the acquisition and the operating cost must be as low as possible.

<u>Example 3:</u> Position of the personal or corporate airplane operator.

To this type of operator the airplane is a tool used to enhance certain aspects of business or to satisfy certain personal needs. Affordability is of prime concern to this type of operator. Affordability usually translates into requirements for low acquisition cost, low operating cost and high resale value.

Government tax policies and tax laws tend to have a significant impact on airplane affordability for this type of operator.

<u>Example 4:</u> Position of the government/military airplane operator.

The government/military airplane operator normally has a limited budget available with which to carry out certain missions. The price paid for an airplane and the cost incurred while operating the airplane all are costs to this operator. What this operator hopes to achieve is to carry out the required missions within the available budget. To make this possible the acquisition cost and the operating cost need to be as low as possible.

Military operators are normally directly involved in the design and development process with the airplane manufacturer. The cost associated with the design and development process can be significant particularly if the production run is small.

<u>Example 5:</u> Position of the consumer/taxpayer.

The consumer/taxpayer is the person who ends up paying for all activities required to design, manufacture, operate and dispose of airplanes. The consumer/taxpayer is ultimately interested in lowering the total cost associated with all aeronautical activities.

At this point it should be observed that all costs mentioned so far are those which can be measured directly in dollars. Such costs are referred to also as economic costs. There are other types of cost as well: social, psychological and environmental, to name a few. Examples of causes of these costs are:

Social Cost: If a company decides to relocate its operations to another part of the country (or the world) then social costs are incurred because of loss of employment, relocation of families, etc. The burden of carrying the economic costs associated with social costs are normally carried by the taxpayer.

Psychological Cost: Airplanes tend to create a significant amount of noise during takeoff and landing. This can cause distress in people and can be thought of as a psychological cost. The psychological distress can become so severe as to lead to anti-noise legislation. Examples are found in FAR 36 which places the burden of the noise problem on the designer and thus, ultimately on the consumer.

Environmental Cost: Pollution caused by certain airplane operations represents an environmental cost. Examples are: atmospheric pollution due to exhaust gasses, fuel dumping in emergencies and chemical discharges resulting from manufacturing and maintenance processes.

Although these types of cost are important, they are difficult to quantify and will not be considered in this text. Many of these costs are ultimately borne by taxpayers although some of them are borne by the consumer (fare paying passenger) through increased operating cost as in the case of costs incurred by meeting FAR 36.

There is much confusion about the relationship between cost and price. If an airplane manufacturer buys certain equipment for installation in an airplane which is being manufactured, he pays a certain price for that equipment. That price paid then becomes part of his cost to produce that airplane!

What a manufacturer charges in the marketplace for a certain airplane becomes the price charged. That price may or may not have a rational relationship to the actual cost. Example: true or not, it was said that Cessna actually 'lost' money on each Cessna 152 it produced. The reason, true or not, was that it was felt that Cessna 152 customers eventually would 'step-up' to a more expensive airplane model on which a profit was actually made. This is the so-called 'loss-leader' policy which is popular in supermarkets as well!

2.2 AIRPLANE PROGRAM, LIFE CYCLE AND LIFE CYCLE COST

The evolution of an airplane from design to manufacturing, operation and finally, disposal is referred to as an airplane program. A typical airplane program can be divided into the following six phases:

Phase 1: Planning and Conceptual Design.

During this phase the activities depicted in Figure 2.1 take place. Planning here consists primarily of mission requirements research. This eventually leads to a mission specification. Conceptual design here consists of the design activities associated with Preliminary Design Sequence I as outlined in Parts I and II of this text.

Some very preliminary cost studies are also conducted during this phase.

Phase 2: Preliminary Design and System Integration.

During this phase the activities depicted in Figure 2.2 take place. The accompanying design activities are those associated with Preliminary Design Sequence II as outlined in Parts I and II of this text. Design trade studies are conducted to find that combination of technology and cost which might result in a viable airplane program.

Phase 3: Detail Design and Development.

During this phase the airplane and system integration design is finalized for certification flight testing and for production.

Phase 4: Manufacturing and Acquisition.

During this phase the airplane is manufactured and delivered to (acquired by) the customer.

Phase 5: Operation and Support.

During this phase the airplane is being acquired by the user and is being operated with the accompanying support activities.

Note: Phases 4 and 5 normally overlap.

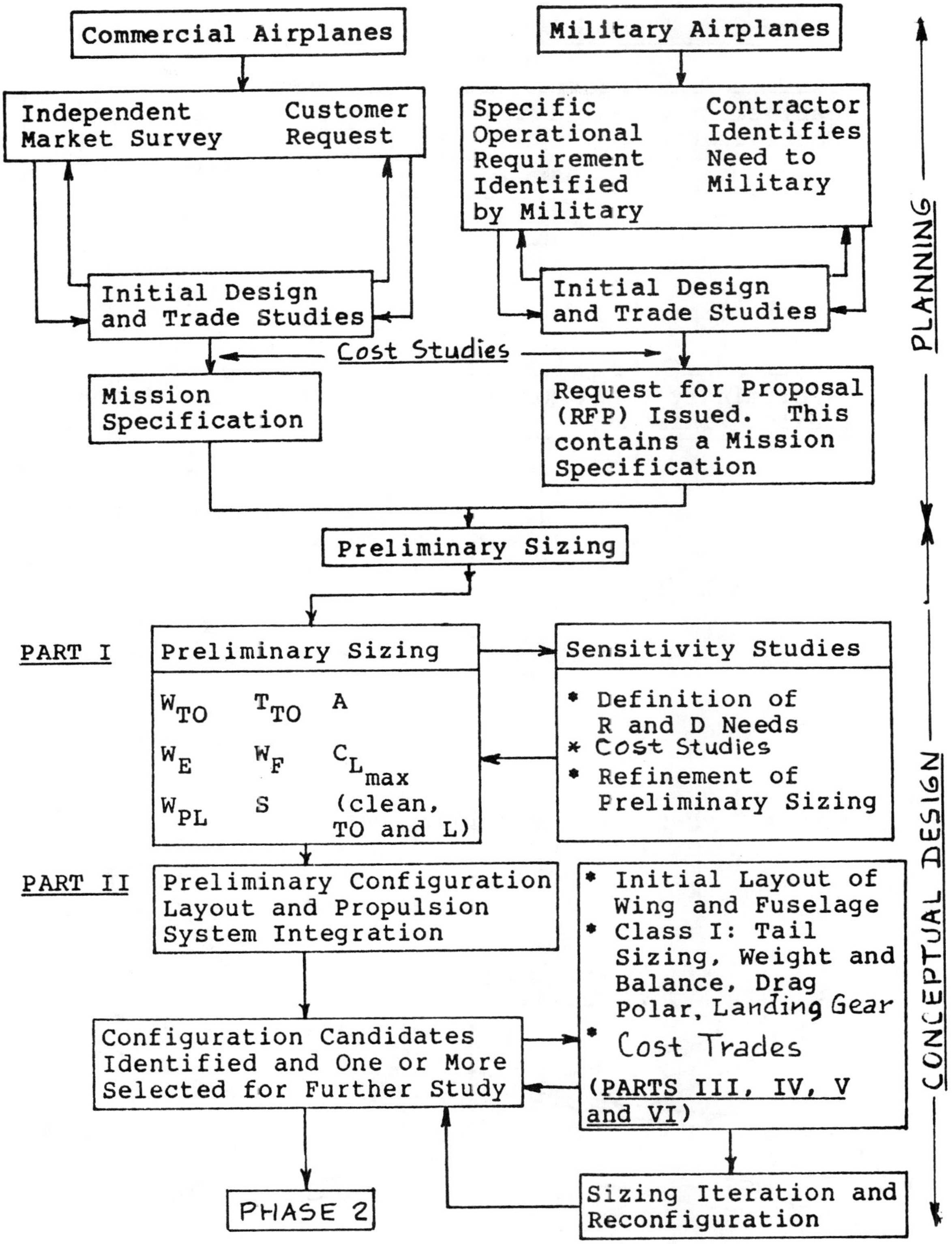

Figure 2.1 Phase 1: Planning and Conceptual Design

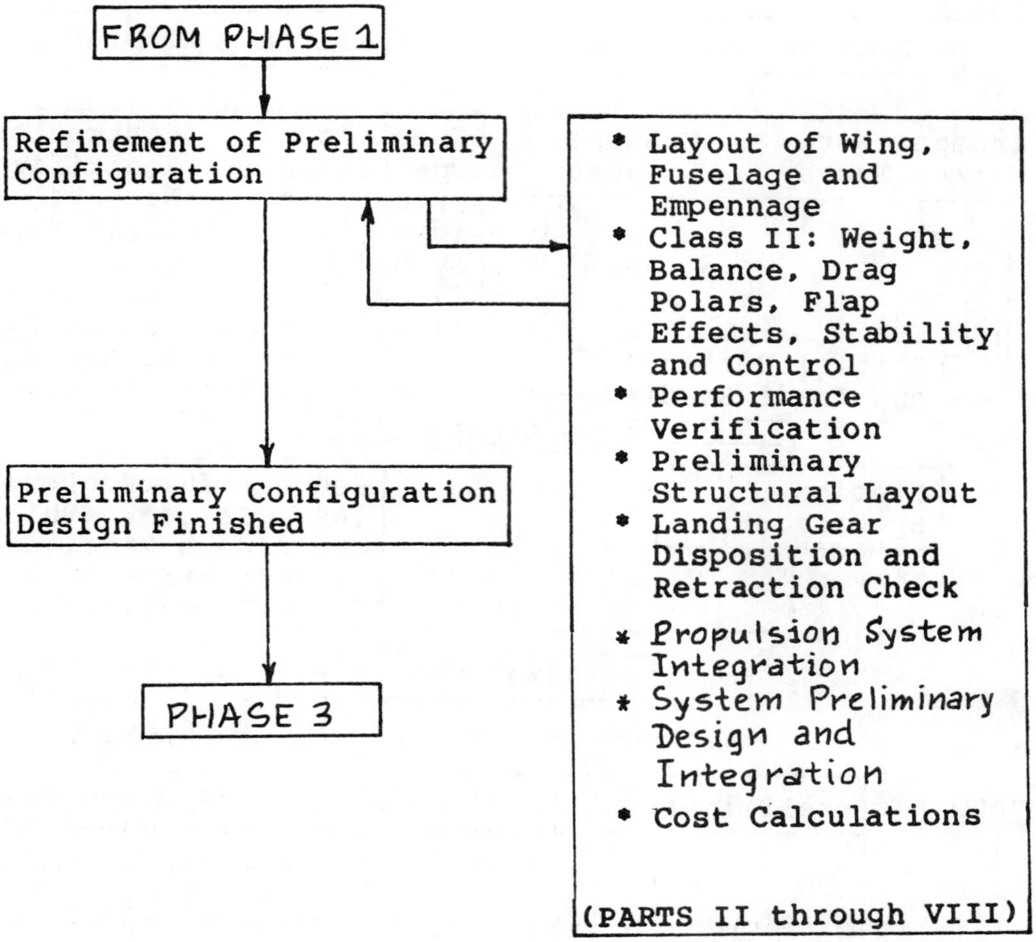

Figure 2.2 Phase 2: Preliminary Design and System Integration

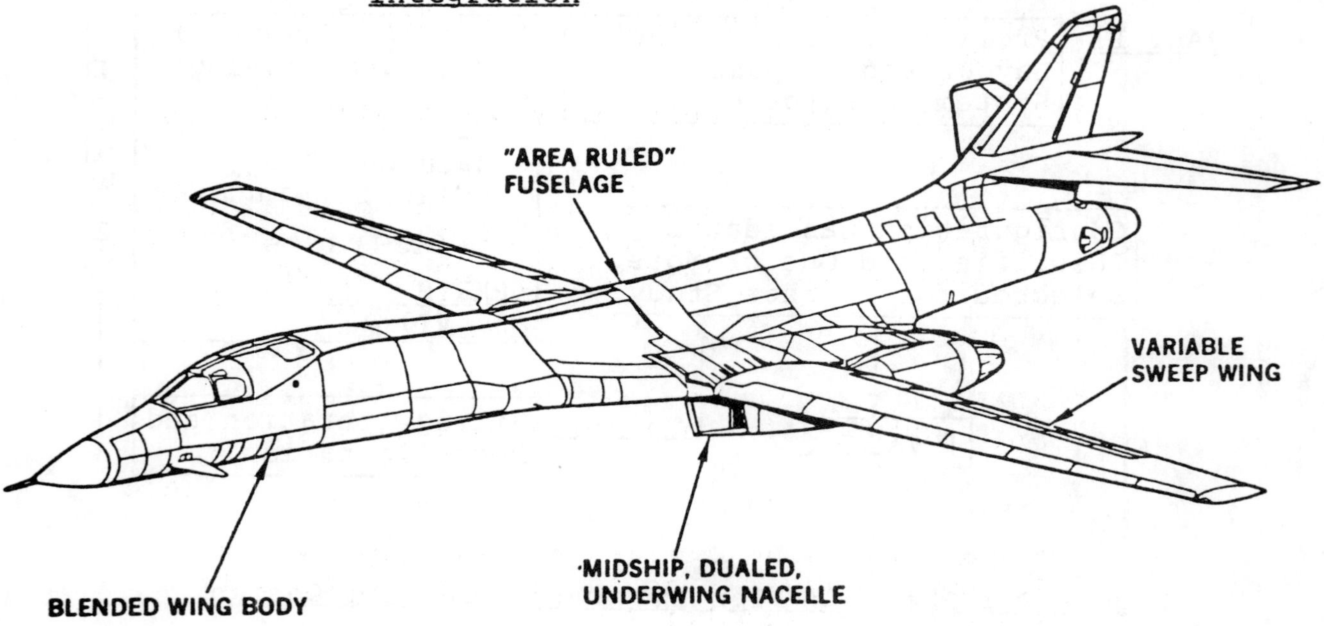

Phase 6: Disposal.

This phase marks the end of the operational life of the airplane. Disposal activities may include destruction of the airplane and disposal of the remaining materials. In case of military airplanes it may also include storage (at Davis-Monthan AFB for example). Disposal becomes necessary when an airplane has reached the limit of its technological or economical life.

Note: Phases 5 and 6 normally overlap.

Figure 2.3 shows these six phases with an indication of their relative impact on costs associated with a typical airplane program. Note the following definitions:

Definition 4: The time elapsed during the six phases of an airplane program (Fig.2.3) is called the Airplane Life Cycle.

Definition 5: The total cost of an airplane program incurred during the airplane life cycle is called the Life Cycle Cost.

Note: the vertical scale in Figure 2.3 indicates the cost associated with each program phase as a percentage of life cycle cost: LCC.

For preliminary cost estimating purposes the life cycle cost of an airplane program is broken down into four cost sources:

1. Research, development, test and evaluation cost: C_{RDTE}.

   This cost source accounts for all costs incurred in phases 1, 2 and 3 in Figure 2.3. A method for estimating C_{RDTE} is presented in Chapter 3.

2. Acquistion cost: C_{ACQ}.

   This cost source includes the manufacturing cost: C_{MAN} and the manufacturer's profit: C_{PRO}. Thus:

   $$C_{ACQ} = C_{MAN} + C_{PRO} \qquad (2.2)$$

   These costs are incurred during Phase 4 in Figure 2.3. A method for estimating C_{ACQ} is presented in Chapter 4.

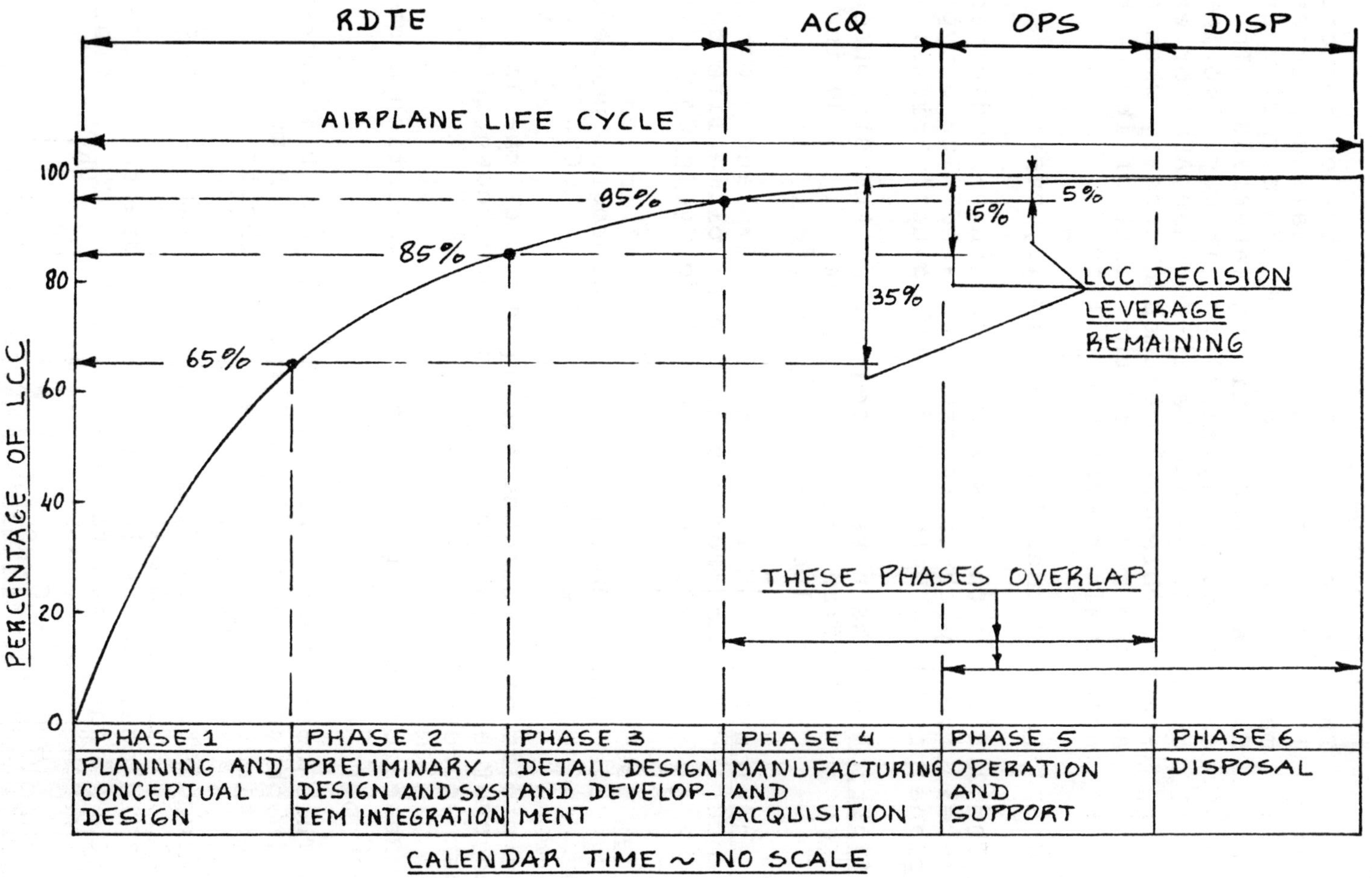

BASED ON AN IDEA IN REF. 8

Figure 2.3 Impact of Airplane Program Phases on Life Cycle Cost

3. Operating cost: C_{OPS}.

   This cost source represents the cost incurred while operating the airplane: see Phase 5 in Figure 2.3. The airplane manufacturer and his suppliers (vendors) usually incur certain support costs during this phase. A method for estimating C_{OPS} is given in Chapter 5 for commercial airplanes and in Chapter 6 for military airplanes.

4. Disposal cost: C_{DISP}.

   This is the cost incurred in disposing of the airplane: see Phase 6 in Figure 2.3. Chapter 7 presents an approach to the estimation of C_{DISP}.

The life cycle cost of an airplane program (LCC) can now be expressed as:

$$LCC = C_{RDTE} + C_{ACQ} + C_{OPS} + C_{DISP} \qquad (2.3)$$

Chapter 7 presents an example of LCC estimation for a military airplane program. An example for a commercial airplane program is given in Section 5.4.

Figure 2.4 shows example cost histories for two airplane programs. Note that the operating cost source is much larger than the acquisition cost source. The latter in turn is much larger than the research, development, test and evaluation cost source. The cost histories of Figure 2.4 are schematically represented in Figure 2.5. It is seen that:

$$C_{OPS} >> C_{ACQ} >> C_{RDTE} \qquad (2.4)$$

What emerges from Figures 2.3 - 2.5 is the following important conclusion:

<u>THE CONCEPTUAL AND PRELIMINARY DESIGN PHASES ARE RESPONSIBLE FOR LOCKING IN MOST OF THE LIFE CYCLE COST OF AN AIRPLANE!!</u>

The implication for airplane design engineers and for airplane program managers is that:

<u>SIGNIFICANT LEVERAGE AFFECTING LIFE CYCLE COST EXISTS ONLY IN THE CONCEPTUAL AND PRELIMINARY DESIGN PHASES!!</u>

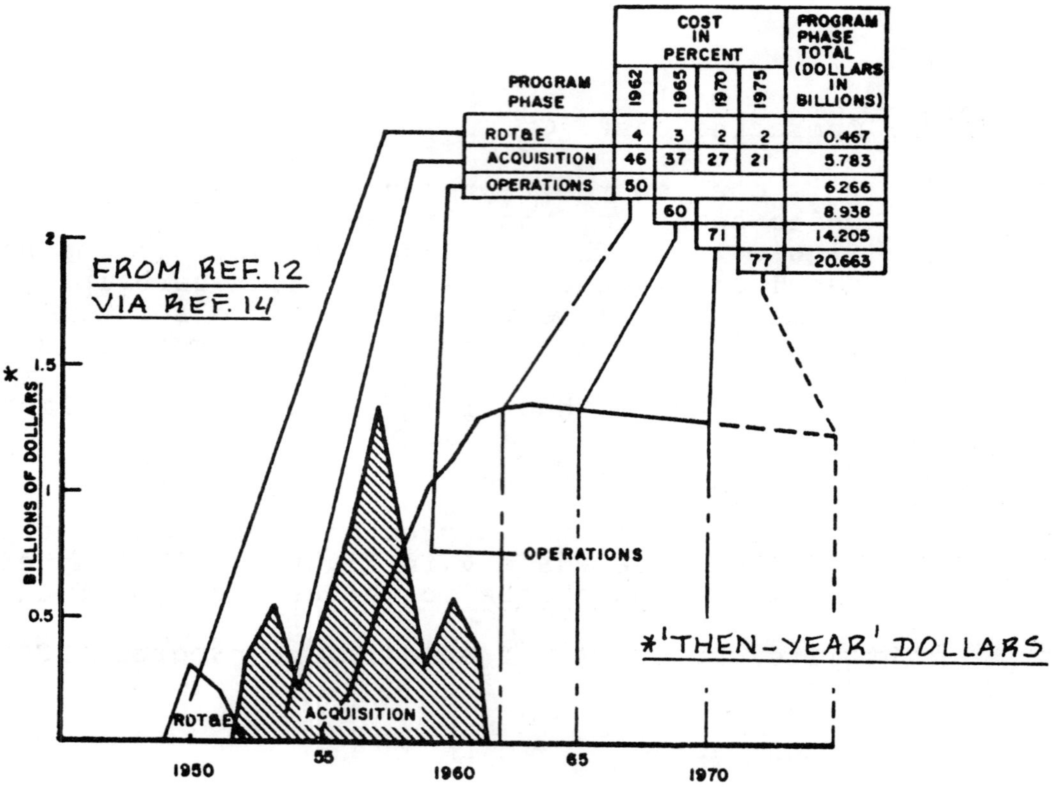

Figure 2.4a Life Cycle Cost History of the B-52 Program

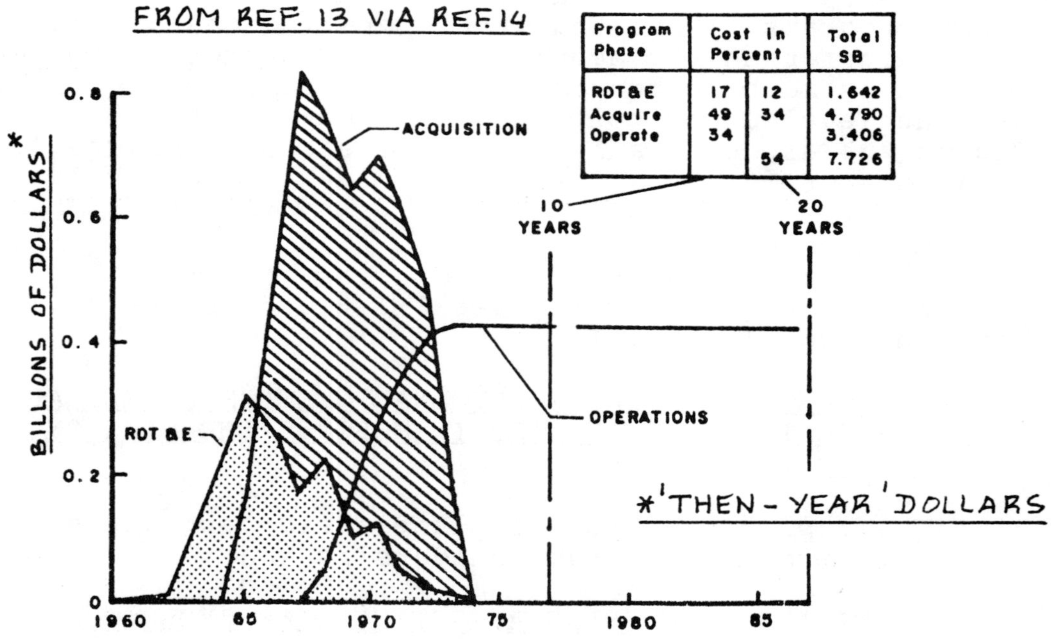

Figure 2.4b Life Cycle Cost History of the F-111 Program

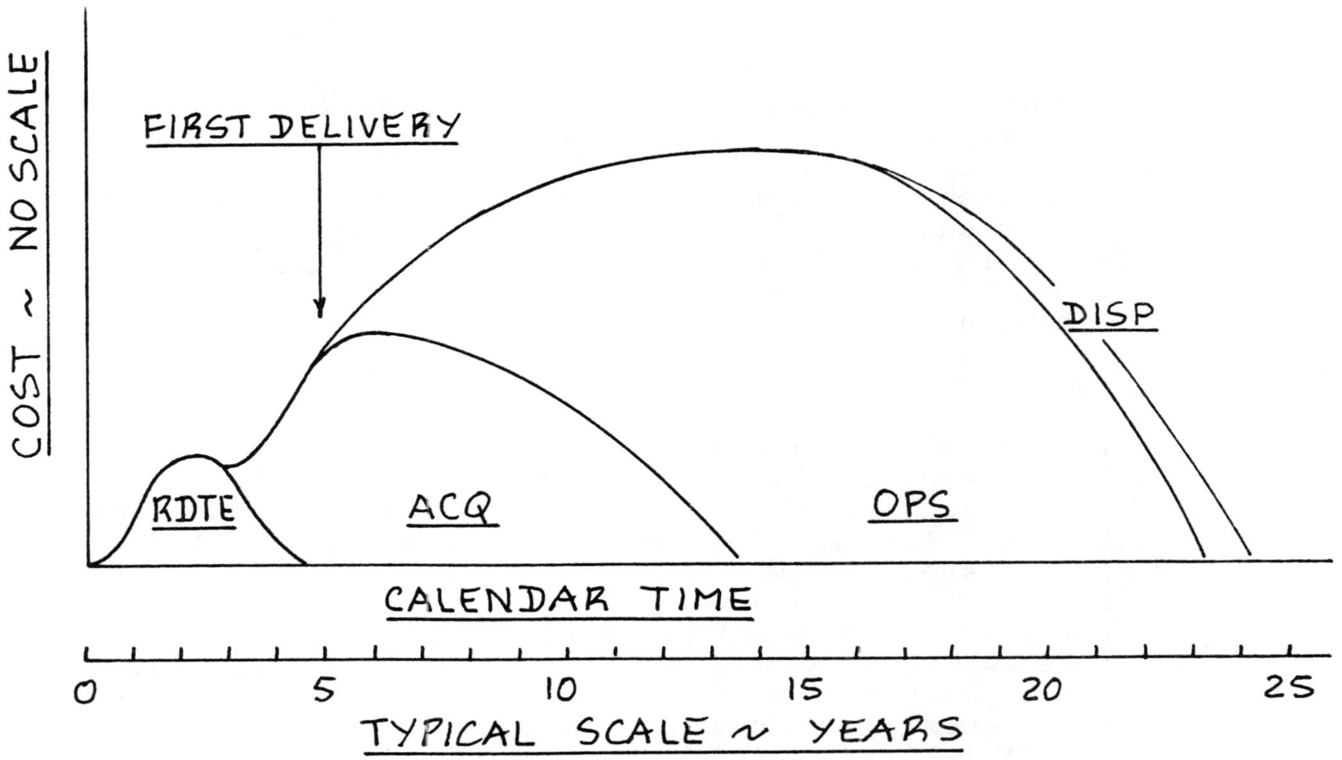

Figure 2.5 Schematic Representation of Life Cycle Cost History of Typical Airplane Programs

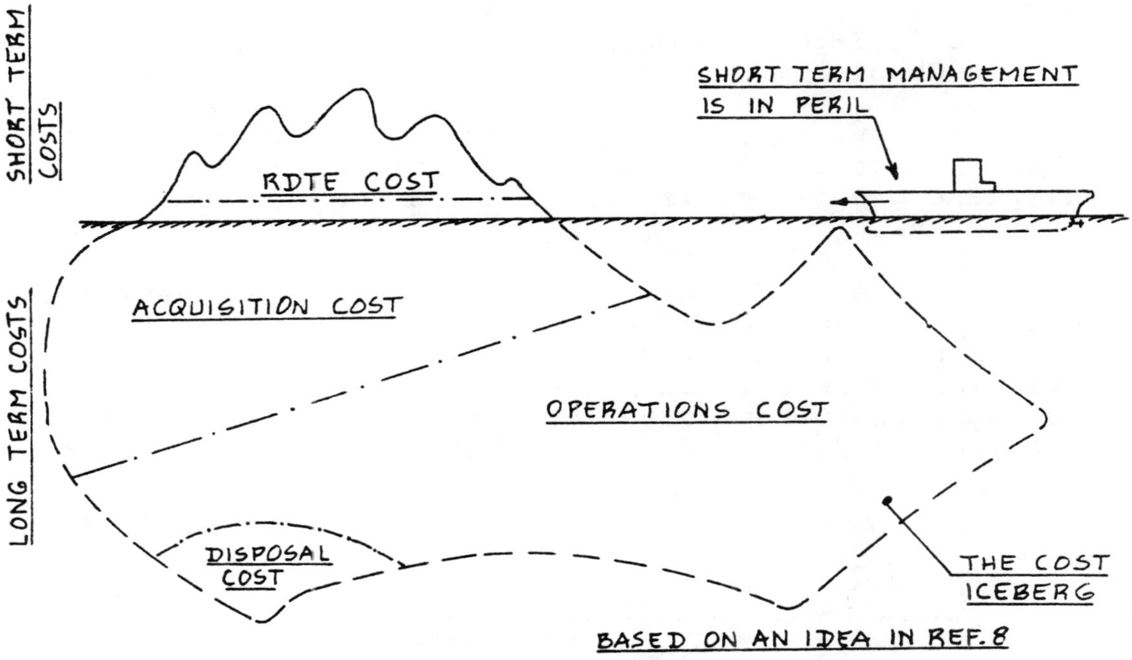

Figure 2.6 The Iceberg Effect in Airplane Program Management

Another way of illustrating these facts is found in Figure 2.6. The airplane program manager is like the captain of a ship navigating close to an iceberg. The program manager has many pressures upon him to lower RDTE costs (the short term outlook). Yielding to these pressures may in fact increase LCC! Performing more RDTE work, when directed at designing to minimize LCC, can result in significant savings to the customer and therefore to society as a whole. It takes courage and wisdom to purposely increase RDTE costs so that LCC will be lower.

During the preliminary airplane design process as described in Parts I through VII many design decisions are made which have a significant effect on airplane life cycle cost. Airplane designers must be aware of these effects if their designs are to be 'cost-effective'.

Step 36 in the preliminary design process (See Part II, page 23) requires the designer to perform a preliminary cost analysis for the new airplane. This in turn means that designers must have available a systematic method for estimating LCC. Chapters 3 through 7 present methods for estimating the major LCC sources of Eqn.2.3.

In addition to having a method for estimating LCC, the designer must have an understanding of how either total LCC or any of its cost sources can be minimized. Several important principles in the areas of DESIGN OPTIMIZATION, DESIGN-TO-COST and DESIGN-TO-PRICE are discussed in Chapter 8.

Finally, it is necessary to have an understanding of various factors (in addition to cost) which may affect an airplane program decision. That subject is taken up in Chapter 9.

Because it is hard for aeronautical engineering students to 'find' hard data on prices, the author has included price data for airplanes, engines, propellers and avionics systems in the form of several appendices. It is hoped that these data (even though they will become obsolete rather rapidly) will be useful to the reader.

2.3 EXAMPLES OF AERONAUTICAL ENTERPRISES

Aeronautical engineering students should be aware of the various types of aeronautical enterprises which exist. After all, these enterprises are most likely to be their future employers.

From an ownership point of view there are three types of aeronautical enterprises:

1. Privately owned.

2. Government owned (country, state, local).

3. Part private and part government owned.

In the following, some selected examples of each are given. There is no implication of importance, relative or absolute, to be attached to these examples. They are merely what came to the author's mind while developing this text. An extensive list of aeronautical enterprises can be found in Reference 9. When consulting Ref.9 the reader will come to the conclusion that aeronautics is very much an international affair! Aeronautics is expected to become even more international in the future.

1. Examples of privately owned aeronautical enterprises are:

Manufacturers of airplanes:

In the USA:

Boeing, McDonnell-Douglas, Lockheed, General Dynamics (including Cessna), Grumman, Raytheon-Beech, Piper and Learjet Corp.

Foreign:

Robin, Grob, Reims-Aviation, Pilatus, Mudry.

Operators of airplanes:

In the USA:

American Airlines, United Airlines, Federal Express, US Air, TWA , Northwest and Southwest Airlines.

Foreign:

British Airways, Martin Air, Cross Air and Tyrolean.

2. Examples of government owned aeronautical enterprises are:

Manufacturers of airplanes:

In the USA: none. But: the U.S. government owns several plants which it leases to private companies.

Foreign:

Fabrica Militar de Aviones, Embraer, CASA, IPTN, Antonov, Ilyushin, PZL, Airbus GmbH and Aeritalia.

Operators of airplanes:

In the USA:

USAF, USNavy, U.S.M.C., US Coast Guard and FAA.

Foreign:

Aeroflot, SAS, Garuda Indonesian and Air France.

3. Examples of part private and part government owned aeronautical enterprises are:

Manufacturers of airplanes:

In the USA: none.

Foreign:

SIAI-Marchetti, Piaggio, Partenavia, Dassault-Breguet, MBB and Fokker.

Operators of airplanes:

In the USA: none.

Foreign:

KLM (Royal Dutch Airlines), Lufthansa, and UTA.

There are many other aeronautical enterprises which supply parts, systems and engines. To save space, these were not included as examples. Nevertheless, these supplier enterprises are important. For an airplane manufacturing industry to be viable, a network of parts and system suppliers (the aeronautics infra-structure) must exist! Airplanes cannot be built without:

* Materials such as: aluminum, steel and titanium alloys, composites, etc.
* Wheels, brakes and tires
* Engines and APU's
* Avionics systems and instruments
* Pumps, lines, sensors, cables and fasteners
* Interiors, including insulating materials

Reference 9 provides an annual listing of airplane operators, manufacturers and suppliers with details about their corporate structure. This reference deals with foreign as well as domestic enterprises.

Last (and not least!) a healthy aeronautics industry requires very long term research capabilities. In most countries such long term research facilities are owned by the government. NASA in the USA is an example.

2.4 SOME OBSERVATIONS ON PROFIT AND COST

In general, the objective of a privately owned aeronautical enterprise is to make a profit. The profit made before taxes is sometimes referred to as the operating margin. Examples of operating margins for several aeronautical enterprises is presented in Table 2.1. Since the tax situation of a company can vary from year to year and from country to country, the operating margin of a company is not the same as its profit. In day-to-day use of these terms they are frequently interchanged.

In the case of military, government owned enterprises, profit is not an objective. Instead, getting the 'best' defense or offense per dollar paid ('biggest bang per buck') is usually the objective.

In the case of civilian, government owned enterprises, profit may not be an immediate objective. Significant participation in some market segment to generate jobs and general prosperity is an important objective of many government owned enterprises.

In the case of part private and part government owned aeronautical enterprises the objective is usually to make a profit albeit not necessarily each year.

Another objective which can play a role in privately owned enterprises is: to 'have fun at building airplanes and not lose money'.

It is expected that the manner in which an aeronautical enterprise is managed will depend on its objective.

Table 2.1 Operating Margins in Percentage of Sales for the Period 1985 - 1988

Aerospace Manufacturers	1985	1986	1987	1988
Boeing Commercial	4.7	4.2	3.6	5.1
Boeing Military	8.0	7.5	1.5	-2.6
Chrysler (Gulfstream)	N/A	9.0	12.9	11.7
General Dynamics Convair	13.3	12.5	10.1	9.3
General Dynamics Cessna	N/A	-82.5	1.7	5.5
General Electric Engines	14.3	14.5	13.9	15.4
G.M. (Hughes)	N/A	7.8	8.6	8.8
Grumman Aerospace	4.4	3.3	1.6	5.3
Honeywell/Sperry	3.4	6.1	6.3	-7.7
Kaman	7.9	7.8	8.2	8.1
Learjet	N/A	N/A	0.0	0.0
Litton Adv. Electronics	11.3	5.7	8.0	7.8
Lockheed Aeronautical	10.6	10.4	8.1	6.3
Loral (Goodyear)	N/A	20.0	18.7	N/A
LTV Aircraft Products	12.3	8.8	9.4	3.4
McDonnell Douglas				
Combat Aircraft	8.7	6.7	5.7	6.6
Transport Aircraft	3.7	3.0	2.7	2.6
Moog (Hydr. Actuators)	15.2	13.8	11.4	-1.0
Northrop Aircraft	9.0	1.6	4.5	1.9
Northrop Electronics	9.0	8.0	6.2	3.6
Raytheon Beech	-2.3	1.4	-1.3	3.0
Rockwell North American	9.3	10.3	13.8	12.4
Teledyne (aviation, electronics)	8.6	6.9	10.2	9.4
United Tech (P and W)	7.6	6.1	12.0	10.8
U.T. (Sikorsky, Hamilton)	9.8	-8.8	3.0	-0.4

Data from: Flight International, June 10, 1989

Airlines	1985	1986	1987	1988
American Airlines	8.7	6.8	6.4	9.1
USAir	9.5	9.2	10.6	7.6
United Airlines	-4.5*	1.3	3.0	7.4
Delta Airlines	7.8	0.8	7.6	7.2
Northwest Airlines	2.9	4.6	3.8	3.6

*Strike

Data from: Aviation Week and Space Techn., June 19, 1989

NOTE: Operating margin is the profit before taxes.

When profit is the primary objective, the temptation is for management decisions to focus on the short term: 'maximizing profit this quarter'. This can lead to the iceberg disaster depicted in Figure 2.6.

When profit is not the immediate objective, management decision making tends to focus on the long term.

The time required for development and certification of an airplane does affect management decision making. Because of the long time it takes to develop and certify airplanes, management decision making in the airplane industry should focus on the long term.

Whatever management objectives are, the cost of developing, certifying, producing and operating an airplane must be known with some certainty before a decision to 'launch' an airplane program is made.

Because airplane programs take many years to evolve through the phases shown in Figure 2.3, inflation (= cost escalation) plays an important role in estimating program cost. So does interest on investment: the cost of money.

Estimates for cost magnitudes are usually given in 'then-year' dollars. It is customary to scale cost data from one 'then-year' to another with a cost escalation factor: CEF. Figure 2.7 shows how this CEF has varied with calendar time since 1965. In Figure 2.7 the CEF has been arbitrarily normalized to a value of 1.0 for 1970. Cost from one year to another may be 'scaled' as follows:

$$Cost_{19XX} = Cost_{19YY}(CEF_{19XX}/CEF_{19YY}) \qquad (2.5)$$

where: the ratio (CEF_{19XX}/CEF_{19YY}) is the CEF-ratio.

The reader is referred to References 8 and 10 for a detailed discussion and development of the ideas of airplane program cost. Chapters 3 through 7 present rapid methods for estimating airplane program costs. These methods are aimed strictly at Phases 1 and 2 as depicted in Figure 2.3. They should be used for preliminary cost estimating purposes only.

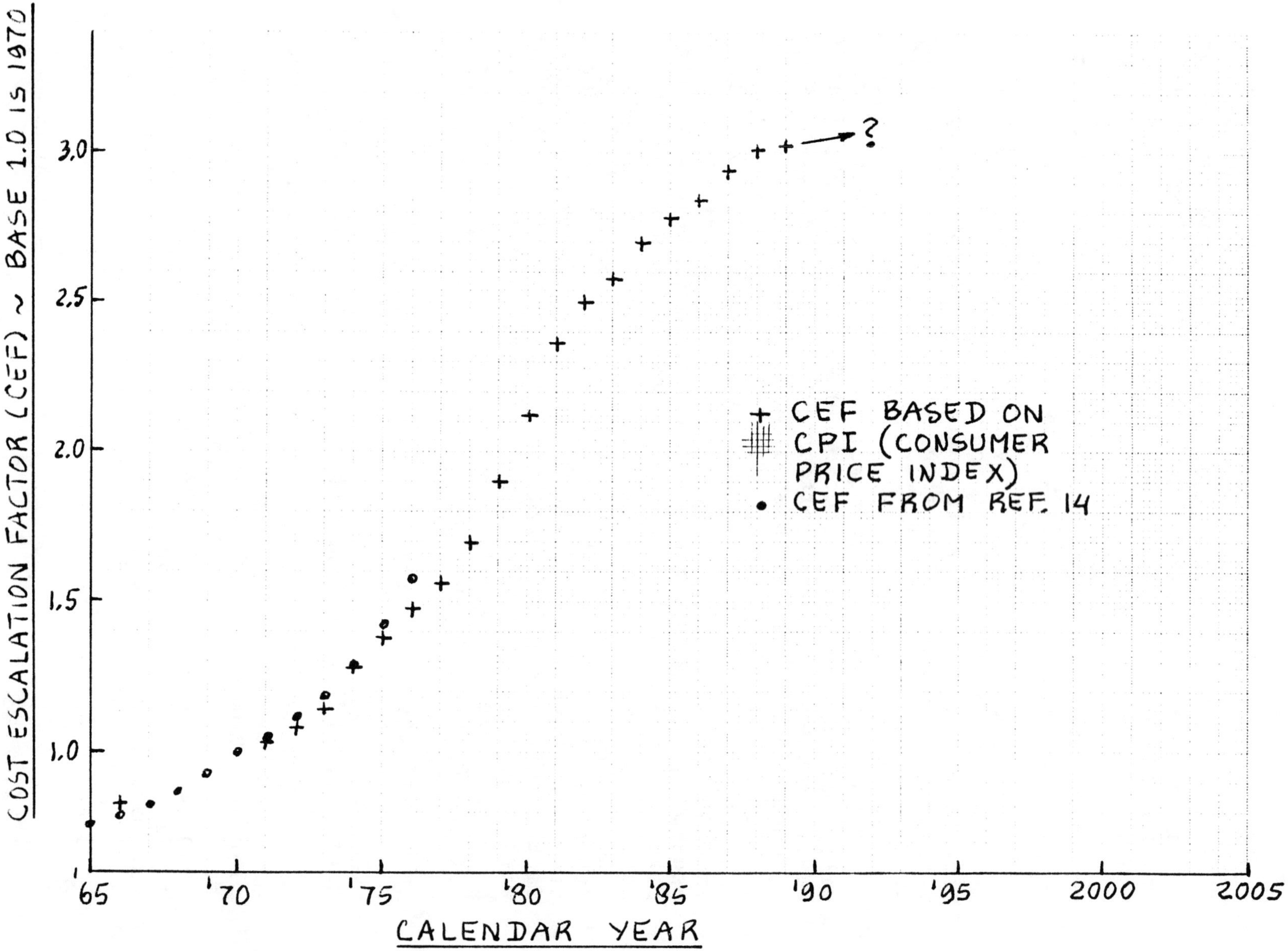

Figure 2.7 Variation of Cost Escalation Factor with Time

3. METHOD FOR ESTIMATING RESEARCH, DEVELOPMENT, TEST AND EVALUATION COST AND A METHOD FOR ESTIMATING PROTOTYPING COST

The purpose of this chapter is to present:

1) A method for estimating research, development, test and evaluation cost (C_{RDTE}) for airplanes.

2) A method for estimating prototyping cost (C_{PROT}) for airplanes.

The RDTE cost, C_{RDTE}, is the first cost source seen in Eqn.(2.3). RDTE cost is accumulated during Phases 1-3 in Figure 2.3. Phases 1-3 involve those activities which take a new airplane all the way from the planning and conceptual design stage to certification. This applies not only to military, but also to commercial airplanes. Phases 1-3 normally include the design, construction, ground and flight testing of a number of static and flight test airplanes. Figure 2.3 shows how the RDTE activities of Phases 1-3 fit into an airplane program.

The method for estimating RDTE cost is based on that of Ref.11. It should be used for preliminary cost estimation purposes only. The method can be applied to military as well as to commercial airplane programs.

RDTE costs are normally broken down into seven cost categories:

3.1 Airframe Engineering and Design Cost: C_{aed_r}

3.2 Development Support and Testing Cost: C_{dst_r}

3.3 Flight Test Airplanes Cost: C_{fta_r}

3.4 Flight Test Operations Cost: C_{fto_r}

3.5 Test and Simulation Facilities Cost: C_{tsf_r}

3.6 RDTE Profit: C_{pro_r}

3.7 Cost to finance the RDTE phases: C_{fin_r}

The total RDTE cost for a new airplane program may be estimated from:

$$C_{RDTE} = C_{aed_r} + C_{dst_r} + C_{fta_r} + C_{fto_r} + C_{tsf_r} +$$

$$+ C_{pro_r} + C_{fin_r} \qquad (3.1)$$

Expressions for estimating these seven categories of RDTE cost are presented in Sections 3.1 through 3.7. An example application is given in Section 3.8.

Not all airplane programs are aimed at eventual production. Some are started for reasons of developing and/or demonstrating some aspect of advanced technology: the X-29 and X-31 programs are typical examples. A method for estimating the cost of such 'experimental' or 'prototyping' programs, together with an application is given in Section 3.9.

3.1 AIRFRAME ENGINEERING AND DESIGN COST: C_{aed_r}

Airframe engineering and design activities typically break down as follows:

1. Planning, conceptual design and associated cost studies (Phase 1 activities of Figures 2.1 and 2.3).

2. Preliminary design and system integration studies, including any associated cost studies (Phase 2 activities of Figures 2.2 and 2.3).

3. Engineering for windtunnel models, mock-ups and engine tests.

4. Design of windtunnel models and mock-ups.

   Note: Figure 3.1 records the number of windtunnel test hours needed in the past for several airplane programs.

5. Design and construction of dedicated test facilities, conduct of developmental tests and static tests including systems tests.

   Note: Items 3-5 are normally part of Phases 1 and 2 of Figures 2.1 through 2.3.

6. Detail design and development (This is represented by Phase 3 in Figure 2.3).

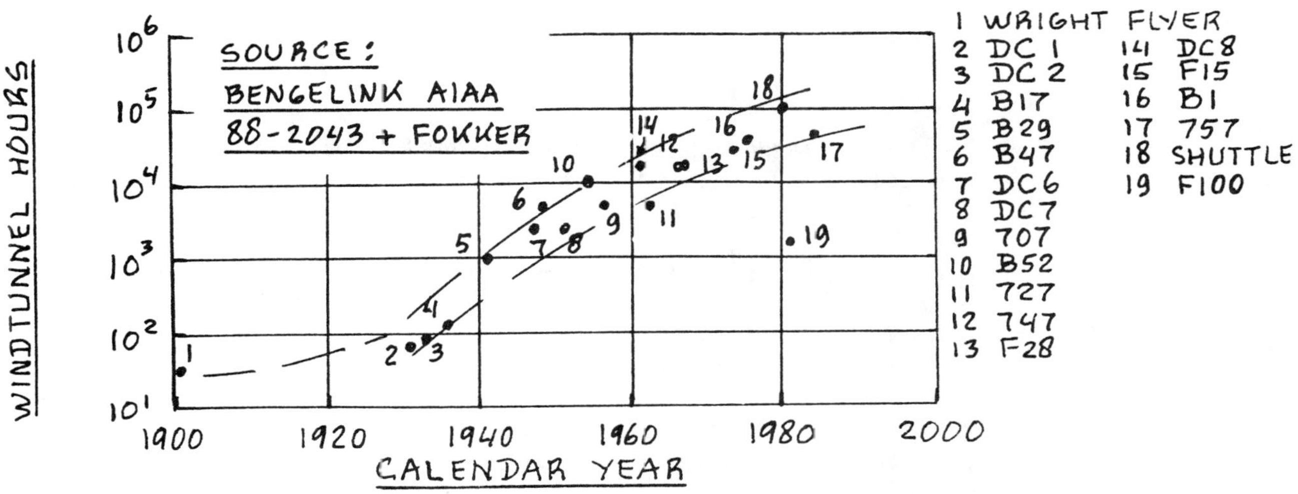

Figure 3.1 Windtunnel Hours Required in Typical Airplane Programs

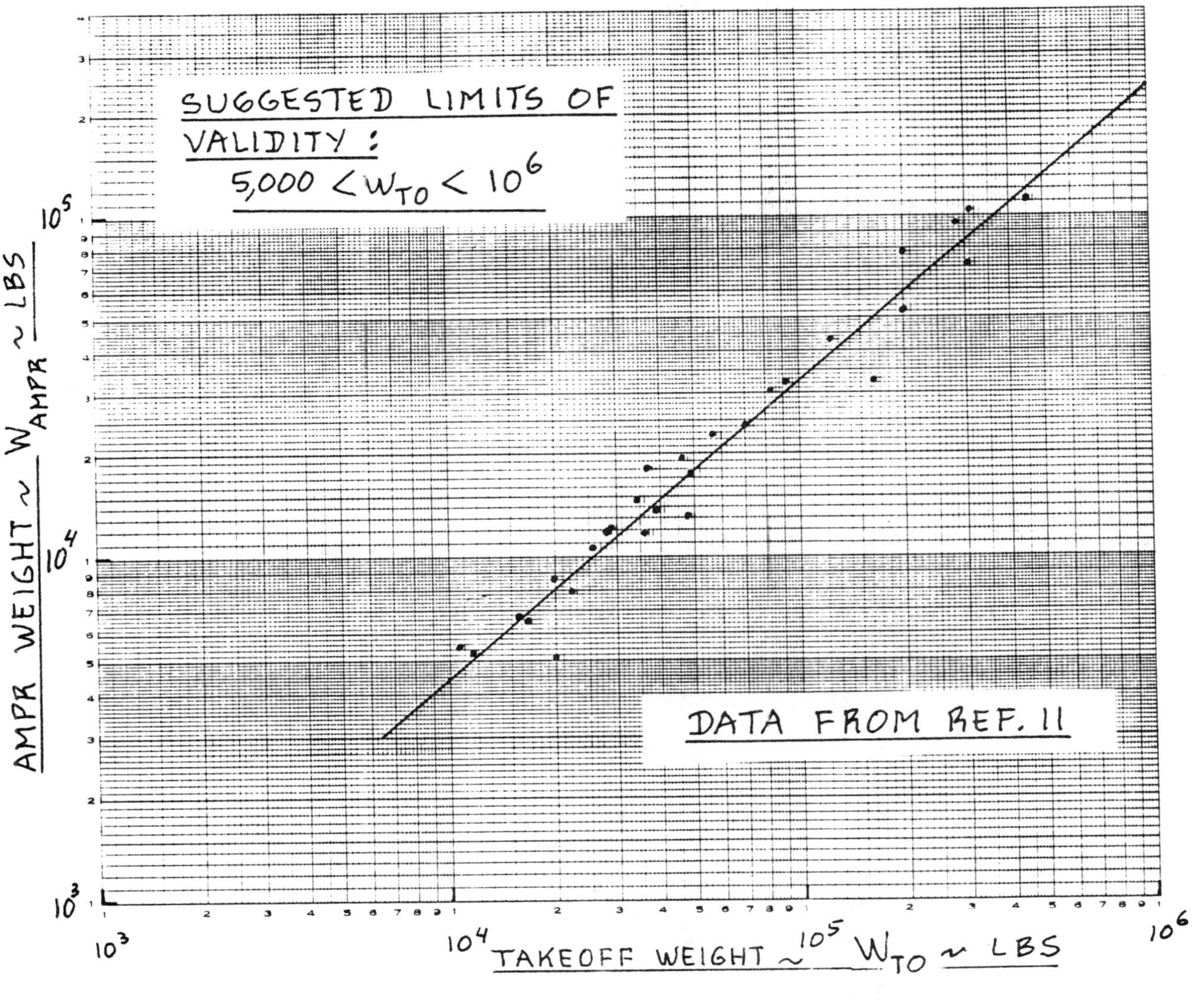

Figure 3.2 Data and Trend Line of AMPR Weight as Related to Takeoff Weight

7. Release and maintenance of drawings and specifications. It is of interest to mention that the first Boeing 747 required 75,000 drawings.

   Note: During the last ten years most companies have switched from 'hand-drafting' to CAD (Computer Aided Drafting). The effect of this on airframe engineering and design cost, C_{aed_r} is to reduce this cost. By how much, varies from one manufacturer to another.

8. Liaison with manufacturing and with vendors.

9. Incorporation and analysis of design changes.

10. Development of specifications for materials, processes and for items purchased from vendors.

11. Analysis of reliability.

12. Analysis of maintainability and accessibility.

    Note: items 6-12 are normally part of Phase 3 of Figure 2.3.

The total engineering manhours required to complete Phases 1-3 of Figure 2.3 may be estimated from:

$$MHR_{aed_r} = \quad (3.2)$$

$$0.0396(W_{ampr})^{0.791}(V_{max})^{1.526}(N_{rdte})^{0.183}(F_{diff})(F_{cad})$$

The airframe engineering and design cost associated with Phases 1-3 of Figure 2.3 can be computed from:

$$C_{aed_r} = (MHR_{aed_r})(R_{e_r}) \quad (3.3)$$

The various terms in Equations (3.2) and (3.3) are defined as follows:

W_{ampr} is the so-called AMPR weight of the airplane. AMPR stands for: Aeronautical Manufacturers Planning Report. This weight is also referred to as the DCPR weight, where DCPR stands for: Defense Contractors Planning Report.
The AMPR weight is defined by:

$$W_{ampr} = W_E - (\text{SUM } W_i)_{i=1}^{i=11} \quad (3.4)$$

with:

W_E = airplane empty weight. A preliminary method for estimating airplane empty weight is given in Part I. A more detailed method for estimating empty weight and airplane component weights is given in Part V.

W_1 = weight of wheels, brakes, tires and tubes. Note: this is NOT the total landing gear weight, W_g of Part V.

W_2 = W_e, weight of engines (See Part V)

W_3 = W_{ess}, weight of starter (See Part V)

W_4 = weight of cooling fluid (Cooling fluid weight should be based on recommendations from the engine manufacturer)

W_5 = weight of rubber or nylon fuel cells (See Part V)

W_6 = W_{els}, weight of batteries, electrical power supply and electrical power conversion equipment (See Pt V)

W_7 = W_{iae}, weight of instruments, avionics and electronic equipment (See Pt V)

W_8 = W_{arm}, weight of armament and fire-control system (See Part V)

W_9 = W_{api}, weight of airconditioning units and fluid (See Part V)

W_{10} = W_{apu}, weight of APU (See Part V)

W_{11} = W_{tfo}, weight of trapped fuel and oil (See Parts I and V)

Many of the weight items 1-11 are often purchased from vendors. These items are NOT normally manufactured by the airframer.

IMPORTANT NOTES:

1. Lacking detailed knowledge of airplane component weights, an alternate (but less accurate) way of finding W_{ampr} is to use Figure 3.2 which translates into:

$$W_{ampr} = \text{invlog}\{0.1936 + 0.8645(\log W_{TO})\} \tag{3.5}$$

where: W_{TO} is the takeoff weight. Methods for determining airplane takeoff weight are given in Parts I and V.

2. The cost of 'engineering' weight items W_i into the airframe is included in Eqn(3.1) despite the fact that W_{ampr} excludes these items!

$V_{max} = V_C$ for civil airplanes (keas)

$= V_H$ for military airplanes (keas)

For a definition of these speeds, see Section 4.2 in Part V of this text.

N_{rdte} is the number of airplanes produced for the RDTE phase. This number includes flight test airplanes as well as airframes used for static testing.

For military programs use: $N_{RDTE} = 6 - 20$

For commercial programs use: $N_{RDTE} = 2 - 8$

F_{diff} is a judgement factor which accounts for the difficulty (i.e. complexity) of a new airplane program. It is suggested to use:

$F_{diff} = 1.0$ for programs involving fairly conventional airplanes. Airplanes of the 1970 era fall in this class of relative difficulty.

$F_{diff} = 1.5$ for programs involving moderately aggressive use of advanced technology.

F_{diff} = 2.0 for programs involving very aggressive use of advanced technology. Examples would be: the X-29 and a laminar flow transport.

NOTE: The reader should not try to 'lowball' this difficulty factor. As a general rule, airplane development tends to be more difficult than anticipated.

F_{cad} is a judgement factor which accounts for the effect of computer aided design capability on airframe engineering and design cost. The following values are suggested:

F_{cad} = 1.2 for manufacturers which are in a CAD learning mode.

F_{cad} = 1.0 for manufacturers which are using 'manual' drafting techniques.

F_{cad} = 0.8 for manufacturers which are experienced in the use of CAD.

R_{e_r} is the engineering dollar rate per hour charged for the airframe engineering activity. It is preferred to use currently hourly rates. However, such information is often not available. Lacking data on current rates, the rates shown in Figure 3.3 may be used. Engineering manhour rates vary with factors such as: local cost-of--living, number of engineers in the job market, union contracts etc.

NOTES: 1. The engineering dollar rate per hour includes the following cost items:

*direct engineering labor
*overhead
*general and administrative costs
*miscellaneous direct charges such as: overtime premiums, travel, per diem, miscellaneous taxes, etc.

2. For airplane programs with a significant security requirement (such as 'black' programs) the hourly rates should be assumed to be 25 percent higher! Security measures add a lot of extra cost to a program.

For future estimates it is suggested to use:

$$(R_{e_r})_{\text{then year}} = \qquad (3.6)$$

$$(R_{e_r})_{1989}\{(CEF_{\text{then year}})/(CEF_{1989})\}$$

where: $(R_{e_r})_{1989}$ is the engineering manhour rate for 1989 as found from Figure 3.3 or from actual data.

$\{(CEF_{\text{then year}})/(CEF_{1989})\}$ is the CEF-ratio as found from Fig.2.7.

<u>NOTE:</u> Equation (3.6), suitably modified, can also be used to adjust the manufacturing and tooling manhour rates of Section 3.3.

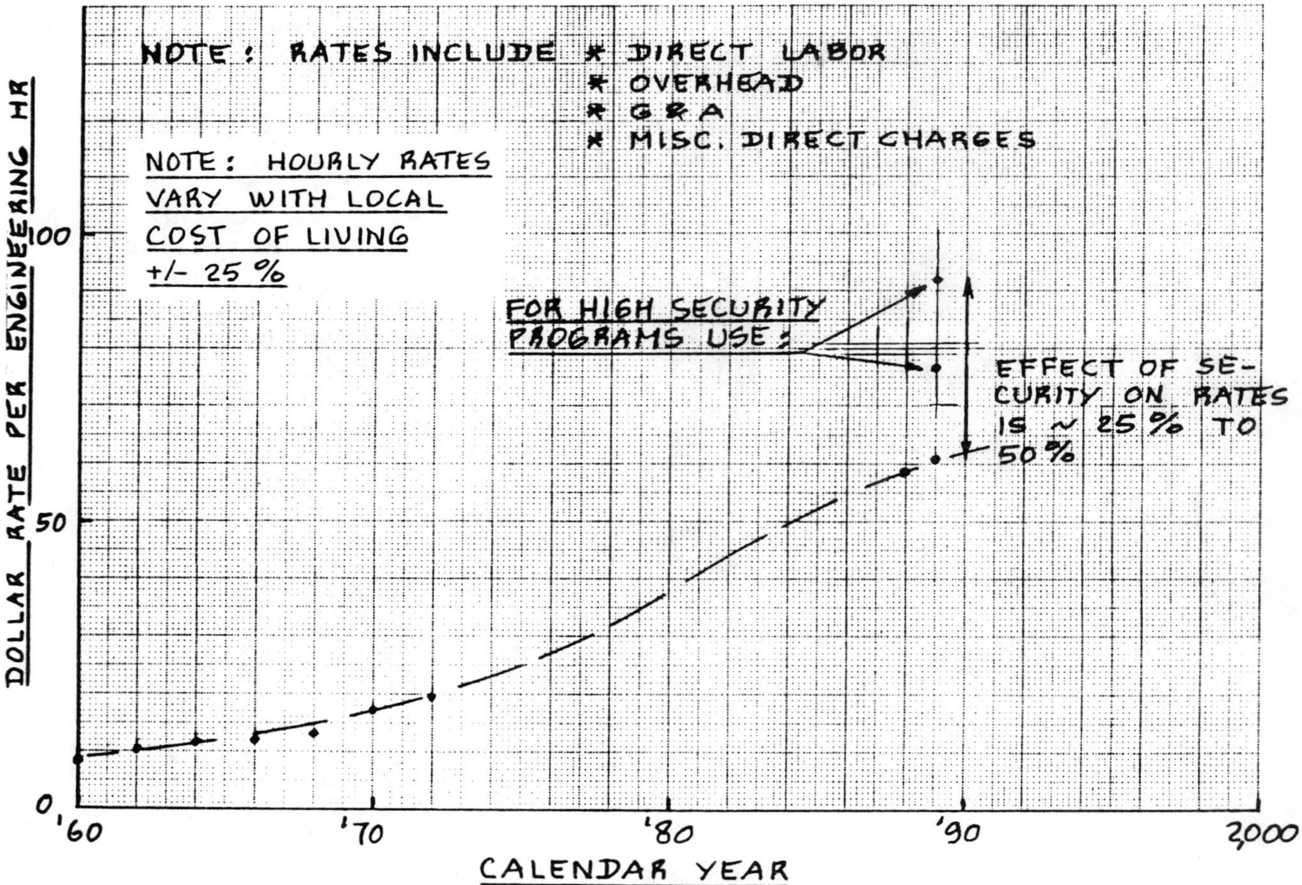

Figure 3.3 Variation of Engineering Manhour Rates with Calendar Time

3.2 DEVELOPMENT SUPPORT AND TESTING COST: C_{dst_r}

Typical activities which are responsible for this cost category are:

1. Windtunnel testing
2. Systems testing (for example the iron bird mentioned in Part IV)
3. Structural testing (strength, fatigue and materials testing)
4. Propulsion testing
5. Simulation for development support testing

The total cost of the above named activities is called C_{dst_r} and may be estimated from:

$$C_{dst_r} = \tag{3.7}$$

$$0.008325(W_{ampr})^{0.873}(V_{max})^{1.890}(N_{rdte})^{0.346}(CEF)(F_{diff})$$

where: CEF is the cost escalation factor of Figure 2.7. Note that the value of CEF is 1.0 for 1970.

All other quantities are defined in Section 3.1.

3.3 FLIGHT TEST AIRPLANES COST: C_{fta_r}

This cost category is normally broken down into the following cost components:

1. Cost of engine and avionics as acquired from vendors: $C_{(e + a)_r}$
2. Manufacturing labor cost: C_{man_r}
3. Manufacturing material cost: C_{mat_r}
4. Tooling cost: C_{tool_r}
5. Quality control cost: C_{qc_r}

To estimate C_{fta_r} it is suggested to use:

$$C_{fta_r} = C_{(e + a)_r} + C_{man_r} + C_{mat_r} + C_{tool_r} +$$

$$+ C_{qc_r} \tag{3.8}$$

where: $C_{(e + a)_r}$ is the cost of engines and avionics for the airplanes and airframes required for the RDTE phases. It can be found from:

$$C_{(e + a)_r} = \qquad (3.9)$$

$$(C_{e_r}N_e + C_{p_r}N_p + C_{avionics_r})(N_{rdte} - N_{st})$$

where: C_{e_r} is the cost per engine. This may be estimated with the help of Appendix B.

N_e is the number of engines per airplane.

C_{p_r} is the cost per propeller. This may be found with the help of Appendix B.

N_p is the number of propellers per airplane. Note that N_p is not necessarily the same as N_e!

$C_{avionics_r}$ is the cost of avionics equipment per airplane. This equipment normally consists of: radar(s), navigation equipment, automatic flight control systems and computers, maintenance and fault detection equipment.

For military airplanes the following may also be included: fire control equipment, missile guidance equipment and other items required for the mission of that particular airplane.

After preparing a listing of all avionics systems required for the mission of the airplane, an individual cost estimate of each such system may be obtained with the help of Appendix C. This appendix also contains a 'short-cut' method for estimating avionics cost.

N_{rdte} is defined in Section 3.1.

N_{st} is the number of static test airplanes. These are NOT normally equipped with engines, propellers or avionics systems.

C_{man_r} is the manufacturing cost of the flight test airplanes. It may be estimated from:

$$C_{man_r} = (MHR_{man_r})(R_{m_r}) \quad (3.10)$$

where: (MHR_{man_r}) is the number of manufacturing manhours required in Phases 1-3 of Figure 2.3. This may be found from:

$$MHR_{man_r} = \quad (3.11)$$

$$28.984(W_{ampr})^{0.740}(V_{max})^{0.543}(N_{rdte})^{0.524}(F_{diff})$$

R_{m_r} is the manufacturing labor rate in dollars per manhour: see Figure 3.4. Also: read the note after Eqn.(3.6).

Other factors are defined in Section 3.1.

C_{mat_r} is the cost of materials to manufacture the flight test airplanes. It is found from:

$$C_{mat_r} = 37.632(F_{mat})(W_{ampr})^{0.689}*$$

$$*(V_{max})^{0.624}(N_{rdte})^{0.792}(CEF) \quad (3.12)$$

where: F_{mat} is a correction factor which depends on the type of materials used in the construction of the airplane:

F_{mat} = 1.0 for airframes made primarily of conventional aluminum alloys

= 1.5 for stainless steel airframes

= 2.0 - 2.5 for airplanes where the primary structure is made with 'conventional' composite materials, Li/Al alloys and/or ARAL

= 3.0 for carbon composite airframes

In most cases, the reader will have to 'estimate' his own 'correction' factor F_{mat}.

All other quantities in Eqn (3.12) are defined in Section 3.1.

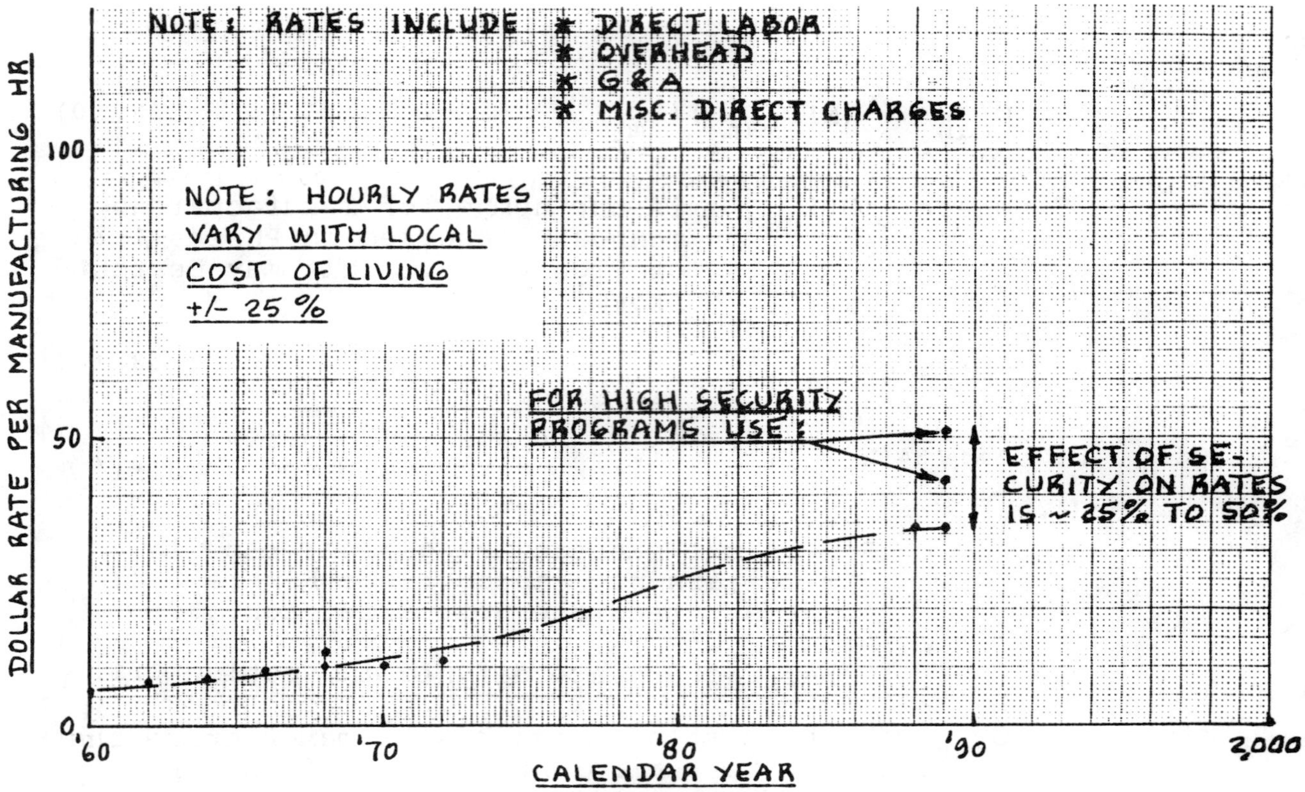

Figure 3.4 Variation of Manufacturing Manhour Rates with Calendar Time

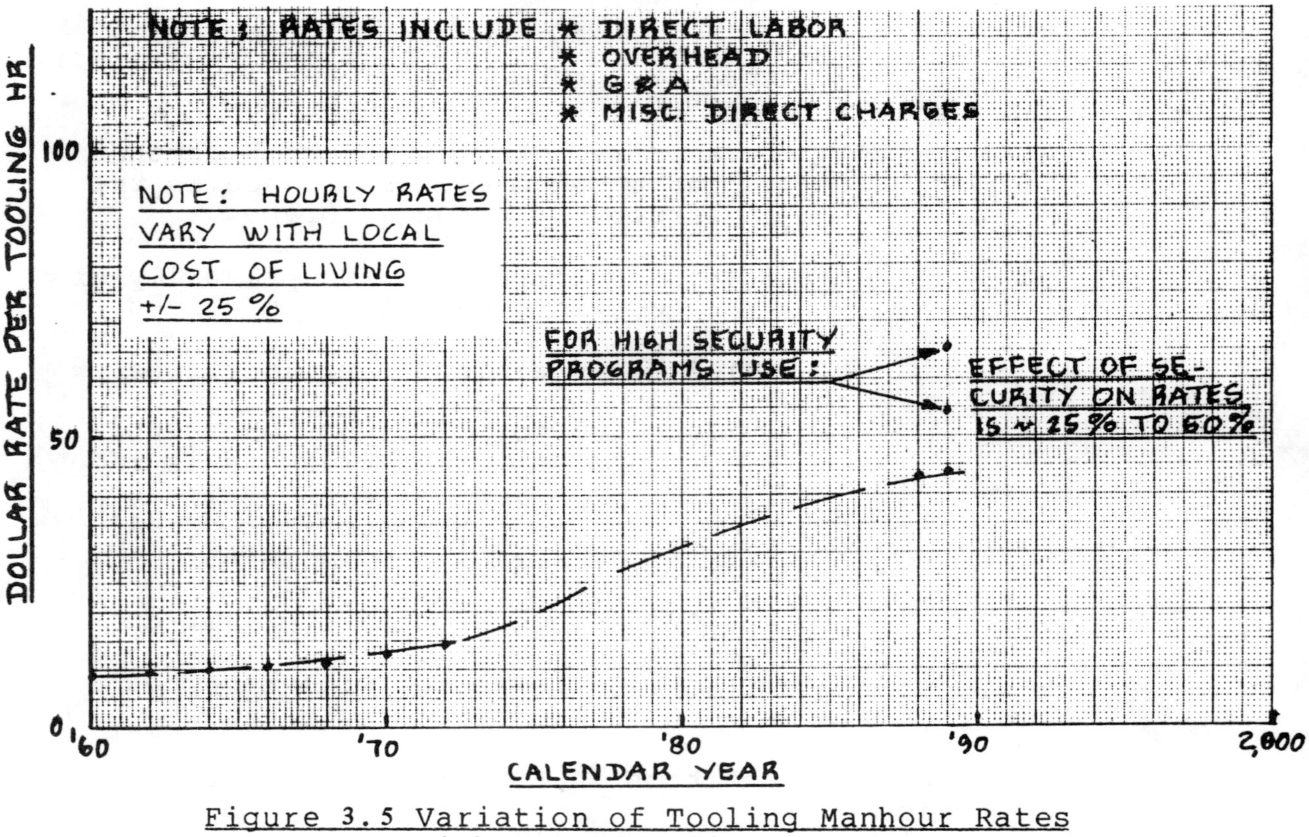

Figure 3.5 Variation of Tooling Manhour Rates with Calendar Time

C_{tool_r} is the tooling cost associated with the manufacturing of flight test airplanes. It can be estimated from:

$$C_{tool_r} = (MHR_{tool_r})(R_{t_r}) \qquad (3.13)$$

where: (MHR_{tool_r}) represents the tooling manhours needed in Phases 1-3 of Fig.2.3. This may be found from:

$$MHR_{tool_r} = 4.0127(W_{ampr})^{0.764}(V_{max})^{0.899}*$$

$$*(N_{rdte})^{0.178}(N_{r_r})^{0.066}(F_{diff}) \qquad (3.14)$$

R_{t_r} is the tooling labor rate in dollars per manhour which can be found from Figure 3.5. Also: read the comment after Eqn.(3.6).

N_{r_r} is the RDTE production rate in units per month. For RDTE a typical rate is 0.33 units per month.

Other factors are defined in Section 3.1.

C_{qc_r} is the quality control cost associated with manufacturing of the flight test airplanes. It may be estimated from:

$$C_{qc_r} = 0.13(C_{man_r}) \qquad (3.15)$$

where: (C_{man_r}) is found from Eqn.(3.10)

NOTE: The manufacturing and tooling manhour rates depend on the same factors listed for the engineering manhour rates on page 27. In addition, if the manufacturing and tooling processes involve toxic materials, special facilities may be required. This would tend to increase the overhead component of the hourly rates. The designer must somehow account for this in any cost estimates.

3.4 FLIGHT TEST OPERATIONS COST: C_{fto_r}

The following activities are normally responsible for this cost category:

1. Flight testing of flight test airplanes. As an example, the Boeing 747 required 1,400 flight test hours before reaching its first 'type' certification.

2. Simulation activities associated with flight testing. This type of activity becomes very significant in the development of inherently unstable airplanes.

3. If the 'observables' of an airplane are important a significant amount of flight test time will have to be dedicated to the determination of the various 'signatures' of the airplane such as:

   * radar cross section

   * infrared signature

   * acoustic signature

   * optical signature

The total cost for these activities may be estimated from:

$$C_{fto_r} = 0.001244(W_{ampr})^{1.160}(V_{max})^{1.371} *$$

$$* (N_{rdte} - N_{st})^{1.281}(CEF)(F_{diff})(F_{obs}) \qquad (3.16)$$

where: F_{obs} is a factor which depends on the importance of having 'low' observables. Based on the author's judgement it is suggested to use:

For commercial airplanes and for military airplanes without a 'stealth' requirement: $F_{obs} = 1.0$

For 'stealthy' airplanes: $F_{obs} = 3.0$

All other terms are as previously defined.

3.5 TEST AND SIMULATION FACILITIES COST: C_{tsf_r}

In many airplane programs it will be found necessary to build new, dedicated test facilities. For example, in the case of a fly-by-wire, inherently unstable airplane a simulation facility will be required which couples the flight control software to the actual flight control hardware and which includes a sophisticated moving base cab with realistic displays. In addition, extensive software validation tests may be required.

There are not sufficient data in the literature to predict this type of cost. It is the responsibility of the designer to understand the facilities available to him during the development of the airplane. If, because of the special nature of the airplane new facilities must be constructed, an estimate of these costs must be made with the help of facilities people.

For preliminary cost estimating purposes it is suggested to use:

$$C_{tsf_r} = (F_{tsf})(C_{RDTE}) \tag{3.17}$$

where: C_{RDTE} is found from Eqn.(3.1)

F_{tsf} is a cost adjustment factor which depends on judgement. The author suggests to use:

F_{tsf} = 0.00 if no extra facilities are required

= 0.20 if extensive test and simulation facilities are required. The X-29, ATF and B-2 programs are examples of this.

3.6 RDTE PROFIT: C_{pro_r}

In most cases a privately held enterprise will want to make a profit on RDTE activities. The total cost of RDTE activities can be a very substantial amount and it can also take many years. The profit cost category may be estimated from:

$$C_{pro_r} = (F_{pro_r})(C_{RDTE}) \tag{3.18}$$

where: F_{pro_r} = 0.10 for a suggested profit of 10 percent.

C_{RDTE} is found from Equation (3.1).

<u>NOTE:</u> The actual profits vary significantly with politics (military), with market conditions (civilian) and with management strategy. Table 2.1 presents examples of profits made by various enterprises during the 1985-1988 time period. The reader should note that these profit figures are for the corporate entities listed. The profit made on a specific RDTE program may be more or it may be less. Many military programs have a fixed profit clause built in.

3.7 COST TO FINANCE THE RDTE PHASES: C_{fin_r}

In many instances a manufacturer will borrow money to finance the RDTE phases depicted in Figure 2.3. Borrowing money in turn costs money. However, even if internally generated cash is used for RDTE financing, there are costs associated with this: the money used could have earned interest!

Methods for estimating these financing costs are judged to be beyond the scope of this text. To estimate financing costs the calendar time element associated with an airplane program becomes of critical importance. A very general discussion of the calendar time factor is included in Chapter 9.

For a fundamental discussion of the cost of money the reader is referred to Reference 8, Section 2.4.4.

Lacking better information on the cost of financing the RDTE phase it is suggested to use:

$$C_{fin_r} = (F_{fin_r})(C_{RDTE}) \qquad (3.19)$$

where: F_{fin_r} = 0.1 to 0.2, depending on the interest rates which are available.

C_{RDTE} is found from Equation (3.1).

3.8 EXAMPLE APPLICATION OF ESTIMATING RDTE COST

In this section an example application of the method for estimating RDTE costs is given. The example applies to the Ourania jet transport of Parts I, II and V of this text. Assume that it is required to estimate RDTE costs for the Ourania program for the year 1992.

The material is organized as follows:

3.8.1 RDTE Cost Input Data
3.8.2 RDTE Cost Calculations

3.8.1 RDTE Cost Input Data

The following input data are required to estimate RDTE costs with the method of Sections 3.1 through 3.7.

W_{TO} = 123,683 lbs from p.11, Part V. This assumes that the airframe will be made mostly from Li/Al alloys.

The AMPR weight, W_{ampr} is determined from Eqn.(3.5):

$$W_{ampr} = \text{invlog}\{0.1936 + 0.8645(\log 123{,}683)\} =$$

$$= 39{,}437 \text{ lbs}$$

$V_{max} = V_C$ = 295 kts from p.43, Part V. Note that this speed is in KEAS, at sea level!

N_{rdte} = 5, consisting of 3 flying prototypes and 2 airframes for structural testing. Because of the decision to use Li/Al alloys, it is prudent to perform a detailed structural evaluation! One airframe will be used for static testing (which includes watertank pressurization testing of the fuselage), the other for dynamic testing to determine the fatigue life. Therefore:

$N_{st} = 2$

CEF = 3.10, extrapolated from Figure 2.7 for 1992.

F_{diff} = 1.8 from page 26/27. This is based on the judgement that the combined use of Li/Al alloys and relaxed static stability (See Part II, p.272) represents a fairly aggressive use of new technology.

F_{cad} = 0.90 from page 27. This assumes moderate experience in the use of computer aided design methods.

N_e = 2 (number of engines per airplane: Pt I, p.55)

C_{e_r} = the cost per engine. This depends on the takeoff thrust of the engine, according to Appendix B. From Part I, p.184: T_{TO} = 47,625 lbs., therefore, the design takeoff thrust per engine is: 47,625/2 = 23,813 lbs.

The engine cost to the manufacturer is the price, EP, charged by the engine manufacturer. From Eqn(B9):

$$EP_{1989} = \text{invlog}\{2.3044 + 0.8858(\log 23,813)\} =$$

$$= \text{USD } 1,518,385$$

This price must be modified with the 1992 to 1989 CEF-ratio according to Eqn (B1). The CEF-ratio is found from Figure 2.7. The result is:

$$C_{e_r} = (3.10/3.02)1,518,385 = \text{USD } 1,558,607$$

N_p = 0 since the Ourania has no propellers.

$C_{avionics_r}$ = the cost of avionics per RDTE airplane.

If a detailed list of required avionics equipment were available, a reliable estimate of this cost contribution could be made. As a shortcut, it will be assumed that the installed avionics cost for the RDTE airplanes will be roughly 10 percent of the airplane market price, AMP_{1992}. The latter is found from Eqns (A14) and (A1):

$$AMP_{1992} = \{(3.10)/(3.02)\}[\text{invlog}\{3.3191 +$$

$$+ 0.8043(\log 123,683)\}] = \text{USD } 26,680,852$$

Therefore, ten percent of this amounts to roughly:

$$C_{avionics_r} = \text{USD } 2,670,000 \text{ per RDTE airplane}$$

R_{e_r} = (3.10/3.02)62.00 = 63.64 USD/mhr for the engineering manhour rate in 1992 as estimated with the help of Eqn.(3.6) and Figure 3.3.

CEF is the cost escalation factor obtained from Figure 2.7. For 1992:

$CEF_{1992} = 3.10$

$R_{m_r} = (3.10/3.02)34.44 = 35.35$ USD/mhr for the manufacturing manhour rate in 1992 as estimated with the help of Eqn.(3.6) and Figure 3.4.

$F_{mat} = 2.3$ from page 31. This is because of the decision to use Li/Al alloys.

$R_{t_r} = (3.10/3.02)43.06 = 44.20$ USD/mhr for the tooling manhour rate in 1992 as estimated with the help of Eqn.(3.6) and Figure 3.5.

$N_{r_r} = 0.33$ units per month (assumed).

$F_{obs} = 1.0$, according to page 34.

F_{tsf} = an adjustment factor which depends on the extra test and simulation facilities required for a new airplane program. Because of the relaxed stability decision made on the Ourania program, page 35 suggests:

$F_{tsf} = 0.2$

$F_{pro_r} = 0.10$, assumed from p.35.

$F_{fin_r} = 0.10$, assumed from p.36.

This completes the list of input requirements for estimating RDTE cost. The RDTE cost calculations are presented in sub-section 3.8.2.

3.8.2 RDTE Cost Calculations

With the input data from sub-section 3.8.1 the following RDTE cost component estimates can be made:

From Eqns (3.3) and (3.2):

$$C_{aed_r} = 0.0396(39{,}437)^{0.791}(295)^{1.526}(5)^{0.183} *$$
$$*(1.8)(0.9)(63.64) = \text{USD } 139{,}046{,}557$$

From Eqn (3.7):

$$C_{dst_r} = 0.008325(39,437)^{0.873}(295)^{1.890}(5)^{0.346} * (3.1)(1.8) = \text{USD } 38,819,483$$

From Eqn (3.9):

$$C_{(e+a)_r} = \{(1,558,607)(2) + (2,670,000)\} * (3) = \text{USD } 17,361,642$$

From Eqns (3.10) and (3.11):

$$C_{man_r} = 28.984(39,437)^{0.740}(295)^{0.543}(5)^{0.524} * (1.8)(35.35) = \text{USD } 236,680,646$$

From Eqn (3.12):

$$C_{mat_r} = 37.632(2.3)(39,437)^{0.689}(295)^{0.624}(5)^{0.792} * (3.1) = \text{USD } 48,975,376$$

From Eqns (3.13) and (3.14):

$$C_{tool_r} = 4.0127(39,437)^{0.764}(295)^{0.899}(5)^{0.178} * (0.33)^{0.066}(1.8)(44.20) = \text{USD } 213,015,201$$

From Eqn (3.15):

$$C_{qc_r} = 0.13(236,680,646) = \text{USD } 30,768,484$$

From Eqn (3.8):

$$C_{fta_r} = 17,361,642 + 236,680,646 + 48,975,376 + 213,015,201 + 30,768,484 = \text{USD } 546,801,349$$

From Eqn (3.16):

$$C_{fto_r} = 0.001244(39,437)^{1.160}(295)^{1.371}(3)^{1.281} * (3.1)(1.8) = \text{USD } 14,791,812$$

From Eqn (3.17):

$$C_{tsf_r} = 0.20(C_{RDTE})$$

From Eqn(3.18):

$C_{pro_r} = 0.10(C_{RDTE})$

From Eqn.(3.19):

$C_{fin_r} = 0.10(C_{RDTE})$

From Eqn (3.1):

$C_{RDTE} = 139{,}046{,}557 + 38{,}819{,}483 + 546{,}801{,}349 +$

$+ 14{,}791{,}812 + 0.20(C_{RDTE}) + 0.10(C_{RDTE}) + 0.10(C_{RDTE})$

from which it follows that:

C_{RDTE} = USD 739,459,201/(0.6) = USD 1,232,432,002 or rounding off to next highest million:

C_{RDTE} = USD 1,232,000,000

This represents the estimated 1992 cost of the entire RDTE part of the Ourania jet transport program.

Figure 3.6 shows the contribution of this RDTE cost to the unit cost of the Ourania transport. Note, that this type of RDTE program makes financial sense only if more than, say 250 transports can be expected to be sold.

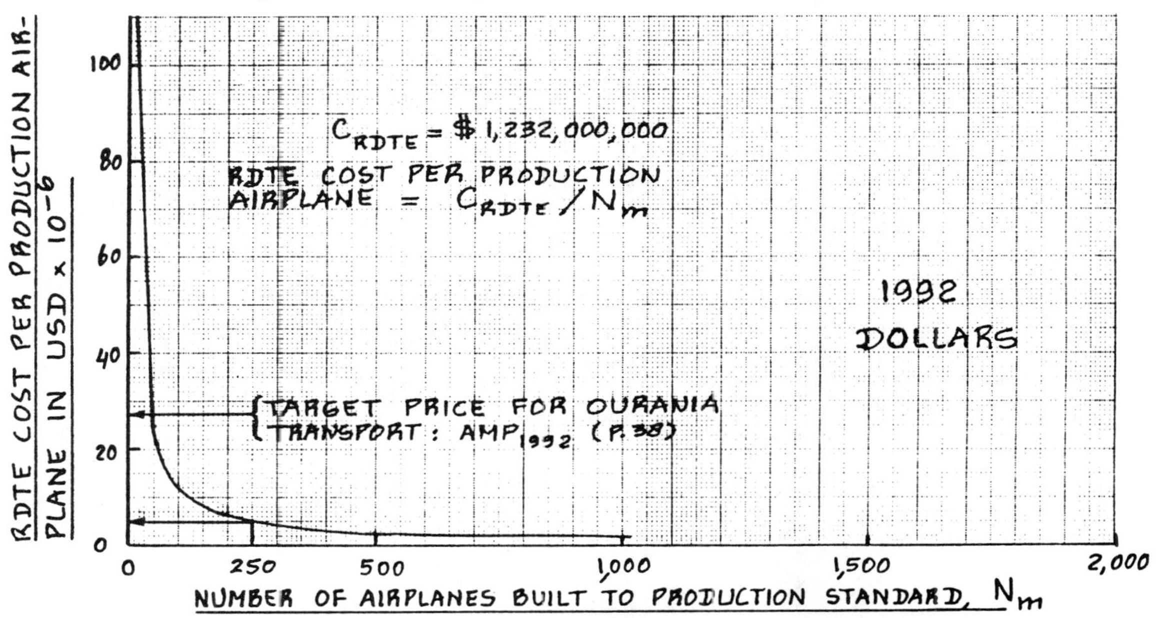

Figure 3.6 Effect of Number of Airplanes Produced on the RDTE Cost Contribution per Airplane

3.9 METHOD FOR ESTIMATING PROTOTYPE PROGRAM COST WITH AN EXAMPLE APPLICATION

The purpose of this section is to present a method for estimating the cost of developing, manufacturing and flight testing prototypes. Prototypes are usually built rather quickly and without provisions required for a follow-on production program.

Nevertheless, many prototype programs do lead to a follow-on production program. Examples are the F-104, T-38, B-47, B-52, C-130 and KC-135 programs.

In many cases a prototype program is conducted mainly for the purpose of evaluating certain technologies: the NASA X-planes, such as the X-29 and the X-31 are good examples of this.

The method presented here should be applied ONLY to prototype programs where 1-4 airplanes are constructed. If more than 4 prototypes are to be constructed the RDTE method of Chapter 3 should be used instead. The method is based on Reference 23.

The prototype program cost may be estimated from:

$$C_{PROT} = (1115.4)(10^3)(W_{ampr})^{0.35}(N_{prot})^{0.99} *$$

$$* (CEF_{then\ year})/(CEF_{1973}) \quad (3.20)$$

where: W_{ampr} is found from Eqn.(3.4) or from Eqn.(3.5).

N_{prot} is the number of prototypes to be built. This number should not exceed 4. When it does, the RDTE method of Chapter 3 should be used.

$CEF_{then\ year}$ is the Cost Escalation Factor (CEF) of Fig. 2.7 for a given 'then year'.

CEF_{1973} is the CEF for 1973 as obtained from Figure 2.7: $CEF_{1973} = 1.14$.

To illustrate the use of the method it will now be applied to the Grumman X-29 prototype program. A three-view of the X-29 is shown in Figure 3.7. The X-29 program was funded by DARPA (Defense Advanced Research Projects Agency). Grumman was awarded the X-29 contract in

December of 1981 for a total of USD 80,000,000. An anonymous source has indicated that Grumman invested a total of USD 20,000,000 of its own money on top of the DARPA funding. The total cost of the X-29 program (until delivery to NASA) was thus estimated at USD 100,000,000.

The following data are found in the 1985-1986 issue of Jane's All the World Aircraft:

$W_{TO} = 17,800$ lbs

$N_{prot} = 2$

Using Eqn.(3.5) it is found that:

$W_{ampr} = 7,380$ lbs

The 1982 CEF value is found from Figure 2.7 as:

$CEF_{1982} = 2.5$

With Eqn.(3.20): $C_{PROT} =$

$(1115.4)(10^3)(7,380)^{0.35}(2)^{0.99}(2.5)/(1.14) =$

$=$ USD 109,724,907

This is in good agreement with the actual cost which was quoted before.

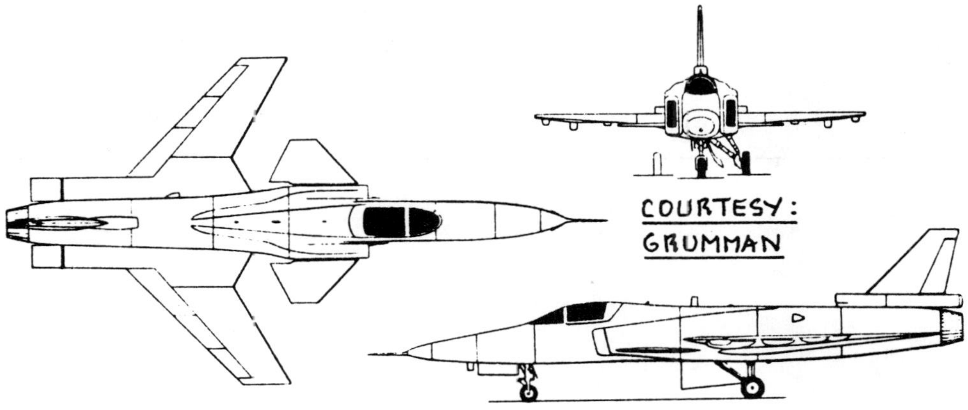

Figure 3.7 Three-view of the Grumman X-29

SIAI-MARCHETTI S.211 AVIONIC SYSTEM

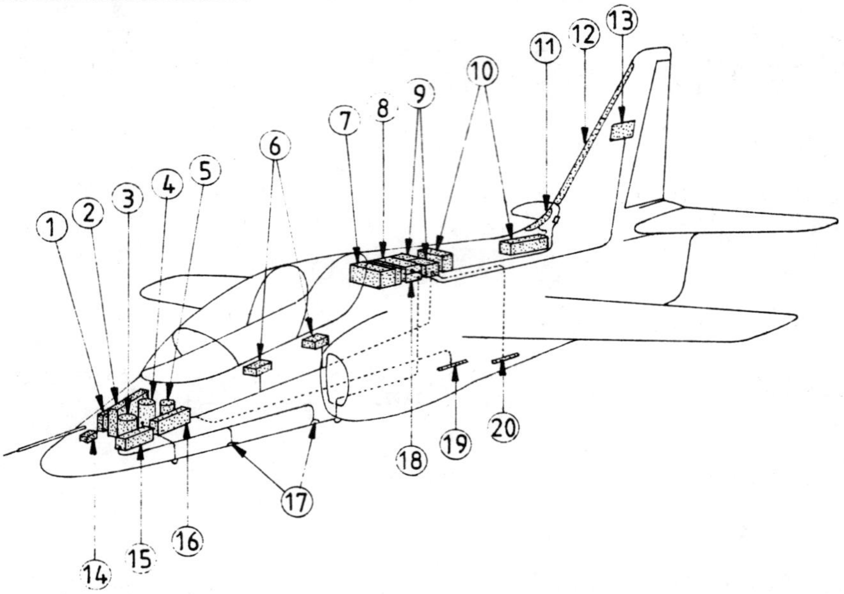

KEY

1. FLIGHT DIRECTOR COMPUTER
2. ADF
3. VERTICAL GYRO
4. DIRECTIONAL GYRO
5. RATE GYRO
6. AVIONIC CONTROL UNIT
7. TACAN
8. DME
9. V/UHF
10. HF
11. UHF ANTENNA
12. V/UHF ANTENNA
13. VOR/ILS ANTENNA
14. ALTITUDE SENSOR
15. R/ALTIMETER
16. IFF
17. R/ALTIMETER ANTENNA
18. VOR/ILS
19. ADF LOOP ANTENNA
20. MARKER BEACON ANTENNA

4. METHOD FOR ESTIMATING MANUFACTURING AND ACQUISITION COST

The purpose of this chapter is to present a method for estimating manufacturing cost: C_{MAN} and acquisition cost: C_{ACQ} for airplane programs. These costs are incurred during Phase 4 of an airplane program as shown in Figure 2.3. The acquisition cost represents the second cost source of the life cycle cost defined in Eqn.(2.3).

The following definitions are important:

<u>Definition 1:</u> The number of airplanes produced during an airplane program is $N_{program}$.

<u>Definition 2:</u> The number of airplanes produced to production standard during an airplane program is N_m.

<u>Definition 3:</u> The number of airplanes produced during the RDTE phases of an airplane program is: N_{rdte}.

Consequently: $$N_{program} = N_m + N_{rdte} \qquad (4.1)$$

Table 4.1 shows examples of the total number of airplanes built in several airplane programs.

<u>Definition 4:</u> The total manufacturing cost associated with an airplane program is: C_{MAN}.

<u>Definition 5:</u> The total acquisition cost associated with an airplane program is: C_{ACQ}.

The difference between acquisition cost and manufacturing cost is the profit made by the manufacturer: C_{PRO}, so that:

$$C_{ACQ} = C_{MAN} + C_{PRO} \qquad (4.2)$$

Again, the costs mentioned sofar represent program costs. The price paid by the user of an airplane (which is his acquisition cost) depends on a number of factors:

Table 4.1 Examples of Airplane Program Production Runs

Note: No. Produced equals $N_{program}$ of Eqn.(4.1).

Fighters

Type	No. Produced
Gen.Dyn. F-111	563
Gen.Dyn. F-16	>3,000
Gloster Meteor	3,545
Gloster Javelin	435
Grumman F9F2-5	1,325
Grumman F9F6-8	1,985
Grumman F11F	201
Grumman F14	> 900
Lockheed F-94	387
Lockheed F-80	1,732
Lockheed T-33	5,691
Lockheed F-104	2,578
McDonnell F-4	>5,000
McDD F-15	>2,000
McDD F-18	>1,500
SAAB JA37	329
SAAB J35A	604

Commercial Transports

Type	No. Produced
Douglas DC-7	338
Douglas DC-8	556
McDD DC-9, MD-80+	>1,000
Boeing 707	870
Boeing 727	1,834
Boeing 737	>2,600
Boeing 747	> 750

General Aviation

Type	No. Produced
Beech Duke	471
Beech Duchess	437
Beech Model 18	8,000
Cessna 140	5,429
Cessna 150	23,839
Cessna 310	5,447
Cessna 421	> 1,900
Cessna 500/501	315

Mil. Trainers/Liaison

Type	No. Produced
Cessna L-19	3,399
Beech T-34B	423
Beech T-34C	> 300
Cessna T-37	1,838
Fokker S-14	20
NAA T2C	521
Northrop T-38	1,189

Mil. Transports/Bombers

Type	No. Produced
Northrop C-125	23
Fairchild C-132	300
Lockheed C-130	>2,000
Lockheed C-141	285
Boeing B-47	2,004
Boeing B-52	742
Boeing KC-135	760
Douglas C-133	50

Regional Transports

Type	No. Produced
Beech M-99	234
Beech 1900	> 100
Fairchild Merlin/ Metroliner	> 800
Embraer-110	> 500
Embraer-120	> 150
DHC-6	> 832
DHC-7	> 115
DHC-8	> 137
Dornier 228	> 133
Fokker F-27	786
Fokker F-28	241
Fokker F-100	> 300
BAC-146	> 100
Pilatus B/N-2B	>1,077
Shorts 330	> 165
Shorts Skyvan	150

1. the total number of airplanes built by the manufacturer: $N_{program}$

2. the number of airplanes acquired: N_{acq}. In commercial programs this number tends to be only a fraction of $N_{program}$. For example, an airplane program may see 1,000 transports built. One airline customer may buy (acquire) 60 airplanes.

   In military programs (discounting foreign sales) N_{acq} is usually the same as N_m.

3. the manufacturers profit which can be negotiated: for large fleet buys a manufacturer often offers a lower profit to enhance his market share.

4. the cost of the RDTE program: C_{RDTE} (See Ch. 3)

An estimate of the unit price per airplane can be obtained from:

$$AEP = \{(C_{MAN} + C_{PRO} + C_{RDTE})/N_m\} \qquad (4.3)$$

This assumes that the RDTE airplanes are not sold during the program. The effect of RDTE cost on the unit cost of an airplane was shown in Figure 3.6.

Another assumption made in Eqn.(4.3) is that no spare parts are bought by the user. In most cases commercial and military customers will want to buy a certain number of spare parts. The cost of spare parts is discussed in Chapters 5 and 6.

Having stated these fundamentals, the remainder of this chapter deals with the development of a method for estimating manufacturing cost: C_{MAN} and manufacturer's profit: C_{PRO}.

The method is a modified version of the methods of References 11 and 14. It should be used for preliminary cost estimation purposes only. The method can be applied to military as well as to commercial airplane programs.

The total airplane program manufacturing cost can be broken down into the following cost categories:

4.1 Airframe Engineering and Design Cost: C_{aed_m}

4.2 Airplane Production Cost: C_{apc_m}

4.3 Production Flight Test Operations Cost: C_{fto_m}

4.4 Cost of financing the manufacturing program: C_{fin_m}

The total manufacturing cost may be estimated from:

$$C_{MAN} = C_{aed_m} + C_{apc_m} + C_{fto_m} + C_{fin_m} \tag{4.4}$$

Expressions for estimating the first three cost categories of the manufacturing cost are presented in Sections 4.1 through 4.3. The cost of financing the manufacturing part of the program is briefly discussed in Section 4.4. In Section 4.5 an equation is presented for estimating the manufacturer's profit. An example application is developed in Section 4.6. A discussion of the effect of several design and production parameters on the estimated price of an airplane is given in Section 4.7.

4.1 AIRFRAME ENGINEERING AND DESIGN COST: C_{aed_m}

Airframe engineering and design typically consists of the following activities:

1. Engineering design work necessitated by problems uncovered during RDTE phases 1-3 (Figure 2.3).
2. Design studies and integration studies associated with special customer requests.
3. Sustaining engineering associated with:
   * errors made in the manufacturing process
   * changes made in manufacturing of components
4. Release and maintenance of drawings and specifications.

   NOTE: during the last ten years most companies have switched from 'hand-drafting' to CAD (Computer Aided Design). The effect of this on C_{aed_m} is to reduce these costs.

   By how much, varies from one manufacturer to another.
5. Liaison engineering with manufacturing and with vendors.
6. Incorporation and analysis of design changes.
7. Development of specifications for materials and processes.
8. Analysis of reliability.
9. Analysis of maintainability and accessibility.

The total (= program) cost of the above named activities is called: C_{aed_m} and may be estimated from:

$$C_{aed_m} = (MHR_{aed_{program}})(R_{e_m}) - (MHR_{aed_r})(R_{e_r}) \qquad (4.5a)$$

or, with Eqn.(3.3):

$$C_{aed_m} = (MHR_{aed_{program}})(R_{e_m}) - C_{aed_r} \qquad (4.5b)$$

where: $MHR_{aed_{program}}$ is the total amount of engineering manhours required for the entire airplane program, that is for <u>all</u> $N_{program}$ airplanes. These manhours may be estimated from:

$$MHR_{aed_{program}} = 0.0396(W_{ampr})^{0.791}(V_{max})^{1.526} *$$

$$*(N_{program})^{0.183}(F_{diff})(F_{cad}) \qquad (4.6)$$

C_{aed_r} is computed with Eqn.(3.3).

R_{e_m} is the engineering manhour rate in dollars per hour for the entire airplane program. It may be assumed equal to R_{e_r} which in turn is obtained from Figure 3.3.

$N_{program}$ is the total number of airplanes produced: See Eqn.(4.1).

Other symbols are defined in Section 3.1.

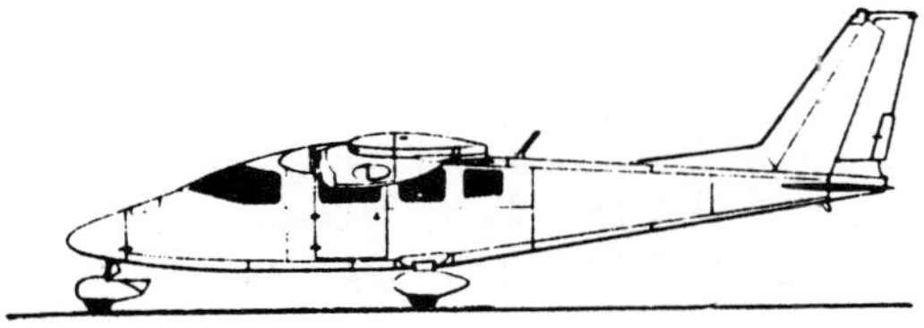

4.2 AIRPLANE PROGRAM PRODUCTION COST: C_{apc_m}

The airplane (program) production cost category, C_{apc_m}, normally consists of the following cost components:

1. Cost of engine and avionics as acquired from vendors: $C_{(e + a)_m}$
2. Cost of interior: C_{int_m}
3. Manufacturing labor cost: C_{man_m}
4. Manufacturing material cost: C_{mat_m}
5. Tooling cost: C_{tool_m}
6. Quality control cost: C_{qc_m}

The following equation is suggested for estimating airplane program production cost, C_{apc_m}:

$$C_{apc_m} = C_{(e + a)_m} + C_{int_m} + C_{man_m} + C_{mat_m} + C_{tool_m} + C_{qc_m} \tag{4.7}$$

where: $C_{(e + a)_m}$ is the cost of engines and avionics equipment as paid to vendors. This can be estimated from:

$$C_{(e + a)_m} = (C_{e_m} N_e + C_{p_m} N_p + C_{avionics_m})(N_m) \tag{4.8}$$

where: C_{e_m} is the cost per engine during the manufacturing phase. This cost can differ from the cost per engine incurred in the RDTE phases: C_{e_r}.

Lacking better information, the method of Appendix B may be used to estimate the cost of production engines.

N_e is the number of engines per airplane.

C_{p_m} is the cost per propeller during the manufacturing phase. This cost can differ from the cost per propeller incurred in the RDTE phases: C_{p_r}.

Lacking better information, the method of Appendix B may be used to estimate the cost of production propellers.

N_p is the number of propellers per airplane. Note that N_p may not be the same as N_e!

$C_{avionics_m}$ is the cost of avionics equipment per airplane. This cost may differ from the avionics cost associated with the RDTE airplanes. Lacking better information, the method of Appendix C may be used.

Note: in civilian airplanes the total cost of avionics typically is 5 to 15 percent of the airplane cost without avionics.

in military airplanes the total cost of avionics, depending on the mission of the airplane, may range from 10 to 100 percent of the airplane cost without avionics.

C_{int_m} is the cost of the airplane interior.

Note: The cost estimating methods of References 11 and 14 do not account for interior cost as a separate cost component. In military airplanes, for which these methods were primarily derived, this is acceptable because their interior cost is included in the statistical equations used for estimating cost. This includes items like passenger seats, ejection seats, insulating materials, etc.

The cost of interiors in commercial airplanes can be substantial and must be accounted for as a separate item. In commercial airplanes the cost of the interior includes items such as:

seats, partitions, cabin sidewall, ceiling and floor materials, galleys, lavatories, closets, storage bins, entertainment systems, lighting systems, insulating materials, etc.

The reader will have observed that interior cost was NOT accounted for in Eqn.(3.8) for the RDTE part of the airplane program. The reason is that flight test airplanes are normally NOT equipped with anything approaching a production interior.

To account for the interior cost component it is suggested to use:

$$C_{int_m} = (F_{int})(N_{pax})(N_m)*$$

$$*(CEF_{then\ year})/(CEF_{1990}) \qquad (4.9)$$

where: F_{int} is the interior cost factor expressed in USD/pax. The following values are suggested for calendar year 1990:

F_{int} = 0 USD/pax for military airplanes

= 500 USD/pax for light general aviation airplanes

= 1,000 USD/pax for regional transport airplanes

= 2,000 USD/pax for jet transports

= 3,000 USD/pax for business jets

N_{pax} is the number of passengers.

CEF values are obtained from Figure 2.7.

C_{man_m} is the labor cost incurred in manufacturing N_m airplanes to production standard. It may be estimated from:

$$C_{man_m} = \qquad (4.10a)$$

$$(MHR_{man_{program}})R_{m_m} - (MHR_{man_r})R_{m_r}$$

or, with Eqn.(3.10):

$$C_{man_m} = (MHR_{man_{program}})R_{m_m} - C_{man_r} \qquad (4.10b)$$

where: $MHR_{man_{program}}$ represents the total number of manhours required for the manufacturing of $N_{program}$ airplanes. It can be found from:

$$MHR_{man_{program}} = 28.984(W_{ampr})^{0.740}(V_{max})^{0.543} * (N_{program})^{0.524}(F_{diff}) \quad (4.11)$$

R_{m_m} is the manufacturing labor rate per hour for the entire program. This may be assumed to be equal to R_{m_r} which in turn is determined from Figure 3.4.

Other symbols are as defined before.

C_{mat_m} is the materials cost incurred while manufacturing N_m airplanes to production standard. It may be found from:

$$C_{mat_m} = C_{mat_{program}} - C_{mat_r} \quad (4.12)$$

where: $C_{mat_{program}}$ is the total materials cost associated with building $N_{program}$ airplanes. It follows from:

$$C_{mat_{program}} = 37.632F_{mat}(W_{ampr})^{0.689}(V_{max})^{0.624} * (N_{program})^{0.792}(CEF) \quad (4.13)$$

C_{mat_r} is found from Eqn.(3.12).

All other symbols are as defined before.

C_{tool_m} is the tooling cost to produce N_m airplanes. It may be estimated from:

$$C_{tool_m} = \qquad (4.14a)$$

$$(MHR_{tool_{program}})R_{t_m} - (MHR_{tool_r})R_{t_r}$$

or, with Eqn.(3.13):

$$C_{tool_m} = (MHR_{tool_{program}})R_{t_m} - C_{tool_r} \qquad (4.14b)$$

where: $MHR_{tool_{program}}$ is the total number of tooling manhours required to build $N_{program}$ airplanes.

It can be estimated from:

$$MHR_{tool_{program}} = 4.0127(W_{ampr})^{0.764}(V_{max})^{0.899} *$$

$$*(N_{program})^{0.178}(N_{r_m})^{0.066}(F_{diff}) \qquad (4.15)$$

R_{t_m} is the tooling labor rate in dollars per manhour as found from Figure 3.5. Note: R_{t_m} may differ from R_{t_r}. It is suggested to ignore this difference in preliminary cost estimations.

N_{r_m} is the airplane manufacturing rate in units per month.

Other symbols are as defined before.

C_{qc_m} is the quality control cost associated with building N_m airplanes. It may be found from:

$$C_{qc_m} = 0.13(C_{man_m}) \qquad (4.16)$$

where: C_{man_m} is found from Eqn.(4.10)

4.3 PRODUCTION FLIGHT TEST OPERATIONS COST: C_{fto_m}

This cost category is normally neglected in preliminary cost estimating calculations. Depending on the airplane program it may not be negligible. The following equation is suggested for estimating C_{fto_m}:

$$C_{fto_m} = N_m(C_{ops/hr})(t_{pft})(F_{ftoh}) \quad (4.17)$$

where: N_m is the number of airplane built to production standard.

$C_{ops/hr}$ is the airplane operating cost per hour as computed with the method of Chapter 5 (commercial airplanes) or Chapter 6 (military airplanes).

t_{pft} is the number of flight test hours flown by the manufacturer before airplane delivery to the customer. It is suggested to use:

t_{pft} = 2 hrs for general aviation airplanes
10 hrs for jet transports
20 hrs for military airplanes

F_{ftoh} is the overhead factor associated with the production flight test activities. Lacking actual overhead data it is suggested to use:

$F_{ftoh} = 4.0$

4.4 COST TO FINANCE THE MANUFACTURING PHASE: C_{fin_m}

Usually a manufacturer will borrow money to finance the manufacturing phase (Phase 4 in Figure 2.3) of an airplane program. Borrowing money in turn costs money. However, even if internally generated cash would be used for production financing, there are costs associated with this: the money used could have been earning interest!

Methods for estimating these financing costs are judged to be beyond the scope of this text. To estimate financing costs, the calendar time element associated with an airplane program becomes critically important. A discussion of the effect of calendar time on interest is also beyond the scope of this text.

For a fundamental discussion of the cost of money the reader should consult Reference 8.

For preliminary cost estimating purposes it is suggested to use:

$$C_{fin_m} = (F_{fin_m})(C_{MAN}) \tag{4.18}$$

where: F_{fin_m} = 0.1 to 0.2, depending on the interest rates which are available.

C_{MAN} is found from Equation (4.4).

4.5 PROFIT: C_{PRO}

As indicated in Chapter 2, privately held aeronautical enterprises will want to make a profit on airplane manufacturing activities. This profit, which is part of the cost of airplane acquisition (Eqn.4.2), may be estimated from:

$$C_{PRO} = (F_{pro_m})(C_{MAN}) \tag{4.19}$$

where: F_{pro_m} = 0.10 for a suggested profit of 10 percent.

C_{MAN} is found from Equation (4.4).

NOTE: Profits can vary significantly with politics (military) and with market conditions (civilian). Table 2.1 presents examples of profits made by various manufacturers during the 1985-1988 time period. The reader must keep in mind that these profit figures are for the corporate entities listed. The profit made on any given manufacturing program may be more or less.

4.6 EXAMPLE APPLICATION OF ESTIMATING MANUFACTURING AND ACQUISITION COST

The purpose of this section is to present an example application of the method for estimating the manufacturing and acquisition costs of an airplane program. In addition, an estimate for the unit price of the airplane will be determined. The example applies to the Ourania jet transport of Parts I, II and V of this text.

Since cost and price are tied to calendar time, it will be assumed that the cost and price estimates are required for the year 1992.

The material is organized as follows:

4.6.1 Manufacturing and Acquisition Cost Input Data

4.6.2 Manufacturing and Acquisition Cost Calculations

4.6.3 Determination of Estimated Unit Price

4.6.1 Manufacturing and Acquisition Cost Input Data

The following input data are required to estimate the manufacturing and acquisition cost for the Ourania program with the method of Sections 4.1 through 4.5.

W_{TO} = 123,683 lbs from p.11, Part V. This assumes that the airframe will be made mostly from Li/Al alloys.

The AMPR weight, W_{ampr}, is determined from Eqn.(3.5):

$$W_{ampr} = \text{invlog}\{0.1936 + 0.8645(\log 123{,}683)\} =$$

$$= 39{,}437 \text{ lbs}$$

$V_{max} = V_C$ = 295 kts from p.43, Part V. Note that this speed is in KEAS, at sealevel!

N_{rdte} = 5 (see p.37)

The number of airplanes produced to production standard, N_m, must be assumed at this point. The assumption is made that a marketing analysis has shown that 500 airplanes can be sold, so that:

N_m = 500

With Eqn.(4.1):

$N_{program}$ = 5 + 500 = 505

F_{diff} = 1.8, according to p.37

F_{cad} = 0.90, according to p.38

R_{e_m} = 63.64 USD/mhr (Set equal to R_{e_r} of p.38)

C_{e_m} = the cost per engine during the production program. Lacking better information at this point, this cost will be assumed equal to the engine price paid in the RDTE program: C_{e_r} of p.38:

C_{e_m} = USD 1,558,607

N_e = 2 (number of engines per airplane: Pt I, p.55)

N_p = 0 since the Ourania has no propellers

$C_{avionics_m}$ = the cost of avionics per airplane in the production phase of the program. At this point it will be assumed that the avionics cost for production airplanes is the same as that for the RDTE airplanes: $C_{avionics_r}$. For the reasons stated on p.38:

$C_{avionics_m}$ = USD 2,670,000 per airplane

F_{int} = 2,000 USD/pax from p.52

N_{pax} = 150 from p.55, Part I.

R_{m_m} = 35.35 USD/mhr (Set equal to R_{m_r} of p.39)

F_{mat} = 2.3 from p.31

CEF_{1992} = 3.10 from p.20

CEF_{1990} = 3.05 from p.20

R_{t_m} = 44.20 USD/mhr (Set equal to R_{t_r} of p.39)

N_{r_m} = the unit production rate per month. It is assumed that the Ourania delivery rate will be 5 per month:

N_{r_m} = 5 units per month

$C_{ops/hr}$ = the total operating cost per block hour for the Ourania. This cost includes the direct and indirect cost of the manufacturer to operate this airplane during production flight testing. Operating cost can be computed with the method of Chapter 5. At this point it will be assumed that this cost for 1992 is USD 2,500 per block hour:

$C_{ops/hr}$ = USD 2,500 per block hour

t_{pft} = the number of flight test hours flown by the manufacturer before delivery to the customer. From Section 4.3:

t_{pft} = 10 hours

The flight test overhead factor (p.55) is set at:

F_{ftoh} = 4.0

The following data from the RDTE cost calculations are also needed:

C_{aed_r} = USD 139,046,557, from p.39

C_{man_r} = USD 236,680,646, from p.40

C_{mat_r} = USD 48,975,376, from p.40

C_{tool_r} = USD 213,015,201, from p.40

F_{fin_m} = 0.10 (assumed from p.56)

F_{pro_m} = 0.10 (assumed from p.56)

C_{RDTE} = USD 1,232,000,000, from p.41

4.6.2 Manufacturing and Acquisition Cost Calculations

With the input from sub-section 4.6.1 the following manufacturing cost categories can now be determined:

From Eqns (4.5) and (4.6):

$$C_{aed_m} = 0.0396(39,437)^{0.791}(295)^{1.526}(505)^{0.183} *$$

$$*(1.8)(0.90)(63.64) - 139,046,557 =$$

$$= 323,557,187 - 139,046,557 = \text{USD } 184,510,630$$

From Eqn.(4.8):

$$C_{(e + a)_m} = \{(1,558,607)(2) + 2,670,000\}(500) =$$

$$= \text{USD } 2,893,607,000$$

From Eqn.(4.9):

$$C_{int_m} = (2,000)(150)(500)(3.10/3.05) =$$

$$= \text{USD } 152,459,016$$

From Eqns (4.10) and (4.11):

$$C_{man_m} = 28.984(39,437)^{0.740}(295)^{0.543}(505)^{0.524} *$$

$$*(35.35)(1.8) - 236,680,646 =$$

$$= 2,657,217,752 - 236,680,646 =$$

$$= \text{USD } 2,420,537,106$$

From Eqns (4.12) and (4.13):

$$C_{mat_m} = 37.632(2.3)(39,437)^{0.689}(295)^{0.624} *$$

$$*(505)^{0.792}(3.1) - 48,975,376 =$$

$$= 1,894,088,669 - 48,975,376 =$$

$$= \text{USD } 1,845,113,293$$

From Eqns (4.14) and (4.15):

$$C_{tool_m} = 4.0127(39,437)^{0.764}(295)^{0.899}(505)^{0.178} *$$

$$*(5)^{0.066}(1.8)(44.20) - 213,015,201 =$$

$$= 579,548,654 - 213,015,201 = \text{USD } 366,533,453$$

From Eqn.(4.16):

$$C_{qc_m} = 0.13(2,420,537,106) = \text{USD } 314,669,824$$

With Eqn.(4.7):

$$C_{apc_m} = 2,893,607,000 + 152,459,016 +$$

$$+ 2,420,537,106 + 1,845,113,293 +$$

$$+ 366,533,453 + 314,669,824 =$$

$$= \text{USD } 7,992,919,692$$

From Eqn.(4.17):

$$C_{fto_m} = 500(2,500)(10)(4) = \text{USD } 50,000,000$$

The financing cost, C_{fin_m}, will be determined from Eqn.(4.18) with $F_{fin_m} = 0.10$ so that:

$$C_{MAN} = (184,510,630 + 7,992,919,692 +$$

$$+ 50,000,000)/(0.9) = \text{USD } 9,141,589,247$$

From Eqns (4.2) and (4.19):

$C_{ACQ} = (1.1)(9,141,589,247) = \text{USD } 10,055,748,000$, or rounding off to the next highest million:

$$C_{ACQ} = \text{USD } 10,056,000,000$$

4.6.3 Determination of Estimated Unit Price

The price per airplane may now be estimated from Eqn.(4.3) as:

$$AEP_{1992} = (10{,}056{,}000{,}000 + 1{,}232{,}000{,}000)/500 =$$

$$= \text{USD } 22{,}576{,}000$$

This price should be checked for realism by comparison to the statistical market price data in Appendix A. Eqn.(A14) with Eqn.(A1) yields:

$$AMP_{1992} = \{(3.10)/(3.02)\}[\text{invlog}\{3.3191 +$$

$$+ 0.8043(\log 123{,}683)\}] = \text{USD } 26{,}680{,}852$$

The estimated price, AEP, is in fair agreement with the market price, AMP. A discussion of various factors which affect the airplane estimated price is presented in section 4.7.

4.7 EFFECT OF SEVERAL DESIGN AND PRODUCTION PARAMETERS ON AIRPLANE ESTIMATED PRICE: AEP

When inspecting the cost estimating equations in Sections 4.1 through 4.5 it is clear that the following parameters play a role in determining the program manufacturing cost and thereby in determining airplane price:

1. Airplane takeoff weight, W_{TO}, because of its effect on W_{ampr} which appears in Eqns.(4.6), (4.11), (4.13) and (4.15).
2. Airplane design cruise speed, V_{max}, which appears in the same equations as W_{ampr}.
3. Total number of airplanes built, $N_{program}$ which appears in the same equations as W_{ampr}.
4. Airplane production rate, N_{r_m}, which appears in Eqn.(4.15) for the tooling cost.
5. Airplane RDTE cost and the number of airplanes (N_m) over which this cost is to be recovered.

The purpose of this section is to show the results of some trade studies conducted to illustrate the effect of these design and production parameters on airplane estimated price. The author gratefully acknowledges the the work of Mr. Larry Bellmard who conducted these trade studies for the Ourania program of Sections 3.8 and 4.6.

Figure 4.1 shows the effect of the following parameters on the airplane estimated price, AEP:

* Total number of airplanes built, $N_{program} = 5 + N_m$

* Takeoff weight, W_{TO}

Observe that for the first 1,000 airplanes the airplane unit price is strongly affected by the number of airplanes built. After the 1,000 mark is reached there is still some price reduction to be seen but the effect is much less dramatic. There are two reasons for the rapid initial unit price reduction:

1) the so-called 'learning curve' effect

2) the RDTE cost contribution per airplane decreases hyperbolically with the total number of airplanes produced. This was already observed in Figure 3.6.

Figure 4.2 illustrates the 'learning curve' effect for three airplane programs: B-727, B-737 and McDD F-15. Note that the number of manhours required per airplane per pound decreases linearly when plotted on a log-log scale. This learning curve effect can be mathematically expressed as follows:

$$(\text{MHRS per unit}) = (\text{MHRS})_1/(N_{program})^n \qquad (4.20)$$

where: $MHRS_1$ is the number of manhours required to produce unit (= airplane) number 1.

$N_{program}$ is the total number of airplanes built.

n is the learning curve exponent. The meaning of n is best illustrated with a graphical example. This is done in Figure 4.3. Note that for a so-called 80 percent learning curve the exponent value is: n = 0.322. For a P percent learning curve the exponent n may be found from:

$$P = 100/(2^n) \qquad (4.21)$$

The learning curves shown in Figure 4.2 represent a 73 percent learning curve. Learning curves of 73 to 80 percent are fairly typical for the airframe manufacturing industry. The 'learning curve' effect noted here is already 'built' into the statistical cost equations of Sections 4.1 through 4.5.

The effect of airplane takeoff weight, W_{TO}, on airplane price is also shown in Figure 4.1. It is seen that the effect of weight on price is important when the number of airplanes produced is small.

Figure 4.4 illustrates the effect of airplane speed, V_{max}, on airplane price. It may be seen that increased speed requirements would increase airplane price rather significantly, when not many airplanes are produced.

Finally, Figure 4.5 shows that airplane production rate has a minor effect on airplane price except for the case where very small numbers are produced.

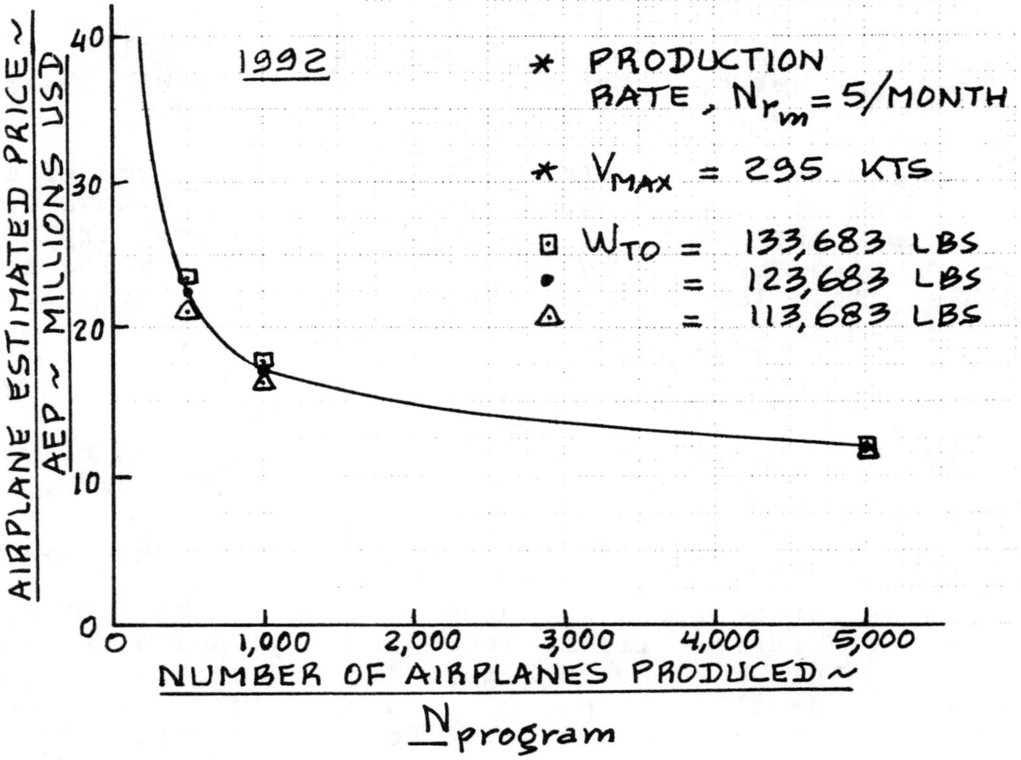

Figure 4.1 Effect of Number of Airplanes Produced and of Takeoff Weight on Airplane Estimated Price

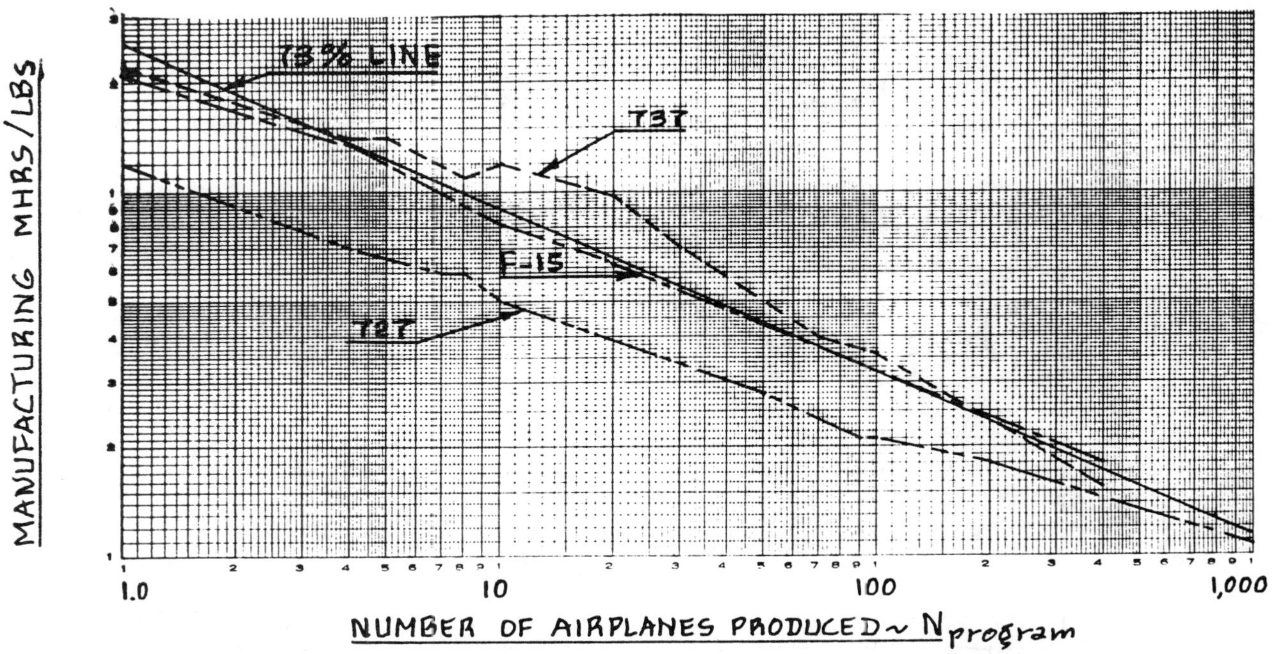

Figure 4.2 Learning Curves for Three Airplane Programs

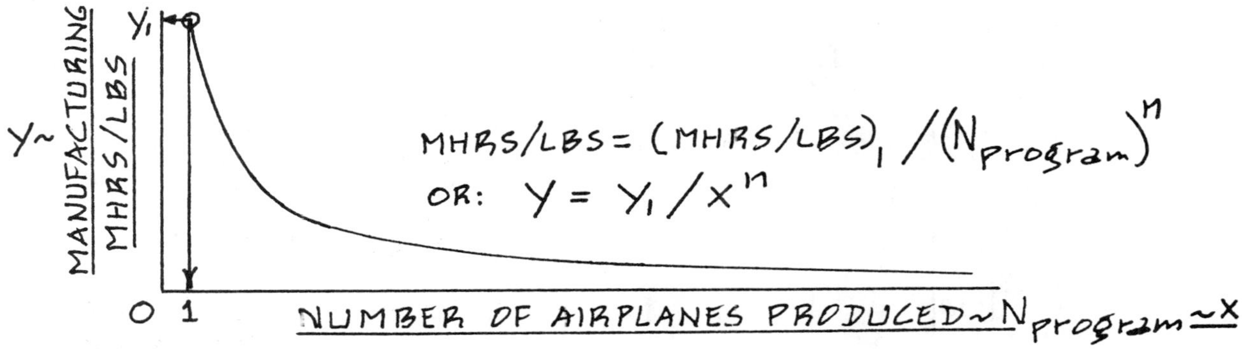

Figure 4.3a Reduction of Manhours with Units Produced

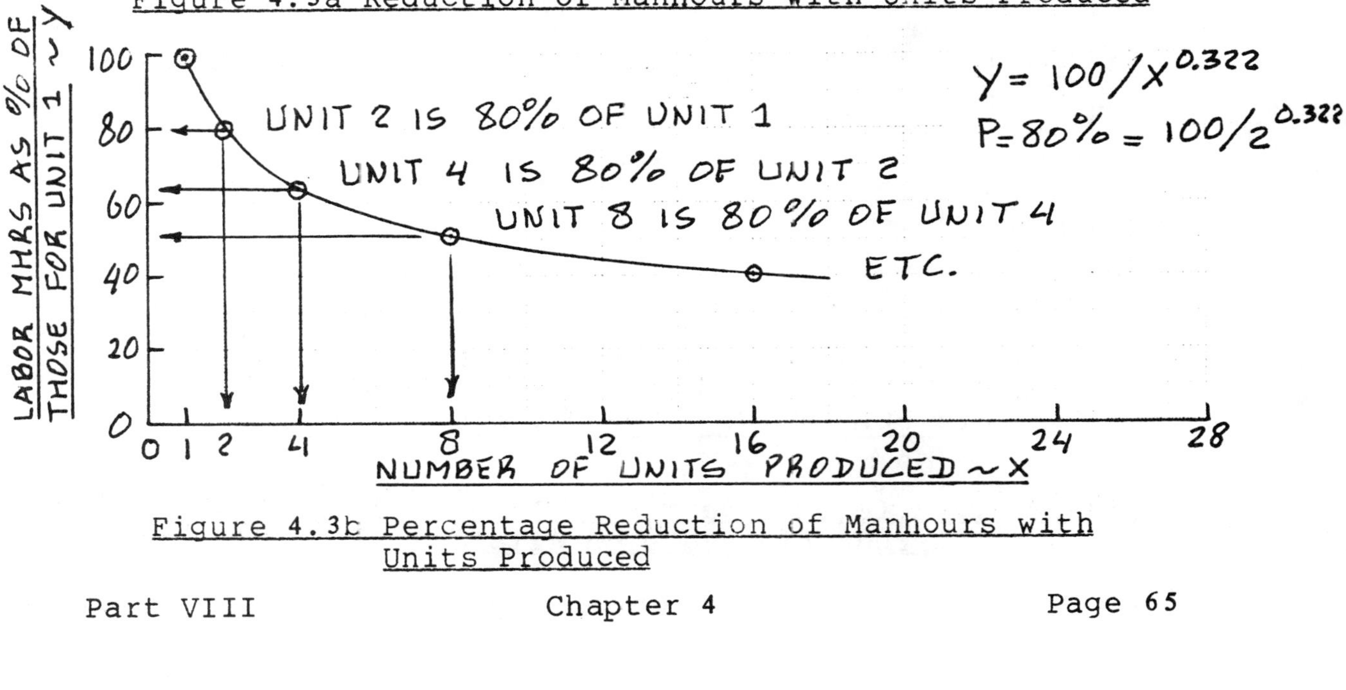

Figure 4.3b Percentage Reduction of Manhours with Units Produced

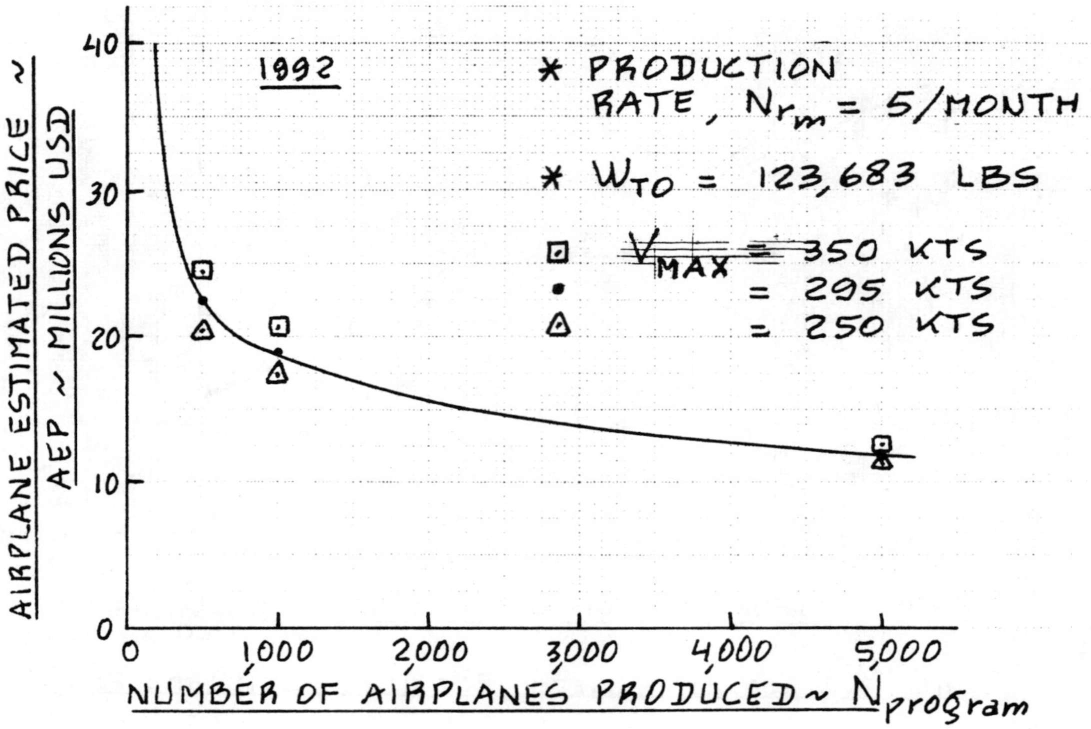

Figure 4.4 Effect of Number of Airplanes Produced and of Cruise Speed on Airplane Estimated Price

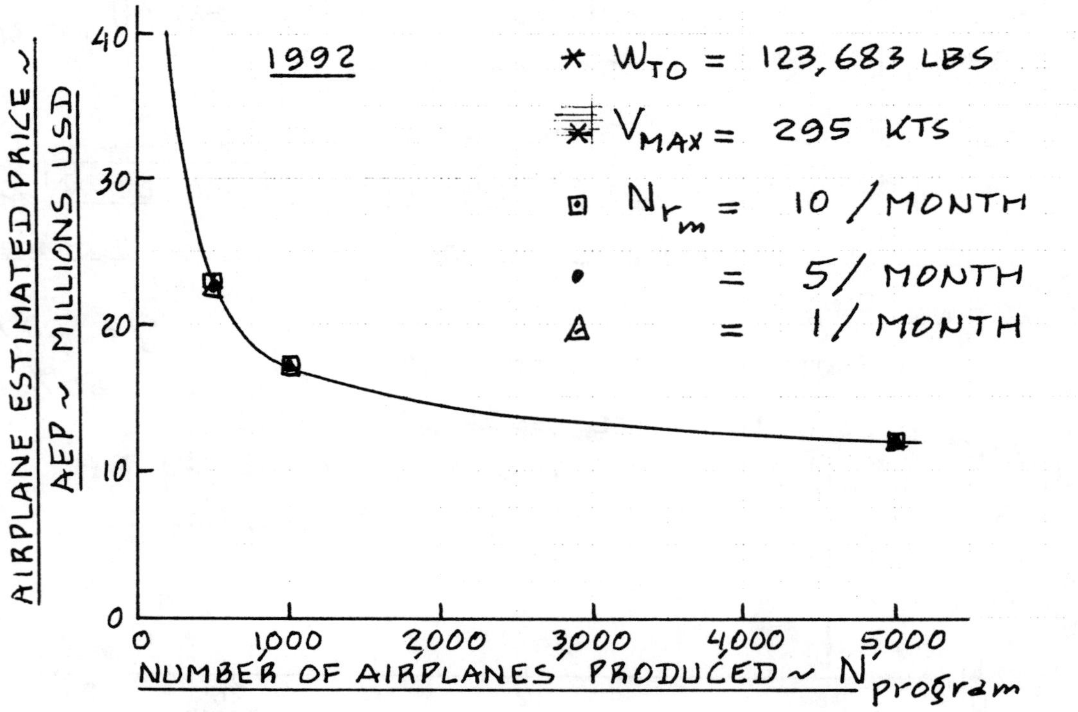

Figure 4.5 Effect of Number of Airplanes Produced and of Production Rate on Airplane Estimated Price

5. METHOD FOR ESTIMATING OPERATING COST OF COMMERCIAL AIRPLANES

The purpose of this chapter is to present a method for estimating the operating cost of commercial airplanes. A commercial airplane is defined as any non-military airplane. The method applies to airliners (all types), corporate airplanes and personal airplanes. A method for estimating the operating cost of military airplanes is presented in Chapter 6.

<u>IMPORTANT ASSUMPTION:</u>

Commercial operators acquire airplanes over many calendar years. As a result, commercial airplane production may vary considerably from year to year. The effect of all this on operating cost is difficult to model. It will therefore be ignored. Instead, the assumption is made that whatever numbers of airplanes are acquired by various operators, are acquired in one particular year and at one particular price.

The methodology presented in this chapter is based on methods and data presented in References 15 - 21. The material in this chapter is organized as follows:

5.1 Method for Estimating the Program Operating Cost of Commercial Airplanes: C_{OPS}

5.2 Method for Estimating the Direct Operating Cost of Commercial Airplanes: DOC

5.3 Method for Estimating the Indirect Operating Cost of Commercial Airplanes: IOC

5.4 Example Application

5.5 Some Important Observations about Commercial Airplane Operating Cost

5.1 METHOD FOR ESTIMATING THE PROGRAM OPERATING COST OF COMMERCIAL AIRPLANES: C_{OPS}

The total (or program) operating cost of commercial airplanes, C_{OPS}, is shown in relationship to other life cycle cost (LCC) sources in Eqn.(2.3) and in Figure 2.3.

The operating cost source is divided into the following two cost categories:

$$C_{OPS} = \sum_{i=1}^{i=n} (C_{ops_{dir}})_i (N_{acq})_i + \sum_{i=1}^{i=n} (C_{ops_{ind}})_i (N_{acq})_i \qquad (5.1)$$

where: C_{OPS} is the program operating cost source of Eqn.(2.3). It is expressed in USD.

$(C_{ops_{dir}})_i$ is the program direct operating cost for the ith airplane customer, expressed in USD/airplane.

$(N_{acq})_i$ is the number of airplanes acquired by the ith customer. Table 5.1 shows examples of fleet sizes acquired by various customers.

Note: because commercial loss rates due to accidents are relatively small, the effect of this on the number of active airplanes will be ignored.

$(C_{ops_{ind}})_i$ is the program indirect operating cost for the ith airplane customer, expressed in USD/airplane.

The airplane program direct and indirect operating costs (per operator and per airplane) as defined by Equation (5.1) may be estimated from:

$$(C_{ops_{dir}})_i = (DOC)_i (R_{bl_{ann}})_i (N_{yr})_i \qquad (5.2)$$

and:

$$(C_{ops_{ind}})_i = (IOC)_i (R_{bl_{ann}})_i (N_{yr})_i \qquad (5.3)$$

where: $(DOC)_i$ is the direct operating cost per nautical mile of the airplane as flown by the ith customer and expressed in USD/nm (per airplane). A method for estimating DOC is presented in Section 5.2.

Table 5.1 Typical Fleet Sizes in 1989

Airline	Boeing 727	Boeing 737	Boeing 747	Boeing 757	Boeing 767	McDonnell Douglas DC-9	McDonnell Douglas DC-10	McDonnell Douglas MD80-82	Lockheed L-1011	Airbus A300-310
Domestic										
American	164	24	2		45		63	153		13
Continental	96	97	8			48	15	65		12
Delta	130	74		50	30	36		40	40	
Eastern	116			25		81	2		20	20
Fed.Express	118		21				24			
Northwest	209		40	33		139	20	8		
Pan Am	70		34							31
TWA	70		19		11	48		33	35	
United	145	134	31		19		56			
US Air	44	201			6	74		31		
Foreign										
KLM		16	21				5			10
SAS					5	60	7	31		
British Airways		49	40	35			8		17	
Air France	24	16	35							22
Lufthansa	24	58	25				11			19
Alitalia			12			43		42		16

$(R_{bl_{ann}})_i$ is the total annual block miles flown by the ith customer, expressed in nautical miles (nm) per airplane. This is related to block speed and to annual utilization as shown in Eqn.(5.4).

$(N_{yr})_i$ is the number of years during which the airplane is operated by the ith customer. It must be recognized that airplanes are sold and resold during their operational lives. For the purpose of program cost estimating it is suggested to use:

For transport airplanes: N_{yr} = 20 years

For corporate airplanes: N_{yr} = 20 years

For personal owner airplanes: N_{yr} = 20 - 40 years

$(IOC)_i$ is the indirect operating cost per nautical mile of the airplane as flown by the ith customer and expressed in USD/nm per airplane. A method for estimating IOC is presented in Section 5.3.

The reason for using the nautical mile (as opposed to the statute mile) as the operating cost basis is an arbitrary one. The reader will find that both mile types are used in the reference literature.

The reason for expressing the program cost, C_{OPS}, as a sum over i customers is the fact that direct and indirect operating costs vary considerably from one customer to another. However, there are no reliable methods for predicting what the cost variations among various customers will be. For the reader who wishes to actually calculate C_{OPS} for any given airplane program it is suggested to use 'reasonable averages' for $(C_{ops_{dir}})_i$ and for $(C_{ops_{ind}})_i$ while setting $(N_{acq})_i$ equal to N_m, the total number of airplanes built to production standard. For the remainder of this chapter the subscript i can therefore be omitted without ambiguity.

The total annual block miles, $R_{bl_{ann}}$, (now without subscript i) in Eqns (5.2) and (5.3) can be expressed as:

$$(R_{bl_{ann}}) = (V_{bl})(U_{ann_{bl}}) \qquad (5.4)$$

where: V_{bl} is the block speed in nm/hr. The concept of block speed is illustrated in Figure 5.1. V_{bl} may be calculated with Eqn.(5.5).

$U_{ann_{bl}}$ is the annual utilization in block hours. This depends on airplane type and on routes flown: see the discussion on page 76.

For zero wind, the block speed, V_{bl}, is found from:

$$V_{bl} = (R_{bl})/t_{bl} \qquad (5.5)$$

where: R_{bl} is the block distance measured in nm. It is defined in Figure 5.1. The block distance depends on the routes flown. It is suggested to use the 'great circle' distance between city pairs for R_{bl}. Table 5.2 gives examples of

block distances between typical city pairs.

t_{bl} is the block time in hours. Figure 5.1 shows how t_{bl} is defined. It may be found from:

$$t_{bl} = t_{gm} + t_{cl} + t_{cr} + t_{de} \qquad (5.6)$$

where: t_{gm} is the time spent in ground maneuvers, such as pulling away from the gate, taxiing to the active runway, takeoff run, landing ground run and taxiing to the gate at the point of final destination. It is expressed in hours. The ground maneuver time may be estimated from:

$$t_{gm} = 0.51(10)^{-6}(W_{TO}) + 0.125 \qquad (5.7)$$

where: W_{TO} is the takeoff weight of the airplane as found with the methods of Part I, Chapter 2 or, more accurately with those of Part V, Chapters 2 and 4.

<u>Note:</u> Eqn.(5.7) includes an allowance of one minute for the takeoff run.

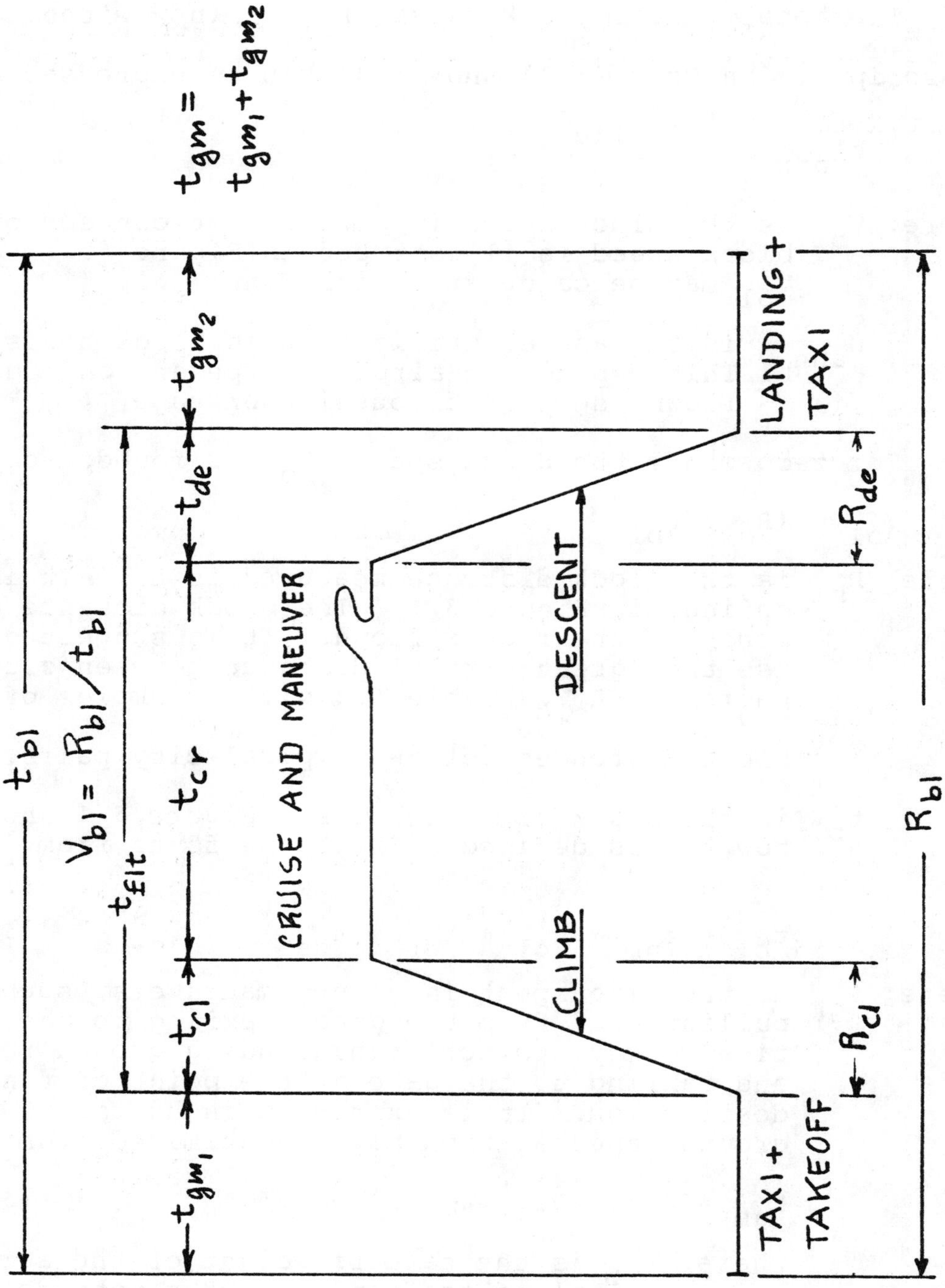

Figure 5.1 Mission Profile Definitions

Table 5.2 Example Block Distances, R_{bl}, Between City Pairs

City-Pairs Under 500 nm	R_{bl} (nm)		R_{bl} (nm)
Detroit - Chicago	220	Wash. DC* - Columbus	301
Buffalo - Wash. DC*	270	Chicago - Pittsburg	380
Chicago - Cleveland	274	Reno - Portland	424
Rochester - Wash. DC*	283	Omaha - Denver	430
Chicago - Des Moines	288	Atlanta - Pittsburg	487

City Pairs From 500 - 1,000 nm	R_{bl} (nm)		R_{bl} (nm)
Oakland - Portland	501	Boise - Denver	572
S.L. City - S.Francisco	551	Seattle - S.Francisco	669
Chicago - Wash. DC**	555	N.Y.*** - Chicago	719
Huntsville- Wash. DC*	558	Cleveland - Ft.Lauderd.	972
Las Vegas - Denver	561	Rochester - Tampa	960

City Pairs From 1,000 - 1,700 nm	R_{bl} (nm)		R_{bl} (nm)
Denver - Cleveland	1,066	Chicago - Seattle	1,522
S.L. City - Chicago	1,104	Los Ang. - Chicago	1,538
Chicago - Boise	1,264	Birmingh. - Los Ang.	1,589
Memphis - Los Ang.	1,430	Huntsville- Los Ang.	1,610
Chicago - Reno	1,469	S.Francis.- Chicago	1,644

City Pairs From 1,700 - 2,300 nm	R_{bl} (nm)		R_{bl} (nm)
Los Ang. - Detroit	1,749	N.Y.*** - Los Ang.	2,190
S.Francis.- Detroit	1,846	N.Y.*** - S.Francis.	2,284
Los Ang. - Wash. DC**	2,016		

* National Airport   ** Dulles   *** JFK

Chicago as listed here is O'Hare.

Data based on United Airlines Information, Ref.20.

t_{cl} is the time required to climb and to accelerate to the cruise speed, expressed in hours. A rapid method for predicting t_{cl} is presented in Part VII, pages 127-128.

t_{cr} is the time spent in cruise, expressed in hrs. It is suggested to use for:

Domestic operations:

$$t_{cr} = (1.06R_{bl} - R_{cl} - R_{de} + R_{man})/V_{cr} \quad (5.8a)$$

International operations:

$$t_{cr} = (1.01R_{bl} - R_{cl} - R_{de} + R_{man})/V_{cr} \quad (5.8b)$$

where: the factors 1.06 and 1.01 account for the fact that great circle distances cannot usually be flown for a variety of reasons.

R_{cl} is the distance to climb and to accelerate to cruise speed, in nm. It can be estimated from:

$$R_{cl} = (V_{cl})(t_{cl}) \quad (5.9)$$

where: V_{cl} is the projected horizontal speed of the airplane during the climb, in kts. For shallow climbs this may be assumed to be the same as the true airspeed in the climb. This speed is a by-product of the climb rate calculations for finding t_{cl}.

The reader is referred to pages 124-128 in Part VII.

R_{de} is the distance covered during the descent, in nm. It can be estimated from:

$$R_{de} = (V_{de})(t_{de}) \quad (5.10)$$

where: V_{de} is the projected horizontal speed of the airplane during the descent, in kts.

Note: For shallow descents this is roughly the same as the true airspeed during the descent. This speed is a by-product of the descent rate calculations of pages 159-161 in Part VII.

t_{de} as used in Eqns (5.6) and (5.10), is the time taken to descend. It may be computed with the method of p.159-161 in Part VII.

R_{man} is the distance covered while maneuvering due to ATC (air traffic control) constraints (in nm), with:

$$R_{man} = (V_{man})(t_{man}) \tag{5.11}$$

where: V_{man} is the speed while maneuvering as required by ATC constraints. It is suggested to use:

below 10,000 ft: $V_{man} = 250$ kts

above 10,000 ft: $V_{man} = V_{cr}$ (kts)

t_{man} is the time spent in ATC maneuvering. It can be found from:

$$t_{man} = 0.25(10)^{-6}W_{TO} + 0.0625 \tag{5.12}$$

V_{cr} is the airplane cruise speed in kts. Different operators cruise airplanes at different speeds. For preliminary costing it is recommended to use:

$$V_{cr} = V_C \tag{5.13}$$

where: V_C is the design cruise speed of an airplane in <u>nautical miles</u> per hour (kts).

<u>NOTE:</u> V_C is normally a 'given' in the mission specification of an airplane. Examples are found on pages 50 and 55 in Part I. The reader should ascertain that V_C of Eqn.(5.13) is consistent with the V_C defined in his V-n diagram calculations of Ch.4, Part V. However, note that V_C in the V-n diagram is expressed in KEAS and not in kts.

The annual utilization in block hours, $U_{ann_{bl}}$, as defined in Equation (5.4) depends on the type of airplane and on the routes flown. The weekly magazine Aviation Week and Space Technology publishes utilization data on a quarterly basis for passenger transports. The monthly magazine Business and Commercial Aviation publishes data on the utilization of other commercial airplanes. The reader will find representative utilization data for several types of airplanes in Table 5.3.

Lacking more detailed information it suggested to determine $U_{ann_{bl}}$ as follows:

For passenger transports:

$$U_{ann_{bl}} = (10^3)[3.4546(t_{bl}) + 2.994 + - \{12.289(t_{bl})^2 - 5.6626(t_{bl}) + 8.964\}^{1/2}] \quad (5.14a)$$

For cargo transports:

$$U_{ann_{bl}} = (10^3)[6.053(t_{bl}) + 5.70 + - \{37.771(t_{bl})^2 + 13.494(t_{bl}) + 32.490\}^{1/2}] \quad (5.14b)$$

where: t_{bl} is determined from Equation (5.6).

Eqns (5.14) are shown graphically in Figure 5.2.

For All Other Types of Airplanes: see Table 5.3.

Observe that annual airplane utilization so far has been expressed in terms of block hours. It is sometimes desirable (for a variety of reasons) to express the annual utilization in terms of flight hours: $U_{ann_{flt}}$. This can be done as follows:

$$U_{ann_{flt}} = (U_{ann_{bl}})(V_{bl})/(V_{flt}) \quad (5.15)$$

where: V_{flt} is the average flight speed at which the airplane is operated, in kts. It is noted that the average flight speed is always greater than the block speed. In turn, the average

Table 5.3 Utilization of Commercial Airplanes

Operator	1988 Airplane Utilization of Narrow-Body Jets in Block Hours per Day				
	727-200	737	757	MD-80	DC-9
American	7.8	9.2	NA	8.8	NA
United	9.2	7.9	NA	NA	NA
Delta	8.9	7.9	8.4	8.0	7.3
US Air/Piedmont	8.2	7.6	NA	NA	7.9
TWA	6.6	NA	NA	8.7	6.0

	1988 Airplane Utilization of Wide-Body Jets in Block Hours per Day				
	747	767	A300	DC-10	L-1011
American	10.2	11.1	NA	9.0	NA
Pan American	11.0	NA	8.4	NA	NA
United	12.0	10.8	NA	10.0	10.2
TWA	11.0	10.8	NA	NA	10.0
Continental	8.0	NA	7.0	10.0	NA

The following data are representative annual-utilization numbers in block hours for various types of airplanes:

Type of Operation:	Long Haul Block Time: $t_{bl}>5$	Medium Haul Block Time: $2<t_{bl}<5$	Short Haul Block Time: $0.5<t_{bl}<2$
Airplane Type			
Jet Transports	3,600-4,400	2,100-3,300	1,000-3,000
Regional Transports	NA	2,000-3,000	1,000-2,500
Corporate Transp.	500-1,500	400-1,200	300-1,000
Personal Transp.	NA	200-800	200-800
Agricultural	NA	NA	500-1,000
Trainers	NA	NA	1,000-2,500

NA = Not Applicable

Data Sources: Various issues of 'Aviation Week and Space Technology' and 'Business and Commercial Aviation'.

* The cargo transport line is based on a 7-day operation and on gate loading times of 0.5 hour or less. If the transport is operated only on a 5-day basis and if the gate loading time is longer, the annual utilization will be less. It is left to the reader to account for these effects!

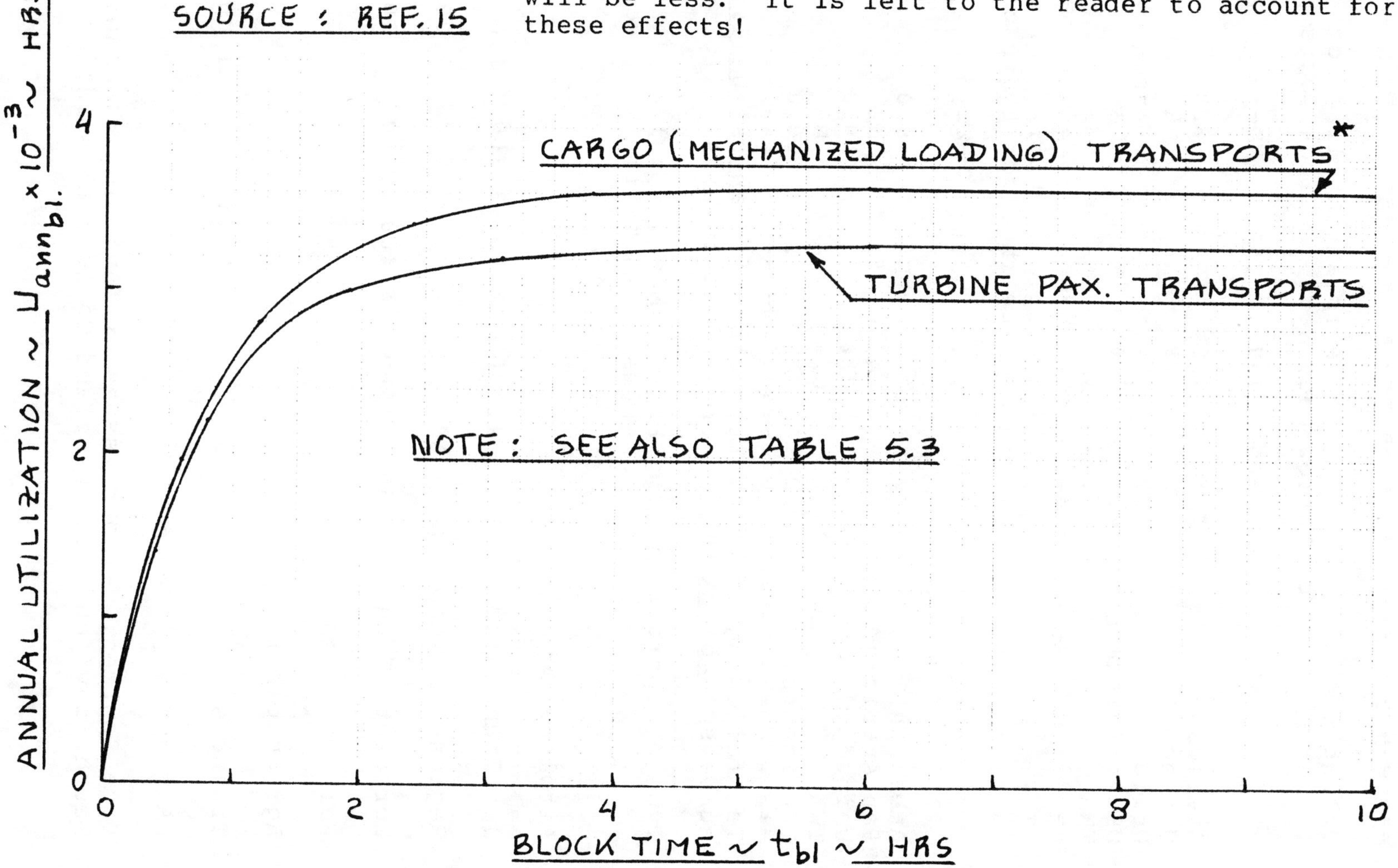

Figure 5.2 Effect of Block Time on Annual Utilization

flight speed is always less than the design cruise speed. This leads to the inequality:

$$V_{cr} > V_{fl} > V_{bl} \qquad (5.16)$$

The average flight speed may be found from:

$$V_{flt} = V_{cr}(t_{cr}/t_{flt}) \qquad (5.17)$$

where: t_{cr} is found from Equations (5.8)

t_{flt} is the average flight time which can be computed from:

$$t_{flt} = t_{cl} + t_{cr} + t_{de} \qquad (5.18)$$

where: t_{cl} and t_{de} are defined on pages 74 and 75, respectively.

Note: it is observed that t_{man} of Eqn.(5.11)

is already included in the definition of t_{cr} via the term R_{man} in Eqn.(5.8)!

Remember, that C_{OPS} in Equation (5.1) represents one of four program cost sources of Life Cycle Cost (LCC) as expressed by Equation (2.3). This cost source as such is normally NOT of any particular interest to a commercial airplane operator. However, the following cost categories are VERY IMPORTANT to typical airplane operators:

1a) Direct operating cost per nautical mile: (DOC)

and:

1b) Direct operating cost per block hour: {(DOC)(V_{bl})}

2a) Indirect operating cost per nautical mile: (IOC)

and:

2b) Indirect operating cost per block hour: {(IOC)(V_{bl})}

Sections 5.2 and 5.3 present methods for estimating DOC and IOC respectively. An example application is provided in Section 5.4. Finally, a number of important observations regarding DOC and IOC are made in Section 5.5.

5.2 METHOD FOR ESTIMATING THE DIRECT OPERATING COST OF COMMERCIAL AIRPLANES: DOC

The purpose of this section is to present a method for estimating the direct operating cost (DOC) in USD/nm incurred while operating commercial airplanes.

The method which is presented here is an adaptation of the so-called ATA-method of References 15 and 16 with various inputs from References 17-21. The ATA-method, as such, applies only to passenger and cargo transport airplanes. With the adaptation given here the method can be applied to any type of commercial airplane.

Table 5.4 presents an overview of cost components and cost elements which contribute to direct operating cost per nautical mile: DOC, which is broken down as:

$$DOC = DOC_{flt} + DOC_{maint} + DOC_{depr} + DOC_{lnr} + DOC_{fin} \tag{5.19}$$

where: DOC_{flt} is the direct operating cost of flying in USD/nm: see Sub-section 5.2.1

DOC_{maint} is the direct operating cost of maintenance in USD/nm: see Sub-section 5.2.2

DOC_{depr} is the direct operating cost of depreciation in USD/nm: see Sub-section 5.2.3

DOC_{lnr} is the direct operating cost of landing fees, navigation fees and registry taxes in USD/nm: see Sub-section 5.2.4

DOC_{fin} is the direct operating cost of financing in USD/nm: see Sub-section 5.2.5

NOTE: The right hand side of Table 5.4 indicates which DOC elements apply to various airplane types.

5.2.1 Direct Operating Cost of Flying: DOC_{flt}

The direct operating cost of flying (in USD/nm) is broken down into the following flying cost elements:

$$DOC_{flt} = C_{crew} + C_{pol} + C_{ins} \tag{5.20}$$

Table 5.4 Overview of Direct Operating Cost (DOC)

DIRECT OPERATING COST ~ DOC ~ USD/nm		AIRLINE	CORPORATE	PERSONAL
DOC COMPONENTS	DOC ELEMENTS			
FLYING ~ DOC_{flt} (5.20)	CREW ~ C_{crew} (5.21)	✓	✓	NO
	FUEL+ ~ C_{pol} (5.25)	✓	✓	✓
	INSURANCE ~ C_{ins} (5.31)	✓	✓	✓
MAINTENANCE ~ DOC_{maint} (5.33)	AIRFRAME LABOR ~ $C_{lab/ap}$ (5.34)	✓	✓	✓
	ENGINE LABOR ~ $C_{lab/eng}$ (5.36)	✓	✓	✓
	AIRFRAME MATERIALS ~ $C_{mat/ap}$ (5.37)	✓	✓	✓
	ENGINE MATERIALS ~ $C_{mat/eng}$ (5.38)	✓	✓	✓
	APPLIED MAINTENANCE BURDEN ~ C_{amb} (5.39)	✓	✓	NO
DEPRECIATION ~ DOC_{depr} (5.40)	AIRFRAME ~ C_{dap} (5.41)	✓	✓	✓
	ENGINE(S) ~ C_{deng} (5.42)	✓	✓	✓
	PROP(S) ~ C_{dprp} (5.43)	✓	✓	✓
	AVIONICS ~ C_{dav} (5.44)	✓	✓	✓
	AIRFRAME SPARE PARTS ~ C_{dapsp} (5.45)	✓	✓	NO
	ENGINE SPARE PARTS ~ C_{dengsp} (5.46)	✓	✓	NO
LANDING & NAVIG. FEES, REGISTRY TAXES ~ DOC_{lnr} (5.47)	LANDING FEES ~ C_{lf} (5.48)	✓	✓	NO
	NAVIGATION FEES ~ C_{nf} (5.52)	✓	✓	NO
	REGISTRY TAXES ~ C_{rt} (5.53)	✓	✓	✓
FINANCE ~ DOC_{fin} (5.55)		✓	✓	✓

where: C_{crew} is the crew cost in USD/nm

C_{pol} is the fuel and oil (pol means: petroleum, oil and lubricants) cost in USD/nm

C_{ins} is the airframe insurance cost in USD/nm

Methods for estimating these cost elements are presented in sub-sub-sections 5.2.1.1 through 5.2.1.3.

5.2.1.1 Crew cost per nautical mile: C_{crew}

The crew cost per nautical mile, C_{crew}, can be found from:

$$C_{crew} = \sum_{j=1}^{j=4} [(n_{c_j})\{(1 + K_j)/V_{bl}\}(SAL_j/AH_j) + (TEF_j/V_{bl})] \quad (5.21)$$

where: For transport airplanes:

n_{c_j} is the number of crew members of each type, j.

j indicates the type of crew member. The following distinctions are normally made:

n_{c_1} stands for captain.

n_{c_2} stands for co-pilot.

n_{c_3} stands for flight engineer.

Note: the cost of the cabin crew (flight attendants) is accounted for in the indirect operating cost, see Section 5.3.

The number of crew members which must be carried depends on government regulations (FARs), company rules and union rules. Lacking precise information the following numbers are suggested:

n_{c_1} = 1 for scheduled block times < 10 hours
= 2 for scheduled block times > 10 hours

n_{c_2} = as for n_{c_1}

n_{c_3} = 0 for airplanes with a 2-man cockpit

= as for n_{c_1} for airplanes with a 3-man cockpit

<u>For personal owner type airplanes:</u>

It is common <u>NOT</u> to count the cost of the pilot, so that:

n_{c_j} = 0 and j=1 only.

<u>For agricultural airplanes:</u>

n_{c_j} = 1 and j=1 only.

K_j is a factor which accounts for such items as: vacation pay, cost of training, crew premium, crew insurance, and payroll tax. This factor varies from one operator to another. Lacking more detailed data on operator work rules and benefits, it is suggested to use: $K_j = 0.26$.

V_{bl} is the airplane block speed in nautical miles per hour. The block speed may be found from:

$$V_{bl} = R_{bl}/t_{bl} \tag{5.22}$$

where: R_{bl} is the block distance in nautical miles. It is defined in Figure 5.1.

t_{bl} is the block time in hours: see Fig.5.1 and Equation (5.6).

SAL_j is the annual salary paid to a crew member of type j. Crew member salaries depend on the following factors:

*equipment flown: salaries tend to increase with airplane weight and with airplane speed
*seniority
*union and company rules

References 15 and 16 contain formulas which relate crew salaries to airplane weight and speed. These formulas applied reasonably well in the sixties because the airline industry was heavily regulated by the CAB:

Civil Aeronautics Board. The author believes that these salary relations can no longer be used. Table 5.5 contains recent data on crew salary ranges. The reader must decide which salaries apply to his airplane.

To estimate crew salaries for future years it is suggested to use:

$$SAL_{then\ year} = \qquad (5.23)$$

$$= SAL_{1989}\{(CEF_{then\ year})/(CEF_{1989})\}$$

where: the CEF values are found from Fig.2.7.

AH_j is the number of flight hours per year for a crew member of type j. Lacking actual data, the following numbers are suggested as applicable to all flight crew members:

<u>For domestic operations:</u>

AH_j = 800 hrs for jets
= 900 hrs for props

<u>For international operations:</u>

AH_j = 750 hrs for jets
= 850 hrs for props

TEF_j is the travel expense factor associated with each type of crew member. Since flight crews normally stay in the same hotel it is not necessary to vary the TEF from one crew member to another. It is suggested to use for calendar year 1990:

<u>Domestic routes:</u> TEF = USD 7.0/bl.hr

<u>International routes:</u> TEF = USD 11.0/bl.hr

For future years use:

$$TEF_{then\ year} = \qquad (5.24)$$

$$= TEF_{1990}(CEF_{then\ year})/(CEF_{1990})$$

where: the CEF values are found from Fig.2.7.

Table 5.5 Salaries and Wages Paid by Operators in 1989

Cockpit Crew: Salary Range in USD/year

Airplane Type:	Captain	First Officer	Flight Engineer
Jet Transport BAC 111 - B747	35,000-144,000	24,000-67,000	20,000-62,000
Corp. Jet Transp. Learjet M23 - Gulfstr.III	30,000-72,000	22,000-52,000	NA
Corp. Turboprops MU-2 - King Air 300	25,000-52,000	20,000-32,000	NA
Regional Turboprops DHC-6 - F-27	20,000-52,000	11,000-21,000	NA
Corporate Recips Single - Twin	20,000-47,000	19,000-26,000	NA

Cabin Crew: Salary Range in USD/year

	Purser	Cabin Attendant
All airplane types	25,000-47,000	19,000-32,000

Maintenance Personnel: Hourly Wage Range, USD/hr

	Fixed Base Operators (FBO)	Airlines
Avionics specialist	6.60 - 15.60	8.00 - 20.60
Dispatcher	5.00 - 10.40	7.00 - 15.50
Inspector	5.00 - 16.60	10.00 - 22.60
Line Personnel	3.35 - 12.80	5.00 - 10.30
Mechanic with A/P	5.00 - 18.40	10.00 - 22.60
Mechanic without A/P	5.00 - 11.30	8.00 - 15.50
Stock Clerk	3.35 - 7.20	5.00 - 10.30

NOTES: 1. For years other than 1989 use:

$$(\text{SAL or Wage})_{\text{then year}} =$$

$$= (\text{SAL or Wage})_{1989}\{(\text{CEF}_{\text{then year}})/(\text{CEF}_{1989})\}$$

where: the CEF data are obtained from Fig.2.7.

2. Salary and wage data do NOT include overhead!
3. Sources: Professional Pilot, Discussions with industry people and CEF data, Fig.2.7.
4. NA = Not Applicable

5.2.1.2 Fuel and oil cost per nautical mile: C_{pol}

The fuel and oil cost per nautical mile, C_{pol}, can be estimated from:

$$C_{pol} = (W_{F_{bl}}/R_{bl})(FP/FD) +$$

$$+ (W_{ol_{bl}}/R_{bl})(OLP/OD) \qquad (5.25)$$

where: $W_{F_{bl}}$ is the block fuel used in lbs. This amount of fuel is the same as the mission fuel used, $W_{F_{used}}$:

$$W_{F_{bl}} = W_{F_{used}} \qquad (5.26)$$

A preliminary method for computing $W_{F_{used}}$ is presented as Eqn.(2.14) of page 16, Part I. More accurate methods to compute $W_{F_{used}}$ are given in Sections 5.4 and 5.5 of Part VII.

NOTE: Reference 20 shows that airline experience indicates that the mission fuel consumption of an airplane tends to increase with calendar time for the following reasons:

* engine deterioration
* airframe surface deterioration (seals, finish)

If the reader wishes to account for this, 0.5 percent increase in fuel consumption per year of service seems to be the industry average.

R_{bl} is the block distance in nautical miles as defined in Figure 5.1.

FP is the price of fuel in USD/gallon

Figure 5.3 shows how fuel prices have varied with calendar time. There is no accurate way to predict how fuel prices will vary in the future. The reader is encouraged to call a local fixed base operator (FBO) to determine actual fuel prices.

NOTE: in this text 'gallon' means 'US gallon'!

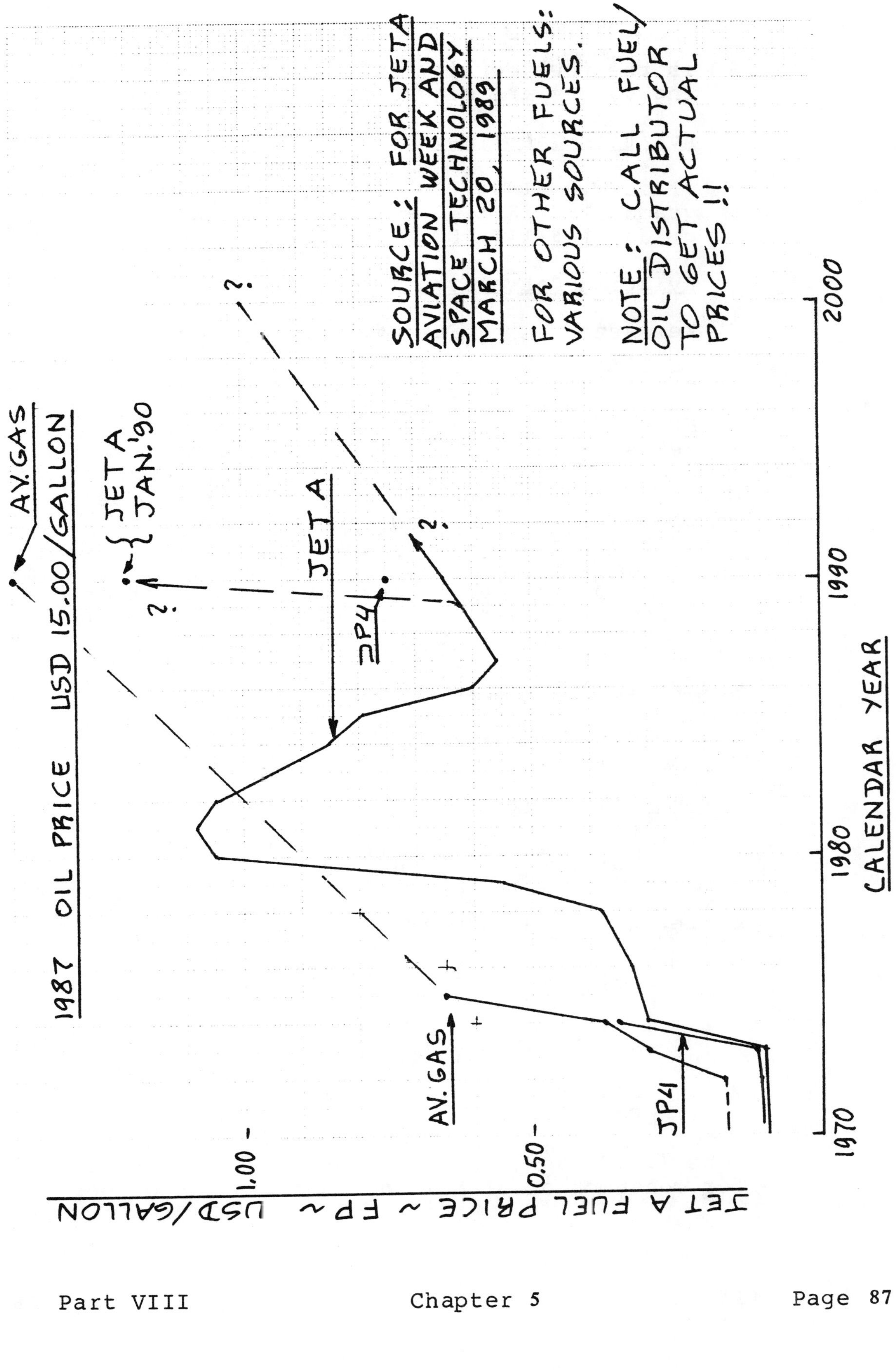

Figure 5.3 Variation of Fuel Prices with Calendar Year

FD is the fuel density in lbs/gallon. The following values are suggested:

<u>Aviation Gasoline:</u>

grades 100/130: FD = 6.0 lbs/gallon
grades 108/135: FD = 5.9 lbs/gallon
grades 115/145: FD = 5.8 lbs/gallon

<u>Aviation Petroleum:</u>

Kerosene:	FD = 6.7 lbs/gallon
JP-1:	FD = 6.65 lbs/gallon
JP-3:	FD = 6.45 lbs/gallon
JP-4:	FD = 6.55 lbs/gallon
JP-5:	FD = 6.82 lbs/gallon
Jet A	FD = 6.74 lbs/gallon

$W_{ol_{bl}}$ is the weight of oil and lubricants used, in lbs. It depends strongly on the type of powerplants and systems in need of lubricants. Manufacturers of engines and systems normally specify the hourly consumption of oil and lubricants. Lacking such specific information, it is suggested to use:

<u>for reciprocating engines:</u>

$$W_{ol_{bl}} = (W_{F_{used}})/70 \tag{5.27}$$

<u>for turbine engines:</u>

$$W_{ol_{bl}} = (0.70)(N_e)(t_{bl}) \tag{5.28}$$

where: N_e is the number of engines

t_{bl} is the block time in hours as found from:

$$t_{bl} = R_{bl}/V_{bl} \tag{5.29}$$

Note that Eqn.(5.29) is a modification of Eqn.(5.5).

Eqns (5.27) and (5.28) may be assumed to account for consumption of oil and lubricants together.

OD is the oil density in lbs/gallon. The following values are suggested:

Oil (MIL-6082B, grade 1100): OD = 7.40 lbs/gallon
Synthetic oil (MIL-L-7808-1): OD = 7.74 lbs/gallon

OLP is the price of oil and lubricants in USD/gallon.

Figure 5.3 also shows the price of oil. Just like is the case for fuel prices, there is no accurate way to predict oil prices in the future.

An alternate method to account for the direct operating cost of oil and lubricants is to assume that it is 5 percent of the direct operating cost of fuel. This leads to:

$$C_{pol} = 1.05(W_{F_{bl}}/R_{bl})(FP/FD) \tag{5.30}$$

5.2.1.3 Cost of airframe insurance per nautical mile: C_{ins}

Airplane operators carry airframe insurance for the following reasons:

1. ground and flight risk of experiencing airframe damage or total loss
2. passenger liability in case of injury or death
3. third party liability in case of injury or death
4. cargo damage risk

Most airplane operators will want to carry airframe damage insurance (to cover item 1) to an amount equal to the replacement value of the airplane. Although some operators will prefer to 'carry' a certain amount of the insurance themselves, there is a cost associated with that as well: a reserve will have to be created and the interest cost on the reserve can be thought of as part of the cost of insurance.

NOTE WELL: In references 15-20 the convention has been adopted to consider item 1 above as part of the DOC while items 2-4 are considered as part of the IOC (Section 5.3). This convention is followed here also!

Insurance rates for hull damage or loss depends on the so-called hull loss rate. Figure 5.4 gives historic insight into the hull losses of jet transports. Note that the annual hull loss rate on the average is about 2.0 per 1,000,000 departures.

For details on insurance rates the reader should consult an aviation underwriter. Lacking such precise information it is suggested to use:

$$C_{ins} = (f_{ins_{hull}})(AMP)/\{(U_{ann_{bl}})(V_{bl})\} \qquad (5.31)$$

where: $f_{ins_{hull}}$ is the annual hull insurance rate in USD per USD airplane price, per airplane and per year. Typical values for this hull insurance rate range from:

$$f_{ins_{hull}} = 0.005 \text{ to } 0.030 \text{ USD/USD/airplane/year}$$

AMP is the airplane market price in USD. AMP may be estimated rapidly with the method given in Appendix A. A more detailed method for computing AMP is presented in Sub-section 4.6.3.

$U_{ann_{bl}}$ is the annual block hour utilization as determined from the discussion on p.76.

V_{bl} is the block speed in nm/hr (kts) as defined by Eqn.(5.5).

An alternate method to account for hull insurance is to set:

$$C_{ins} = 0.02(DOC) \qquad (5.32)$$

where: DOC is found from Eqn.(5.19).

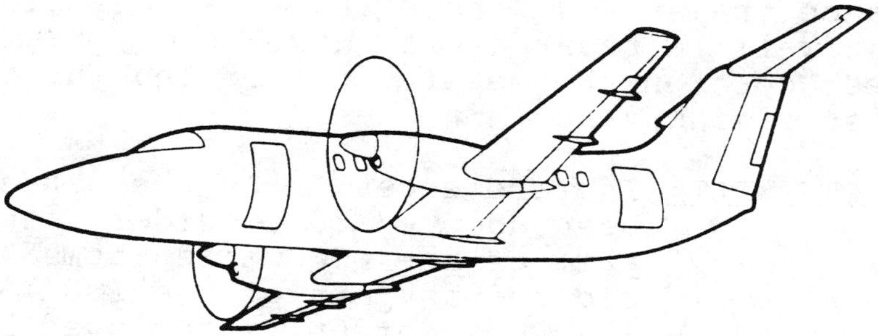

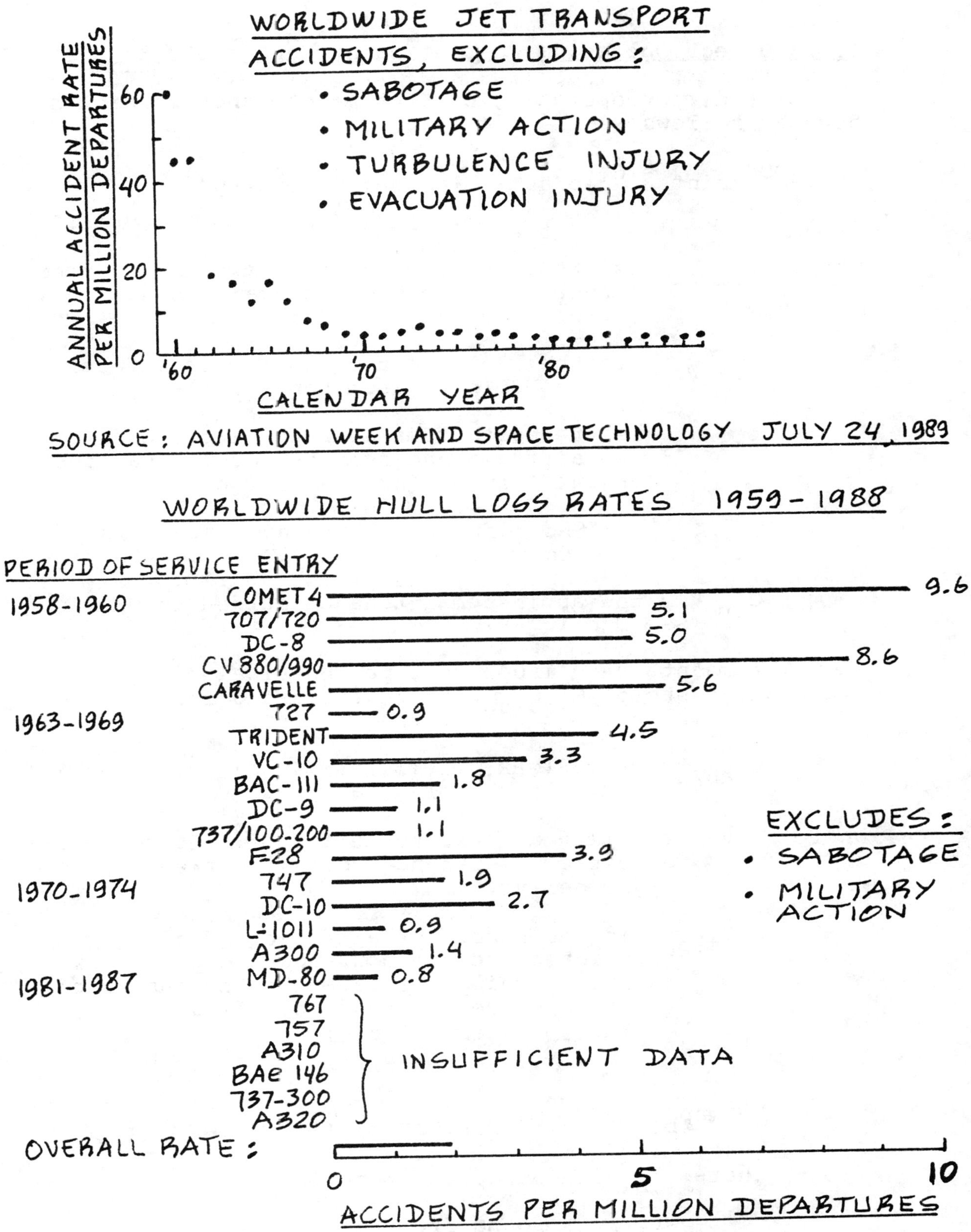

Figure 5.4 Worldwide Accident and Hull Loss Rates

5.2.2 Direct Operating Cost of Maintenance: DOC_{maint}

The direct operating cost of maintenance is broken down as follows:

$$DOC_{maint} = C_{lab/ap} + C_{lab/eng} + C_{mat/ap} + + C_{mat/eng} + C_{amb} \qquad (5.33)$$

where: $C_{lab/ap}$ is the labor cost of airframe and systems (other than the engines) maintenance in USD/nm. See: Eqn.(5.34).

$C_{lab/eng}$ is the labor cost of engine(s) maintenance in USD/nm. See: Eqn.(5.36).

$C_{mat/ap}$ is the cost of maintenance materials for the airframe and systems (other than the engines) in USD/nm. See: Eqn.(5.37).

$C_{mat/eng}$ is the cost of maintenance materials for the engines in USD/nm. See Eqn.(5.38).

C_{amb} is the applied maintenance burden in USD/nm. See Eqn.(5.39).

The maintenance labor cost per nautical mile of airframe and systems, $C_{lab/ap}$, can be estimated from:

$$C_{lab/ap} = 1.03(MHR_{map_{bl}})(R_{l_{ap}})/(V_{bl}) \qquad (5.34)$$

where: 1.03 is called a non-revenue factor. It accounts for extra maintenance costs incurred due to flight delays.

$MHR_{map_{bl}}$ is the number of <u>airframe and systems</u> maintenance man hours needed per block hour. This is related to the number of maintenance man hours per flight hour, $MHR_{map_{flt}}$ as follows:

$$MHR_{map_{bl}} = (MHR_{map_{flt}})(t_{flt}/t_{bl}) \qquad (5.35)$$

where: t_{bl} is found from Eqn.(5.6)

t_{flt} is found from Eqn.(5.18)

It is desirable to use actual man hour data where possible. Lacking such information the data of Table 5.6 (given per flight hour) and/or Fig. 5.5 (given per block hour) may be used as a guide.

<u>NOTE WELL:</u> Aviation Week and Space Technology publishes data on airframe and systems maintenance on a regular basis. The reader should check the latest issues for updated information. Ref.20 contains detailed maintenance data for United Airlines airplanes for the year 1975.

$R_{l_{ap}}$ is the airplane maintenance labor rate per manhour in USD/hr. It is desirable to use actual labor rate data. Lacking such information, Table 5.5 may be used: use the 'mechanic with A/P' data.

V_{bl} is the block speed as found from Eqn.(5.5).

The maintenance labor cost per nautical mile of engines, $C_{lab/eng}$ can be estimated from:

$$C_{lab/eng} = 1.03(1.3)(N_e)(MHR_{meng_{bl}})(R_{l_{eng}})/(V_{bl}) \qquad (5.36)$$

where: the factor 1.3 accounts for cycle-dependent labor as opposed to cycle-independent labor.

N_e is the number of engines per airplane.

$MHR_{meng_{bl}}$ is the number of engine maintenance hours needed per block hour per engine. Whenever possible, it is desirable to use actual manhour data. Lacking such information the data of Table 5.6 and/or Figures 5.6 and 5.7 may be used.

$R_{l_{eng}}$ is the engine maintenance labor rate per man hour in USD/hr. It may be found from Table 5.5: use the 'mechanic with A/P'data.

The cost of maintenance materials for airframe and systems (other than the engines) per nautical mile may be estimated from:

$$C_{mat/ap} = 1.03(C_{mat/apblhr})/V_{bl} \qquad (5.37)$$

where: $C_{mat/apblhr}$ is the airframe and systems maintenance materials cost per airplane block

Table 5.6 Maintenance Manhour Data for Commercial Airplanes

Airplane Type	Annual Utilization in flight hours, $U_{ann_{flt}}$	Maintenance Manhours* per flight hour, MHR_{flthr}	Data from year
Cessna 150	250	0.3	1974
Cessna Skywagon	250	0.5	1974
Beech Kingair 90	350	1.0	1974
Cessna Citation I	400	3.0	1974
McDD DC-9-30	2,900	6.4	1973
B-707-300	3,196	8.4	1973
B-727-100	2,670	7.9	1973
B-727-200	2,800	6.5	1973
B-737-200	2,200	6.6	1973
B-747-100	3,525	14.5	1973
L-1011	1,870	14.1	1974
McDD DC-10-10	2,450	10.9	1973

Data Source: Nicolai, Ref.14

* sum of airframe (+ systems) and engine maintenance hrs:

$$MHR_{flthr} = MHR_{map_{flt}} + MHR_{meng_{flt}}$$

For additional data, see Ref.20 (1976). The distribution between airframe (+ systems) and engine maintenance manhours varies <u>considerably</u> with airplane and engine type, and in particular with engine 'maturity'.

Engine maintenance manhours have been reported anywhere from 10 to 50 percent of the total maintenance manhours per flight hour. For modern jet transports with highly mature engines, the trend is for the engine maintenance manhours to constitute about 10 percent of the total maintenance manhours per flight hour.

Lacking the specific information given in Ref.20 it is suggested to split the total maintenance manhours per flight hour evenly between the airframe (+ systems) and the engine(s).

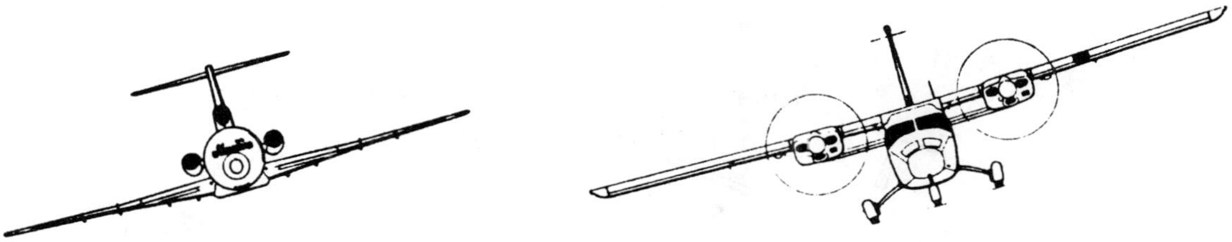

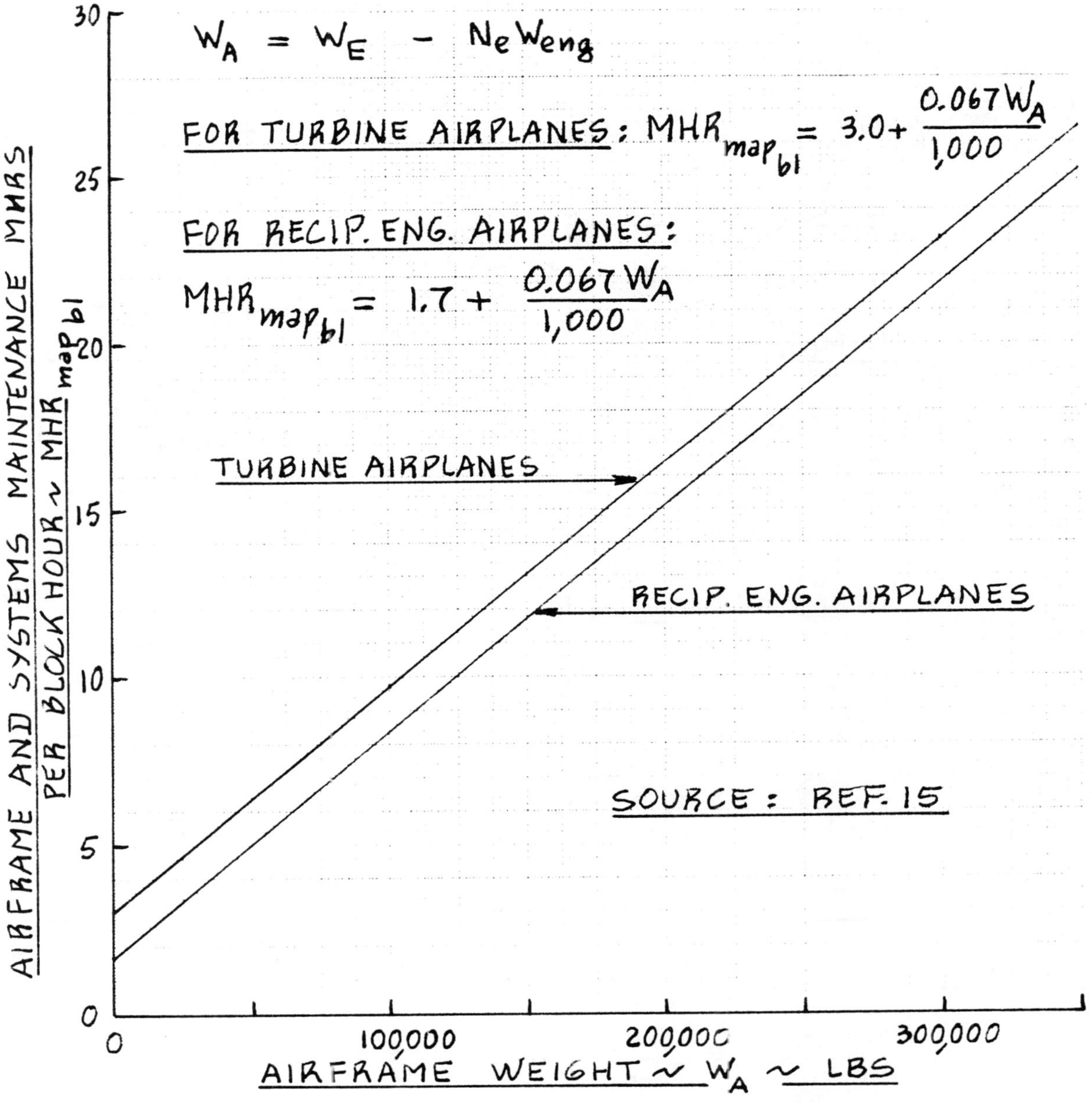

Figure 5.5 Effect of Airframe Weight on Airframe and Systems Maintenance Man Hours

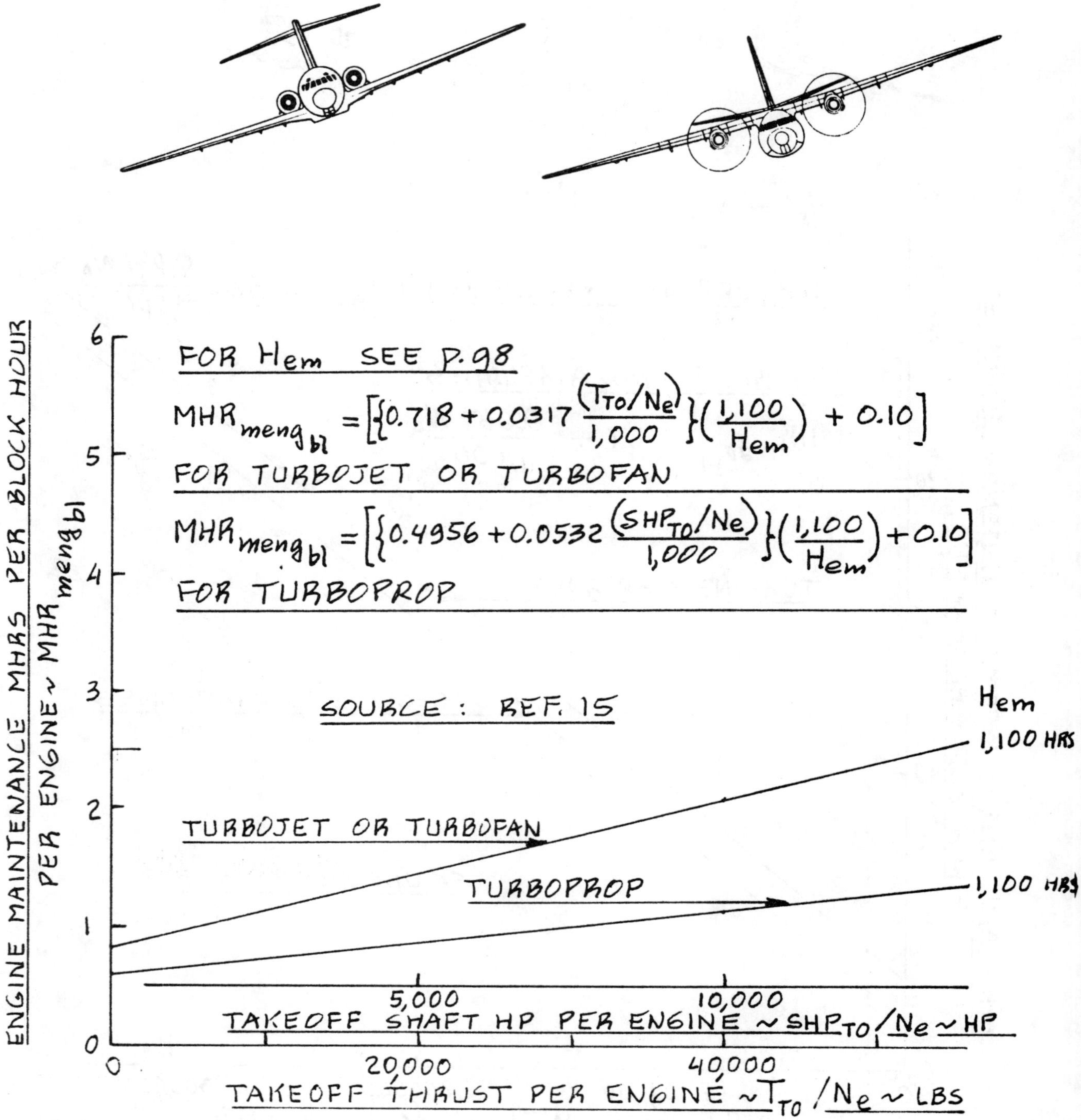

Figure 5.6 Effect of Takeoff Thrust and Takeoff Shaft Horsepower on Turbine Engine Maintenance Manhours

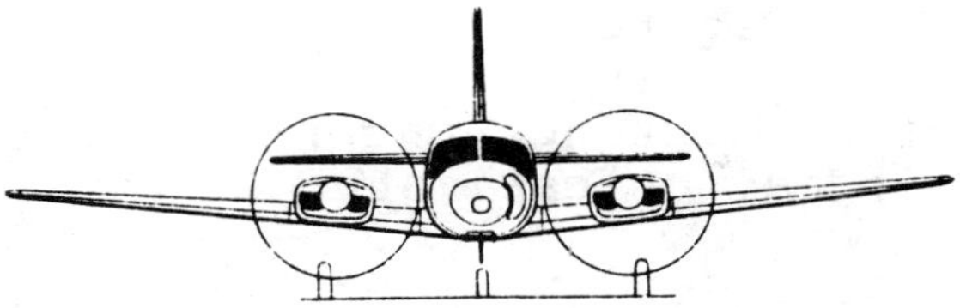

$$MHR_{meng_{bl}} = \left\{ 0.0765 \left(\frac{W_{eng}}{1{,}000} \right)^2 + 0.2495 \left(\frac{W_{eng}}{1{,}000} \right) \right\} \times$$

$$\times \left\{ \frac{0.70}{K_{H_{em}}} + 0.30 \right\}$$

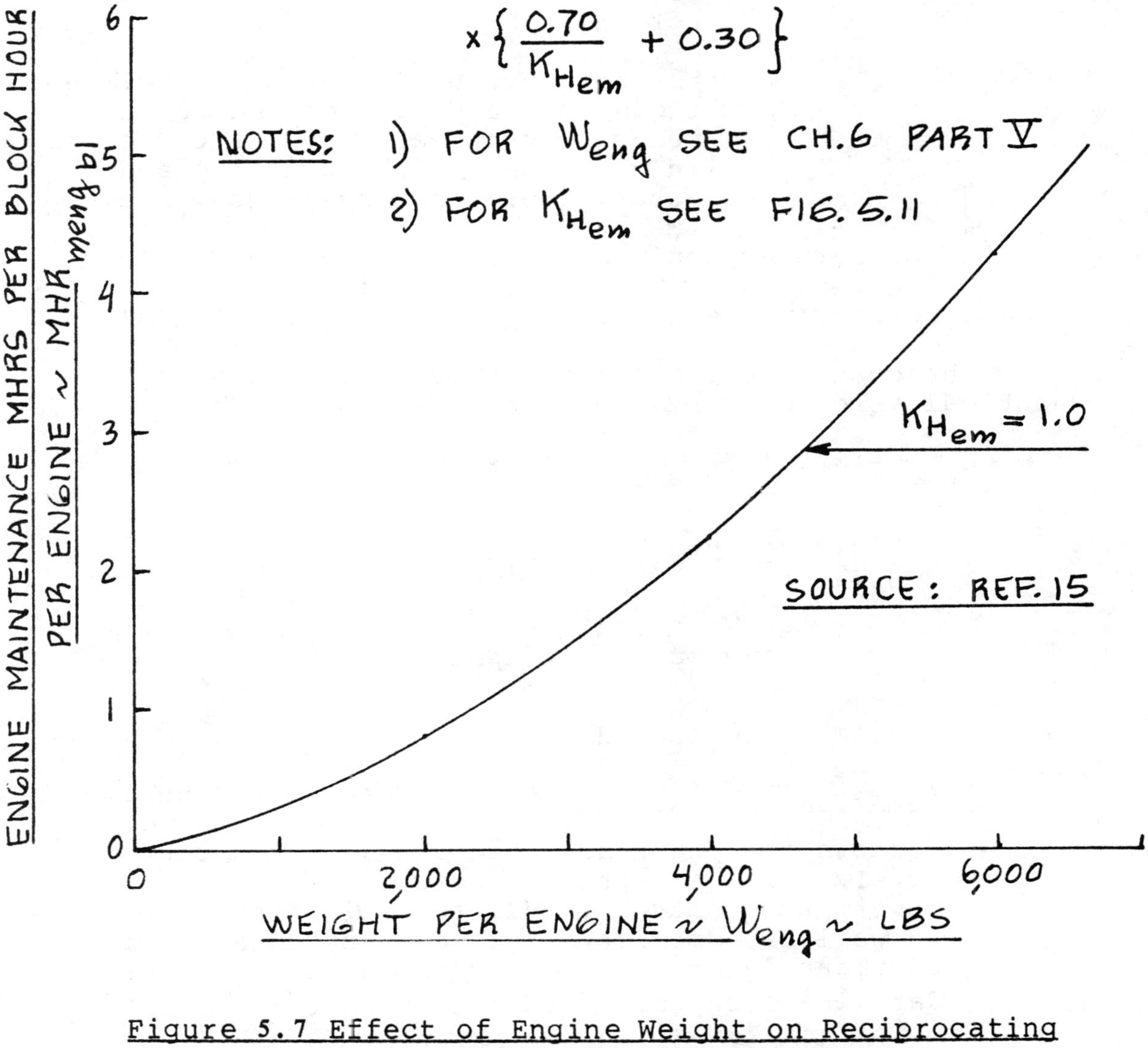

Figure 5.7 Effect of Engine Weight on Reciprocating Engine Maintenance Man Hours

hour in USD/hr. It may be found from Figure 5.8.

The cost of maintenance materials for the engine(s) per nautical mile may be estimated from:

$$C_{mat/eng} = 1.03(1.3)(N_e)(C_{mat/engblhr})/V_{bl} \qquad (5.38)$$

where: the factor 1.3 accounts for cycle-dependent as opposed to cycle-independent material.

$C_{mat/engblhr}$ is the engine maintenance materials cost per engine per airplane block hour in USD/hr. It may be found from Figures 5.9-5.11. Note that one factor which must be decided on is H_{em},

the attained number of hours between engine overhauls. That number depends on the 'maturity' of the engine in question. Lacking data based on actual operating experience, use:

For turbine engines: H_{em} = 3,000 - 5,000 hrs

For recipr. engines: H_{em} = 1,000 - 2,000 hrs

The cost of the applied maintenance burden per nautical mile, C_{amb}, may be estimated from:

$$C_{amb} = 1.03[(f_{amb/lab})\{(MHR_{map_{bl}})(R_{l_{ap}}) + $$

$$+ (N_e)(MHR_{eng_{bl}})(R_{l_{eng}})\} + $$

$$+ (f_{amb/mat})\{C_{mat/apblhr} + $$

$$+ (N_e)C_{mat/engblhr}\}]/(V_{bl}) \qquad (5.39)$$

where: $f_{amb/lab}$ and $f_{amb/mat}$ are overhead distribution

factors for labor and for material cost, respectively. These overhead factors are intended to cover such costs as: building, lighting, heating as well as administrative costs associated with airplane maintenance. Lacking actual overhead data it is suggested to use (see top of p.102):

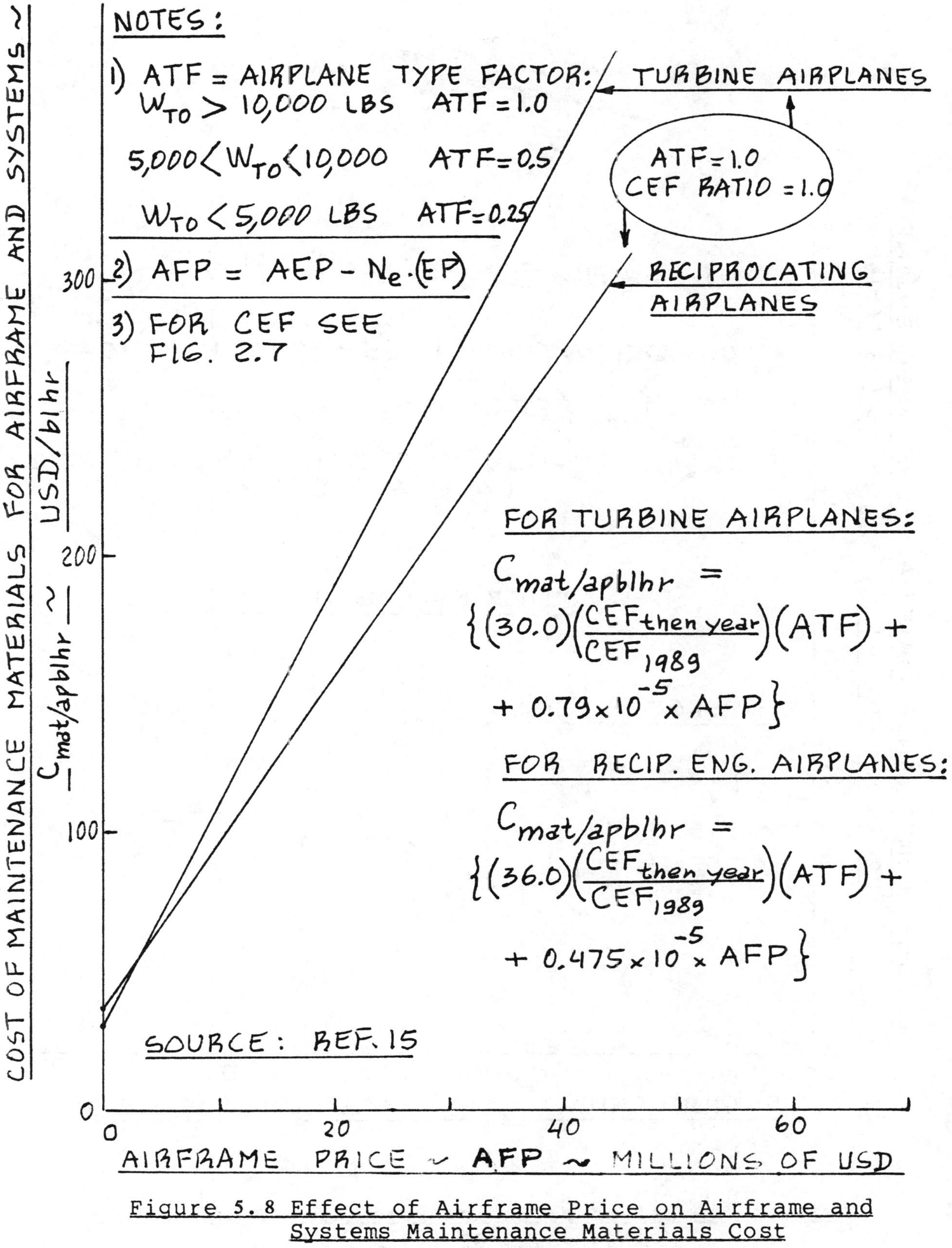

Figure 5.8 Effect of Airframe Price on Airframe and Systems Maintenance Materials Cost

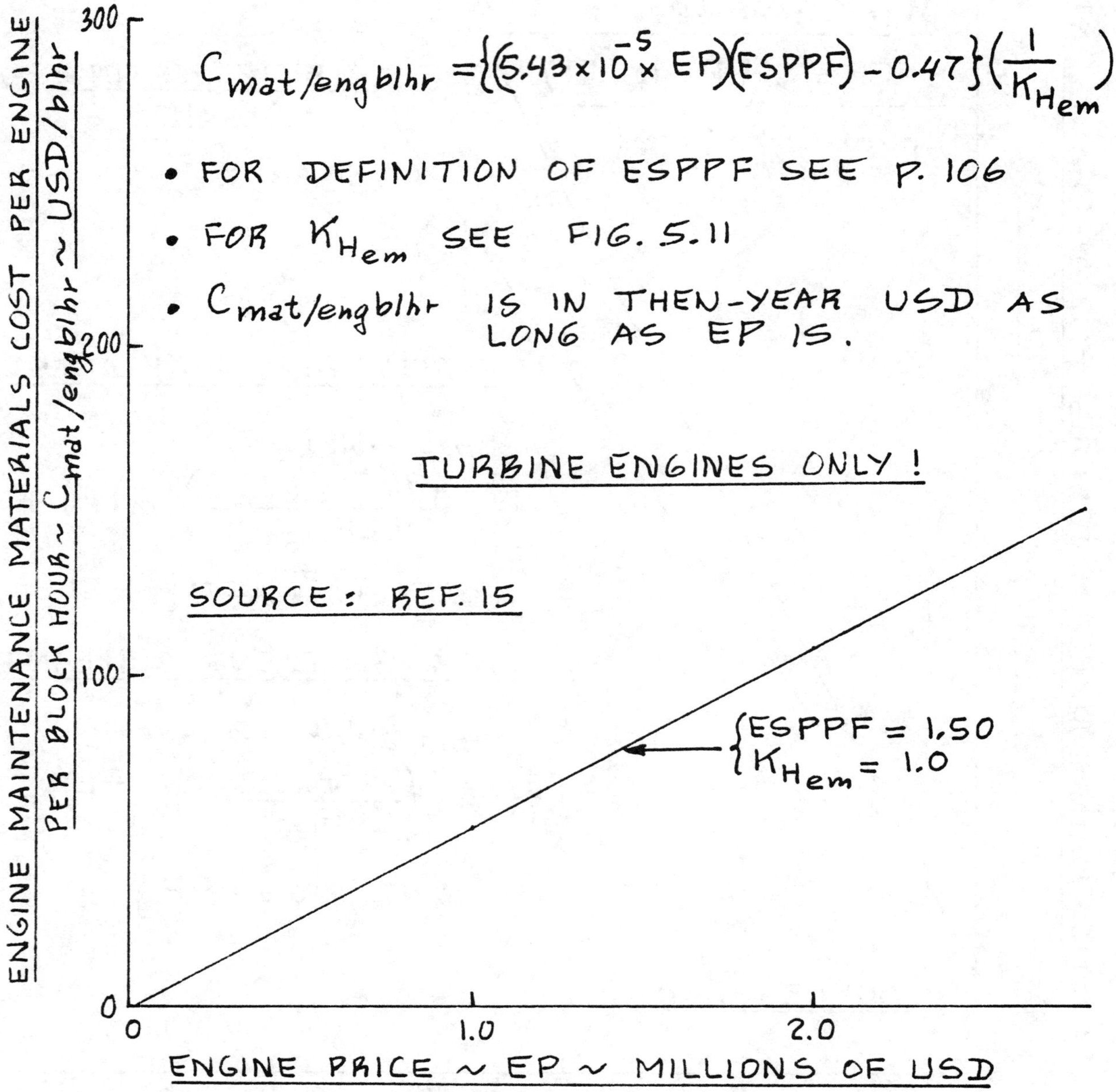

Figure 5.9 Effect of Engine Price on Turbine Engine Maintenance Materials Cost

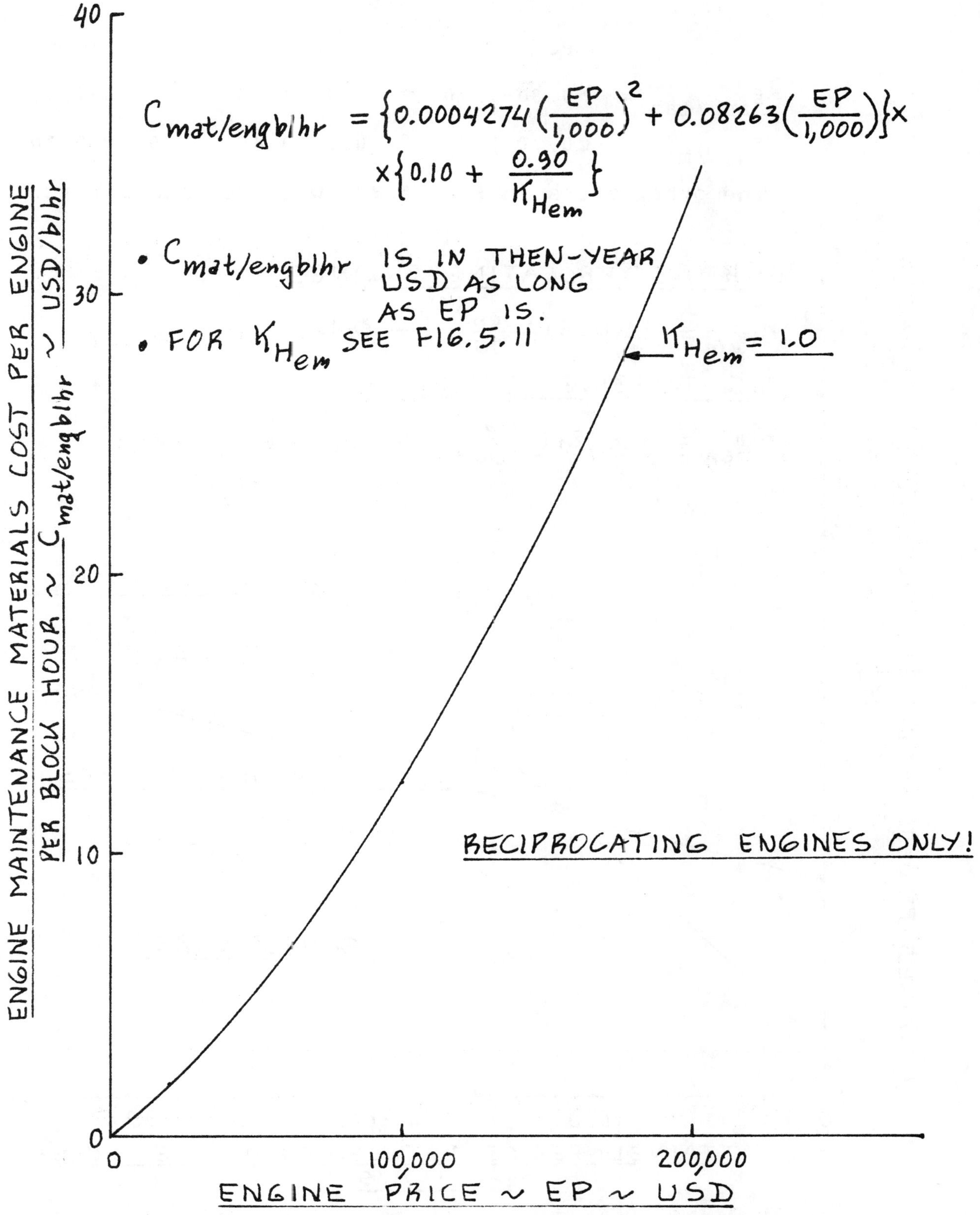

Figure 5.10 Effect of Engine Price on Reciprocating Engine Maintenance Materials Cost

	Personal	Corporate	Airlines
$f_{amb/lab}$ =	0.80-0.90	0.90-1.00	1.00-1.40
$f_{amb/mat}$ =	0.20-0.30	0.30-0.40	0.40-0.70

The other terms in Eqn.(5.39) were defined before.

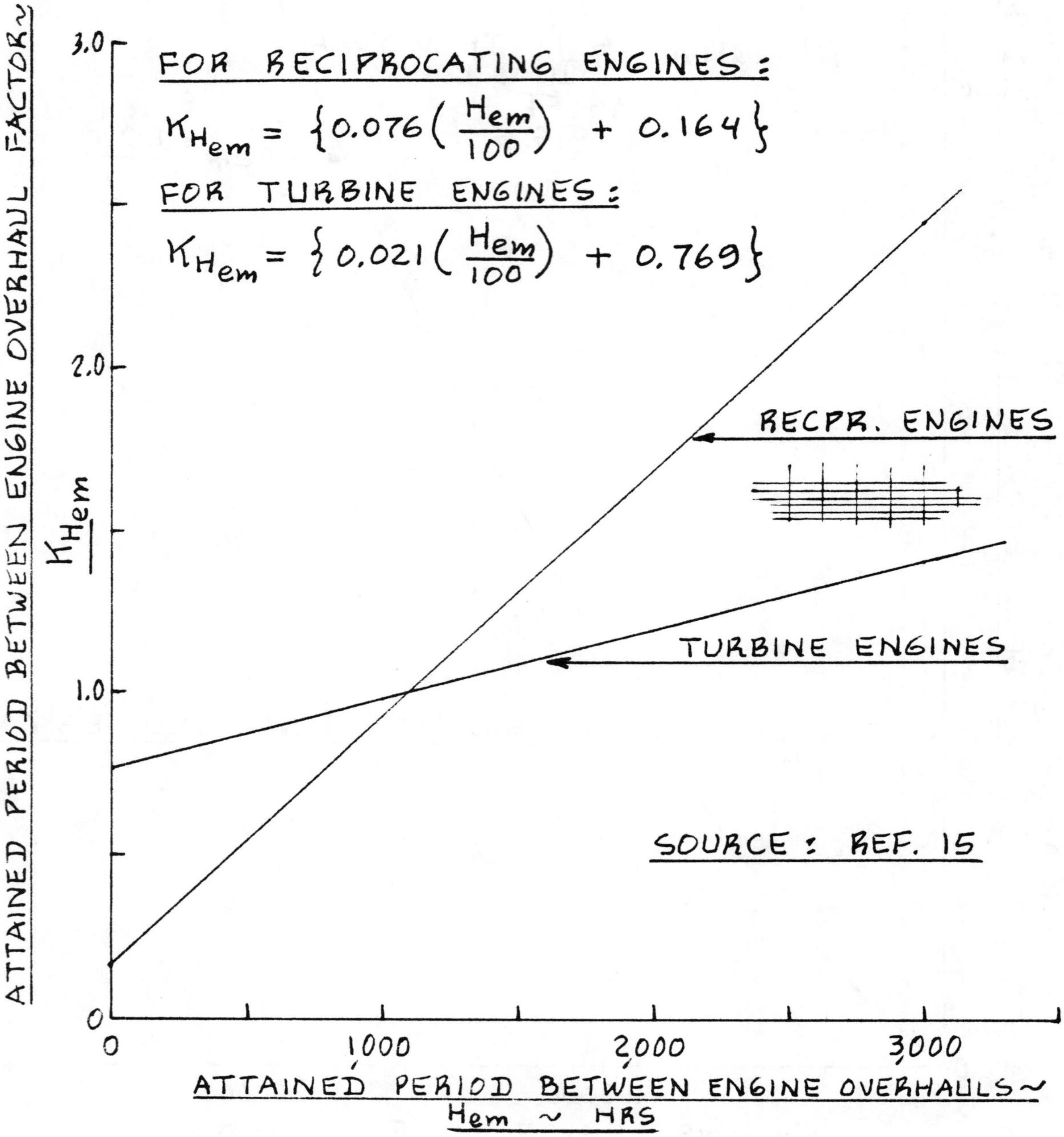

Figure 5.11 Factor Depending on Attained Engine Period Between Engine Overhauls

5.2.3 Direct Operating Cost of Depreciation: DOC_{depr}

The direct operating cost of depreciation is broken down as follows:

$$DOC_{depr} = C_{dap} + C_{deng} + C_{dprp} + C_{dav} + C_{dapsp} + C_{dengsp} \tag{5.40}$$

where: C_{dap} is the cost of airplane depreciation without engines and without propellers, avionics systems and spares (= airframe), in USD/nm. See Eqn.(5.41).

C_{deng} is the cost of engine depreciation (as mounted on the airplane but without propellers), in USD/nm. See Eqn.(5.42).

C_{dprp} is the cost of depreciation of propellers, in USD/nm. See Eqn.(5.43).

C_{dav} is the cost of depreciation of avionics systems in USD/nm. See Eqn.(5.44).

C_{dapsp} is the cost of depreciation of airplane spare parts in USD/nm. See Eqn.(5.45).

C_{dengsp} is the cost of depreciation of engine spare parts in USD/nm. See Eqn.(5.46).

The cost of airframe depreciation per nautical mile, C_{dap}, can be estimated from:

$$C_{dap} = \frac{(F_{dap})\{(AEP) - (N_e)(EP) - (N_p)(PP) - (ASP)\}}{(DP_{ap})(U_{ann_{bl}})(V_{bl})} \tag{5.41}$$

where: F_{dap} is an airframe depreciation factor. This factor depends on the perceived resale value of the airplane: see Table 5.7, p.107.

AEP is the airplane estimated price in USD. It can be estimated with the methods of Ch.4 or Appendix A.

N_e is the number of engines per airplane.

EP is the engine price (per engine) in USD. It can be found from Appendix B.

N_p is the number of propellers per airplane.

PP is the price per propeller in USD. It can be determined from Appendix B.

ASP is the avionics systems price per airplane in USD. This can be found from Appendix C.

DP_{ap} is the airplane depreciation period. This depends on the operator's business strategy. Table 5.7 may be used as a guide.

$U_{ann_{bl}}$ is the annual utilization in block hours. It may be found from page 76.

V_{bl} is the block speed in nm/hr as found from Equation (5.5).

The cost of engine depreciation per nautical mile, C_{deng}, may be determined from:

$$C_{deng} = \frac{(F_{deng})(N_e)(EP)}{(DP_{eng})(U_{ann_{bl}})(V_{bl})} \tag{5.42}$$

where: F_{deng} is the engine depreciation factor. This factor depends on the perceived resale value of each engine: see Table 5.7, p.107.

DP_{eng} is the engine depreciation period. This depends on the operator's business strategy. The data of Table 5.7 may serve as a guide.

The cost of depreciation of propellers, C_{dprp}, in USD/nm can be found from:

$$C_{dprp} = \frac{(F_{dprp})(N_p)(PP)}{(DP_{prp})(U_{ann_{bl}})(V_{bl})} \tag{5.43}$$

where: F_{dprp} is the propeller depreciation factor. This factor depends on the perceived resale value of each propeller: see Table 5.7, p.107.

DP_{prp} is the propeller depreciation period. This depends on the operator's business strategy. The data of Table 5.7 may serve as a guide.

The cost of depreciation of avionics systems, C_{dav}, in USD/nm may be estimated from:

$$C_{dav} = \frac{(F_{dav})(ASP)}{(DP_{av})(U_{ann_{bl}})(V_{bl})} \quad (5.44)$$

where: F_{dav} is the avionics systems depreciation factor. This factor depends on the perceived resale value of the avionics systems. Experience has indicated little or no resale value as shown in Table 5.7, p.107.

DP_{av} is the avionics systems depreciation period. This depends on the operator's business strategy AND on federal regulations (in terms of requiring new types of equipment: TCAS is an example). The data of Table 5.7, p.107 may serve as a guide.

The cost of depreciation of airplane spare parts, C_{dapsp}, in USD/nm may be estimated from:

$$C_{dapsp} = \frac{(F_{dapsp})(F_{apsp})\{(AEP) - (N_e)(EP)\}}{(DP_{apsp})(U_{ann_{bl}})(V_{bl})} \quad (5.45)$$

where: F_{dapsp} is the airplane spare parts depreciation factor. This factor depends on the perceived resale value of the airplane spare parts. The data of Table 5.7, p.107 may serve as a guide.

F_{apsp} is the airplane spare parts factor. It is equal to the ratio of airplane spare parts cost to the cost of the airplane minus engines. This factor depends on repair and maintenance operating experience with the airplane. Lacking actual data it is suggested to use: $F_{apsp} = 0.10$.

DP_{apsp} is the airplane spare parts depreciation period. This depends on the operator's business strategy. The data in Table 5.7 may serve as a guide.

Note: the airplane spare parts factor may be as-med to account also for spare parts needs for propellers and avionics systems.

The cost of engine spare parts depreciation, C_{dengsp}, in USD/nm can be estimated from:

$$C_{dengsp} = \frac{(F_{dengsp})(F_{engsp})(N_e)(EP)(ESPPF)}{(DP_{engsp})(U_{ann_{bl}})(V_{bl})} \tag{5.46}$$

where: F_{dengsp} is the engine spare parts depreciation factor. It depends on the perceived re-sale value of the engine spare parts. The data of Table 5.7 should serve as a guide.

F_{engsp} is the engine spare parts factor. It is equal to the ratio of the cost of engine spare parts to the cost of one engine. The engine manufacturer usually recommends a ratio for each given engine depending on experience in the field. For preliminary cost estimating purposes it is suggested to use: $F_{engsp} = 0.50$.

ESPPF is the engine spare parts price factor. It depends on the engine manufacturer's policy in pricing spare parts. If all engine com-ponents could be purchased at the same price as a fully assembled engine, the value of ESPPF would be 1.0. It is suggested to use: ESPPF = 1.50.

DP_{engsp} is the depreciation period for engine spare parts. It depends on the operator's business strategy. The data of Table 5.7 may be used as a guide.

Table 5.7 Suggested Depreciation Periods and Depreciation Factors for Airframe and Equipment

Item	Suggested Depreciation Period	Residual Value in Percent	Depreciation Factor*
Airframe	$DP_{ap} = 10$	15	$F_{dap} = 0.85$
Engines	$DP_{eng} = 7$	15	$F_{deng} = 0.85$
Propellers	$DP_{prp} = 7$	15	$F_{dprp} = 0.85$
Avionics	$DP_{av} = 5$	0	$F_{dav} = 1.00$
Airplane Spares	$DP_{apsp} = 10$	15	$F_{dapsp} = 0.85$
Engine Spares	$DP_{engsp} = 7$	15	$F_{dengsp} = 0.85$

* Depreciation factor =

= {1 - (Residual Value)/(Original Price)}

5.2.4 Direct Operating Cost of Landing Fees, Navigation Fees and Registry Taxes: DOC_{lnr}

There is no uniform agreement whether the operating costs caused by landing fees, navigation fees and registry taxes are to be considered direct or indirect operating costs. In this text the decision has been made to include them as direct operating costs.

The direct operating cost of landing fees, navigation fees and various taxes can be broken down in the following manner:

$$DOC_{lnr} = C_{lf} + C_{nf} + C_{rt} \tag{5.47}$$

where: C_{lf} is the direct operating cost due to landing fees in USD/nm. This can be found from:

$$C_{lf} = (C_{aplf})/\{(V_{bl})(t_{bl})\} \tag{5.48}$$

where: C_{aplf} is the airplane landing fee per landing. This is a function of airplane takeoff weight (it is assumed, that after landing, it will take off again!) AND of local airport authority policy. Lacking actual landing fee data it is suggested to use:

$$C_{aplf} = 0.002W_{TO} \text{ USD/lbs} \quad (5.49)$$

V_{bl} is the block speed in nm/hr as found from Eqn.(5.5).

t_{bl} is the block time as defined by Equation (5.6).

An alternate method is to consider C_{lf} to be a simple fraction of the DOC. Based on data from Ref.20 it is suggested to use:

$$C_{lf} = (f_{lf})(DOC) \quad (5.50)$$

where: f_{lf} is a factor which depends on airplane size. This factor depends on local airport authorities which set these landing fees. Lacking actual data it is suggested to use:

$$f_{lf} = 0.036 + 4(10^{-8})W_{TO} \quad (5.51)$$

where: W_{TO} is the airplane takeoff weight in lbs.

C_{nf} is the cost of navigation fees in USD/nm. It may be estimated from:

$$C_{nf} = (C_{apnf})/\{(V_{bl})(t_{bl})\} \quad (5.52)$$

where: C_{apnf} is the navigation fee charged per airplane per flight, expressed in USD/flight. This fee varies from route to route and from country to country.

Lacking actual data it is suggested to use:

For operations in the USA:

C_{apnf} = 0 USD/flight

For operations outside the USA:

C_{apnf} = 10.00 USD/flight

C_{rt} is the direct cost of registry taxes expressed in USD/nm. Based on data from Ref.20 it is suggested to use:

$$C_{rt} = (f_{rt})(DOC) \qquad (5.53)$$

where: f_{rt} is a factor which depends on airplane size. This factor depends on state and country governments which set these registry taxes. Lacking actual data it is suggested to use:

$$f_{rt} = 0.001 + (10^{-8})W_{TO} \qquad (5.54)$$

where: W_{TO} is the airplane takeoff weight in lbs.

5.2.5 Direct Operating Cost of Financing: DOC_{fin}

The direct operating cost of financing the airplane in USD/nm, DOC_{fin}, depends on how an operator is financing his fleet of airplanes. Operators can elect to borrow money to pay for airplane(s) and spare parts as well as to finance their operation. They frequently elect to lease part or all of their flying equipment.

Even when an operator decides to use his own money to finance the airplanes and their operation, there is a cost associated with that: interest.

Methods for estimating financing cost are judged to be beyond the scope of this text. However, the reader may wish to account for financing costs by the following simple 'rule-of-thumb':

$$DOC_{fin} = 0.07(DOC) \qquad (5.55)$$

This 'rule-of-thumb' is based on the observation that financing costs typically run to 7 percent of the total DOC as found from Eqn.(5.19).

5.3 METHOD FOR ESTIMATING THE INDIRECT OPERATING COST OF COMMERCIAL AIRPLANES: IOC

The purpose of this section is to provide a method for estimating the indirect operating cost (IOC) in USD/nm, incurred while operating commercial airplanes.

Table 5.8 presents an overview of cost components and cost elements which contribute to the indirect operating cost per nautical mile.

NOTE: The right hand side of Table 5.8 indicates which IOC elements apply to various airplane types.

The indirect cost associated with airplane operations vary significantly from one operator to another. Also, as will be seen, the airplane designer has very little influence over this cost category. The method proposed here to estimate the indirect operating cost assumes that the IOC can be expressed as a simple fraction of DOC:

$$IOC = f_{ioc}(DOC) \tag{5.56}$$

where: f_{ioc} is the fraction of DOC by which IOC is estimated. When reviewing published data on the relationship between IOC and DOC, it is found that f_{ioc} is strongly dependent on block distance, R_{bl}, as well as on the type of airplane. Figure 5.12 provides some data for this fraction.

Lacking actual IOC data it is suggested to use:

For transport operations: f_{ioc} as in Fig.5.12

For personal owner and for corporate airplanes: $f_{ioc} = 0$.

As will be seen, airplane designers have very little design leverage over the magnitude of the indirect operating costs of an airplane operator. Despite this fact, the author believes that it is desirable for aeronautical engineering students to develop at least a basic understanding of those factors which determine IOC. For that reason a discussion of IOC breakdown is presented next. For a discussion of IOC trends, see Section 5.5.

Table 5.8 Overview of Indirect Operating Cost (IOC)

INDIRECT OPERATING COST ~ IOC ~ USD/nm		AIRLINE	CORPORATE	PERSONAL
IOC COMPONENTS	IOC ELEMENTS			
PASSENGERS ~ IOC_{pax} (5.58)	MEALS ~ C_{mls}	✓	✓	NO
	INSURANCE ~ $C_{pax/ins}$	✓	✓	✓
	CABIN ATTENDANTS ~ C_{cat}	✓	✓	NO
	PASSENGER HANDLING, BAGGAGE HANDLING, SALES & RESERVATIONS, SECURITY, MISC ~ $C_{pax/gen}$	✓	NO	NO
STATION ~ IOC_{sta} (5.59)	MAINTENANCE OF GROUND EQUIPM'T & FACILITIES ~ C_{mgef}	✓	✓	NO
	DEPRECIATION OF GROUND EQUIPM'T & FACILITIES ~ C_{dgef}	✓	✓	NO
AIRPLANE SERVICE, CONTROL AND FREIGHT ~ IOC_{ascf} (5.60)	AIRPLANE SERVICE ~ C_{aps}	✓	✓	✓
	AIRPLANE CONTROL ~ C_{apc}	✓	✓	NO
	FREIGHT HANDLING ~ C_{frt}	✓	NO	NO
PROMOTION, SALES AND ENTERTAIN-MENT ~ IOC_{pse} (SUB-SECT. 5.3.4)	COMMISSIONS TO TRAVEL AGENCIES,	✓	NO	NO
	PUBLICITY AND ADVER-TISING,	✓	NO	NO
	ENTERTAINMENT	✓	NO	NO
GENERAL AND ADMINISTRATIVE ~ IOC_{gaa} (SUB-SECT. 5.3.5)	ADMINISTRATIVE, ACCOUNTING AND CORPORATE STAFF + FACILITIES	✓	✓	NO

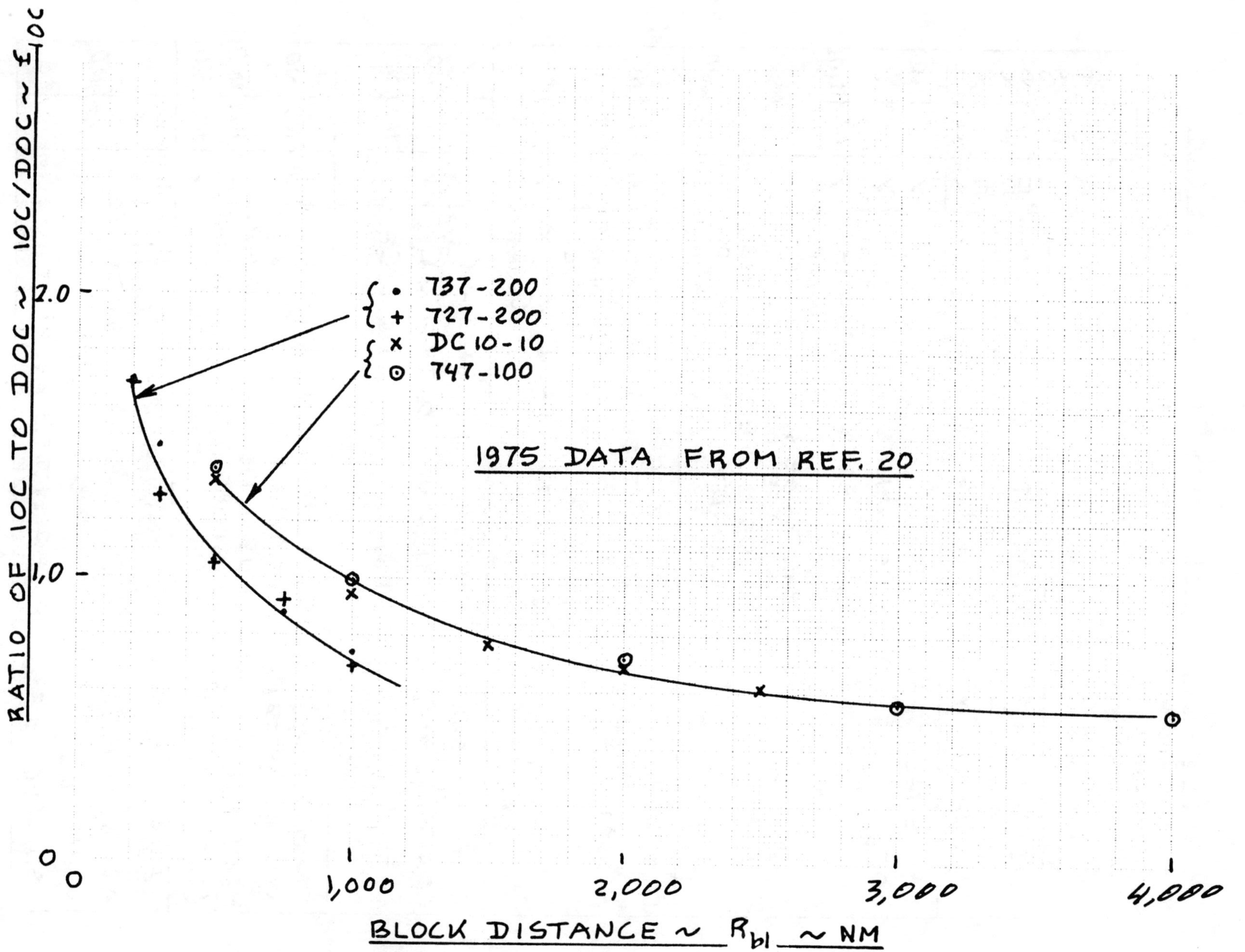

Figure 5.12 Effect of Block Distance on the Ratio of IOC to DOC

The indirect operating cost per nautical mile for a commercial airplane is broken down into the following IOC cost components:

$$IOC = IOC_{pax} + IOC_{sta} + IOC_{ascf} + \\ + IOC_{pse} + IOC_{gaa} \qquad (5.57)$$

where: IOC_{pax} is the indirect operating cost for passenger services, expressed in USD/nm: see Sub-section 5.3.1.

IOC_{sta} is the indirect cost for maintaining and depreciating ground equipment and ground facilities, expressed in USD/nm: see Sub-section 5.3.2.

IOC_{ascf} is the indirect operating cost for airplane and traffic servicing, control and for freight, expressed in USD/nm: see Sub-section 5.3.3.

IOC_{pse} is the indirect operating cost for promotion, sales and entertainment, expressed in USD/nm: see Sub-section 5.3.4.

IOC_{gaa} is the indirect operating cost for general administrative expenses, expressed in USD/nm: see Sub-section 5.3.5.

Sub-sections 5.3.1 through 5.3.5 contain breakdowns for these indirect operating cost components.

5.3.1 Indirect Operating Cost for Passenger Service: IOC_{pax}

The indirect operating cost for passenger services may be broken down as follows:

$$IOC_{pax} = C_{mls} + C_{pax/ins} + C_{cat} + C_{pax/gen} \qquad (5.58)$$

where: C_{mls} is the cost for meal service. This depends on the routes flown and on the level of service which the operator wants to provide for his passengers.

The airplane designer has minor leverage over this cost item. The design of the airplane galleys constitutes this leverage. No quantitative method for estimating this leverage is available.

$C_{pax/ins}$ is the cost for passenger insurance. There exists an international agreement between airline operators on the amount of passenger insurance that must be carried: the so-called Warsaw Convention. The cost of this insurance to an airline varies with the underwriter of that insurance and with the accident history of the airline.

Other than designing the airplane to be safer than previous airplanes, the airplane designer has no design leverage over this cost element.

C_{cat} is the cost for cabin attendants. The number of cabin attendants which must be carried aboard an airplane is regulated by the FAA. Part III, p.59 shows how many cabin attendants are required. By using salary data from Table 5.5 it is possible to determine the actual cost of flight attendants.

The designer has no leverage over this cost element.

$C_{pax/gen}$ is the cost associated with the following items:

* Passenger handling
* Passenger baggage handling
* Sales and reservations
* Security
* Miscellaneous passenger costs

No reliable quantitative method is available for estimating this cost element. The designer of the airplane does have some design leverage over the following items:

* design of baggage space and the ease of access to baggage handling personnel or equipment

* boarding and unboarding of passengers is affected significantly by the interior cabin layout.

5.3.2 Indirect Operating Costs for Station Operation: IOC_{sta}

The indirect operating cost for station operation can be broken down as follows:

$$IOC_{sta} = C_{mgef} + C_{dgef} \tag{5.59}$$

where: C_{mgef} is the cost of maintaining ground equipment and facilities

C_{dgef} is the cost of depreciation of ground equipment and facilities

The designer has some leverage over these cost elements: requirements for ground facilities and for ground equipment are dependent on the type of systems which need to be accessed for servicing while the airplane is at the gate. No quantative method is available to estimate this cost element.

5.3.3 Indirect Operating Cost for Airplane Service, Airplane Control and Freight: IOC_{ascf}

The indirect operating cost for aircraft service, control and freight is broken down into the following cost elements:

$$IOC_{ascf} = C_{aps} + C_{apctrl} + C_{frt} \tag{5.60}$$

where: C_{aps} is the cost of airplane service. It depends on how well the airplane was designed from a gate servicing viewpoint: see Part III, pages 82-83.

If all required service vehicles cannot have simultaneous access to the airplane then the cost of servicing will increase.

C_{apctrl} is the cost of airplane control. It depends on the amount of equipment and personnel needed to push (or maneuver) the airplane away from the gate or help it approach the gate.

C_{frt} is the cost associated with handling freight. This freight is NOT passenger baggage. Many airlines will try to carry extra freight when possible to bring in additional revenue.

Depending on how this freight must be handled (people, containers, handling equipment) this cost component can be high or low. The airplane designer has definite design leverage over this cost element. Sub-section 3.3.5 in Part III provides details on the layout of cargo/baggage holds.

5.3.4 Indirect Operating Cost for Promotion, Sales and Entertainment: IOC_{pse}

The following activities contribute to this indirect cost component:

* Commissions to travel agencies

* Publicity and advertising campaigns

* Entertainment

The airplane designer has no leverage over this item unless there is something about the appearance or public appeal of the airplane that causes this cost to increase.

No quantitative method for estimating this cost component is avalable.

5.3.5 Indirect Operating Cost for General and Administrative Cost: IOC_{gaa}

The following activities contribute to this indirect cost component:

* requirements for administrative and accounting personnel as well as for their facilities

* requirements for corporate staffers and their facilities

The mentality of upper level management has a major effect on the magnitude of this cost item: the degree of fanciness in the corporate offices and the various perks associated with company officers all influence this cost item. No quantitative method is available for estimating this cost component. The airplane designer certainly has no design leverage over this cost component.

5.4 EXAMPLE APPLICATION

The purpose of this section is to present an example application of the methods for estimating program operating cost, C_{OPS}, direct operating cost, DOC, and indirect operating cost, IOC, of Sections 5.1, 5.2 and 5.3 respectively. The example has been constructed around the Ourania jet transport, to be consistent with the examples given in Sections 3.8 and 4.6. The data will be computed for calendar year 1990.

The material is organized as follows:

5.4.1 Ourania Mission Data

5.4.2 Calculation of Direct Operating Cost: DOC

5.4.3 Calculation of Indirect Operating Cost: IOC

5.4.4 Calculation of Program Operating Cost: C_{OPS}

5.4.5 Discussion of Results

5.4.1 Ourania Mission Data

Before operating costs can be determined, some baseline data about the airplane mission are needed. These baseline data are essentially those dealing with block distance, block speed and block times. Definitions and formulas for these factors were given in Section 5.1.

The block distance for the Ourania will be assumed to be equal to the 1,500 nm range requirement specified in Table 2.18 of Part I: R_{bl} = 1,500 nm.

Block time, t_{bl}, is determined from Eqn.(5.6). This in turn requires the ground maneuvering time, t_{gm}, the time-to-climb, t_{cl}, the time to cruise, t_{cr}, and the time to descend, t_{de}. These times will be found next.

The ground maneuvering time, t_{gm}, is obtained from Eqn.(5.7). This in turn requires the takeoff weight, W_{TO}, which is: W_{TO} = 123,683 from p.11, Part V.

Therefore: t_{gm} = 0.51(0.123683) + 0.125 = 0.19 hrs

The time-to-climb, t_{cl}, is found from p.56 of Part I as: t_{cl} = 14 min. = 0.23 hrs

The time-to-cruise, t_{cr}, is found from Eqn.(5.8a). This in turn requires the block distance, R_{bl}, the distance-to-climb, R_{cl}, the distance-to-descend, R_{de}, the maneuvering distance, R_{man}, and the cruise speed, V_{cr}. These distances will be found next.

The distance-to-climb is found from p.56, Part I as: R_{cl} = 64 nm.

The distance-to-descend is found from Eqn.(5.10). The average speed during the descent will be assumed to be 250 kts. For the Ourania it will also be assumed that the average descent rate from 35,000 ft is 2,500 ft/min so that:

t_{de} = 35,000/2,500 = 14 min = 0.23 hrs. Therefore:

R_{de} = 250(0.23) = 58 nm

The maneuvering distance, R_{man}, is determined from Eqn.(5.11). The maneuvering speed is set equal to the cruise speed: V_{man} = V_{cr} = 473 kts (p.56, Part I).

The maneuvering time, t_{man}, follows from Eqn.(5.12):

t_{man} = 0.25(0.123683) + 0.0625 = 0.093 hrs

Therefore:

R_{man} = 473(0.093) = 44 nm

The cruise time now follows from Eqn.(5.8a):

t_{cr} = {1.06(1,500) - 64 - 58 + 44}/473 =

= 1,512/473 = 3.20 hrs

The block time now follows from Eqn.(5.6):

t_{bl} = 0.19 + 0.23 + 3.20 + 0.23 = 3.85 hrs

The block speed can now be found from Eqn.(5.5) as:

$$V_{bl} = 1{,}500/3.85 = 390 \text{ kts}$$

The annual utilization in block hours may be computed from Eqn.(5.14a) as:

$$U_{ann_{bl}} = (10)^3[3.4546(3.85) + 2.994 +$$

$$- \{12.289(3.85)^2 - 5.6626(3.85) + 8.964\}^{1/2}] = 3{,}282 \text{ hrs}$$

NOTE: this annual utilization amounts to an average daily utilization of about 9 hours. For a well run airline, on 1,500 nm stage lengths, this is a reasonable number. Table 5.3 also substantiates this.

5.4.2 Calculation of Direct Operating Cost: DOC

The Ourania direct operating cost in USD/nm is computed with Eqn.(5.19). The five cost components of DOC are detailed in sub-sections 5.2.1 through 5.2.5. The calculations for these cost components are presented in sub-sub-sections 5.4.2.1 through 5.4.2.5. The DOC is computed in sub-sub-section 5.4.2.6.

5.4.2.1 Direct operating cost of flying: DOC_{flt}

This cost is found with Eqn.(5.20).

Crew Cost: C_{crew}

This cost may be estimated with Eqn.(5.21).

The mission specification for the Ourania jet transport of Table 2.18, Part I, p.55 specifies a crew of two. Therefore, in Eqn.(5.21):

$n_{c_1} = 1$ and also: $n_{c_2} = 1$

From page 83: $K_j = 0.26$

The crew salaries for the captain (n_{c_1}) and for the first officer (n_{c_2}) are determined from Table 5.5. The annual flight hours for each are found from p.84:

$SAL_1 = 85{,}000$ USD/yr $\qquad AH_1 = 800$ hrs

$SAL_2 = 50{,}000$ USD/yr $\qquad AH_2 = 800$ hrs

These data are assumed to be applicable for 1990.

The travel expense factor, TEF, according to p.84 is:

$$TEF_1 = TEF_2 = 7 \text{ USD/blhr}$$

The crew cost per nautical mile can now be estimated from Eqn.(5.21) as:

$$C_{crew} = \{(1.26/390)(85{,}000/800) + 7/390\} +$$

$$+ \{(1.26/390)(50{,}000/800) + 7/390\} =$$

$$= 0.361 + 0.220 = 0.581 \text{ USD/nm}$$

Fuel and oil cost: C_{pol}

This cost is determined with Eqn.(5.30)

The block fuel used, $W_{F_{bl}}$, is the same as the mission fuel used, $W_{F_{used}}$, which in turn is given by Eqn.(2.14) of Part I. Reference is made to the example calculation for fuel used in sub-section 2.6.2 of Part I. Note that mission phases 6 and 8 as depicted in the mission profile of Table 2.18 (Part I) should not be counted as normal operational fuel used. Therefore, the M_{ff} calculation of p.58 of Part I must be modified to:

$$M_{ff} = (0.992)(1)(0.990)(1)(0.909)(0.980)(0.995)*$$

Mission Phase: 9 8 7 6 5 4 3

$$*(0.990)(0.990) = 0.853$$

Mission Phase: 2 1

The mission phase numbers refer to Table 2.18 in Part I. This yields for the operational mission fuel used, $W_{F_{used}}$, with Eqn.(2.14) of Part I:

$$W_{F_{bl}} = (1 - 0.853)(123{,}683) = 18{,}181 \text{ lbs}$$

The 1990 price of Jet A fuel will be assumed to be 0.65 USD/gallon (Figure 5.3), while the density (p.88) is

6.74 lbs/gallon. Therefore, with Eqn.(5.30):

$$C_{pol} = (1.05)(18,181/1,500)(0.65/6.74) =$$

$$= 1.227 \text{ USD/nm}$$

Insurance cost: C_{ins}

This cost is found from Eqn.(5.32):

$$C_{ins} = 0.02(DOC)$$

The direct operating cost of flying may now be computed from Eqn.(5.20) as:

$$DOC_{flt} = 0.581 + 1.227 + 0.02(DOC) =$$

$$= 1.808 + 0.02(DOC) \text{ USD/nm}$$

5.4.2.2 Direct operating cost of maintenance: DOC_{maint}

This cost is found from Eqn.(5.33).

Maintenance labor cost for airframe and systems: $C_{lab/ap}$

This cost may be calculated from Eqn.(5.34). The number of maintenance manhours per blockhour for airframe and systems, $MHR_{map_{bl}}$, may be found from Figure 5.5. This is seen to depend on the airframe weight, W_A, defined as:

$$W_A = W_E - W_{pwrplt} = 65,133 - 9,891 = 55,242 \text{ lbs},$$

where the weight data were found on p.11 of Part V.

Figure 5.5 now yields:

$$MHR_{map_{bl}} = 3.0 + 0.067(55.242) = 6.7 \text{ mhrs/blhr}$$

The airplane maintenance labor rate per manhour is assumed from Table 5.5 to be: $R_{l_{ap}}$ = 16 USD/mhr.

Therefore, with Eqn.(5.34):

$$C_{lab/ap} = 1.03(6.7)(16)/390 = 0.283 \text{ USD/nm}$$

Maintenance labor cost for engines: $C_{lab/eng}$

This cost is found from Eqn.(5.36). The number of engine maintenance hours per engine, per block hour, $MHR_{meng_{bl}}$, may be determined from Figure 5.6. It is seen to depend on the takeoff thrust per engine, (T_{TO}/N_e), and on the attained period between engine overhauls, H_{em}.

The takeoff thrust per engine is found from page 184 in Part I as:

$$(T_{TO}/N_e) = 47,625/2 = 23,813 \text{ lbs}$$

The attained overhaul period for the engine, H_{em}, is assumed to be 3,000 hrs (see p.98). For a new engine, this is not an unreasonably high number. Therefore, from the equation in Figure 5.6:

$$MHR_{meng_{bl}} = [\{0.718 + 0.0317(23.813)\}*$$

$$*(1,100/3,000) + 0.10] = 0.64 \text{ mhrs/blhr}$$

The hourly labor rate for engine maintenance will be assumed to be the same as that for the airframe:

$R_{l_{eng}}$ = 16 USD/hr so that with Eqn.(5.36):

$$C_{lab/eng} = 1.03(1.3)(2)(0.64)(16)/390 = 0.070 \text{ USD/nm}$$

<u>NOTE:</u> it is of interest to 'check' whether the total manhours for maintenance are reasonable. This total is: 6.7 + 2(0.64) = 7.98 mhrs/blhr. This can be translated into mhrs/flthr with Eqn.(5.35). The value for t_{flt} is found from Eqn.(5.18) as:

$$t_{flt} = 0.23 + 3.20 + 0.23 = 3.66 \text{ hrs}$$

The block time, t_{bl}, was found to be 3.85 hours, so that the maintenance manhours per flight hour follow from Eqn.(5.35) as:

$$\text{MHR per flt hour} = (\text{MHR per bl hour})(t_{bl})/(t_{flt}) =$$

$$= 7.98(3.85/3.66) = 8.4 \text{ mhrs/flthr.}$$

This number appears reasonable when compared with the data in Table 5.6.

Maintenance materials cost for airplane and systems: $C_{mat/ap}$

This cost can be found from Eqn.(5.37). The cost of airplane maintenance materials per blockhour, $C_{mat/apblhr}$, is determined from Figure 5.8. It depends on the airframe price, AFP:

$$AFP = AEP - N_e(EP) =$$

$$= 22,576,000 - 2(1,558,607) = \text{USD } 19,458,786$$

where: the airplane and engine prices are found from pages 62 and 58 (Section 4.6), respectively.

With Figure 5.8:

$$C_{mat/apblhr} = 30.0 + 0.79(194.58786) =$$

$$= 183.72 \text{ USD/blhr}$$

where: the 1990 to 1989 CEF ratio has been assumed to be equal to 1.0.

Therefore, with Eqn.(5.37):

$$C_{mat/ap} = 1.03(183.72)/390 = 0.485 \text{ USD/nm}$$

Maintenance materials cost for engines: $C_{mat/eng}$

This cost is computed from Eqn.(5.38). The cost of engine maintenance materials per blockhour, $C_{mat/engblhr}$, may be found from Figures 5.9 and 5.11. It was already assumed that the attained overhaul period for the engines is H_{em} = 3,000 hrs (See p.122). Thus, from Figure 5.11:

$K_{H_{em}} = 1.40.$

Then, with the equation in Figure 5.9:

$$C_{mat/engblhr} = \{0.0543(1,558.607)(1.5) +$$

$$- 0.47\}/1.40 = 90.34 \text{ USD/blhr}$$

Thus, with Eqn.(5.38):

$$C_{mat/eng} = 1.03(1.3)(2)(90.34)/390 = 0.620 \text{ USD/nm}$$

Applied maintenance burden cost: C_{amb}

This cost may be found from Eqn.(5.39). The overhead distribution factors are selected from the top of page 102 as:

$f_{amb/lab} = 1.30$ $\qquad$ $f_{amb/mat} = 0.60$

Therefore from Eqn.(5.39):

$$C_{amb} = 1.03[1.30\{(6.7)(16) + 2(0.64)(16)\} + $$
$$+ 0.60\{183.72 + 2(90.34)\}]/390 = 1.016 \text{ USD/nm}$$

Total direct maintenance cost: DOC_{maint}

The total direct cost for maintenance now follows with Eqn.(5.33) as:

$$DOC_{maint} = 0.283 + 0.070 + 0.485 + 0.620 + 1.016 =$$
$$= 2.474 \text{ USD/nm}$$

5.4.2.3 Direct operating cost of depreciation: DOC_{depr}

This cost is found with Eqn.(5.40).

Airplane Depreciation Cost: C_{dap}

This cost may be computed with Eqn.(5.41).

The following terms are found in Table 5.7:

Airframe depreciation factor: $F_{dap} = 0.85$

Airplane depreciation period: $DP_{ap} = 10$ yrs

The following price terms were found in Section 4.6:

Airplane estimated price:	AEP =	USD 22,576,000
Engine price:	EP =	USD 1,558,607
Avionics systems price:	ASP =	USD 2,670,000

Number of propellers: $N_p = 0$. Now, with Eqn.(5.41):

$$C_{dap} = \frac{0.85\{22,576,000 - 2(1,558,607) - 2,670,000\}}{10(3,282)(390)} =$$

$$= 1.115 \text{ USD/nm}$$

Engine Depreciation Cost: C_{deng}

This cost is determined with Eqn.(5.42).

The following terms are found in Table 5.7:

Engine price depreciation factor: $F_{deng} = 0.85$

Engine depreciation period: $DP_{eng} = 7$ yrs

With Eqn.(5.42):

$C_{deng} = \{0.85(2)(1,558,607)\}/\{7(3,282)(390)\} =$

$= 0.296$ USD/nm

Propeller Depreciation Cost: C_{dprp}

This cost is zero: the Ourania has no propellers.

Avionics Systems Depreciation Cost: C_{dav}

This cost may be found with Eqn.(5.44).

The following terms are found in Table 5.7:

Avionics systems depreciation factor: $F_{dav} = 1.00$

Avionics systems depreciation period: $DP_{av} = 5$ yrs

With Eqn.(5.44):

$C_{dav} = \{1.0(2,670,000)\}/\{5(3,282)(390)\} =$

$= 0.417$ USD/nm

Airplane Spare Parts Depreciation Cost: C_{dapsp}

This cost may be found with Eqn.(5.45).

The following terms are found from Table 5.7:

Airplane spare parts depreciation factor:

$F_{dapsp} = 0.85$

Airplane spare parts depreciation period:

$DP_{apsp} = 10$ yrs

The airplane spare parts factor, F_{apsp}, is assumed to have the value 0.10: see p.105.

With Eqn.(5.45):

$$C_{dapsp} = \{0.85(0.1)(19,458,786)\}/\{10(3,282)(390)\} =$$
$$= 0.129 \text{ USD/nm}$$

Engine Spare Parts Depreciation Cost: C_{dengsp}

This cost may be estimated from Eqn.(5.46).

The following terms are found in Table 5.7:

Engine spare parts depreciation factor:

$$F_{dengsp} = 0.85$$

Engine spare parts depreciation period:

$$DP_{engsp} = 7 \text{ yrs}$$

The engine spare parts factor, F_{engsp}, will be set equal to 0.50 as suggested on p.106. The engine spare parts price factor, ESPPF will be set equal to 1.50 as suggested on page 106. Therefore, with Eqn.(5.46):

$$C_{dengsp} = \{0.85(0.50)(2)(1,558,607)*$$
$$*(1.5)\}/\{7(3,282)(390)\} = 0.222 \text{ USD/nm}$$

Total Direct Operating Cost of Depreciation: DOC_{depr}

From Eqn.(5.40):

$$DOC_{depr} = 1.115 + 0.296 + 0 + 0.417 + 0.129 +$$
$$+ 0.222 = 2.181 \text{ USD/nm}$$

5.4.2.4 Direct operating cost of landing fees, navigation fees and registry taxes: DOC_{lnr}

This cost may be determined with Eqn.(5.47).

Cost of Landing Fees: C_{lf}

This cost is estimated with Eqn.(5.48).

The landing fee per landing, C_{aplf}, is determined from Eqn.(5.49) as:

$$C_{aplf} = 0.002(123,683) = USD\ 247.37$$

Therefore, with Eqn.(5.48):

$$C_{lf} = 247.37/\{(390)(3.85)\} = 0.165\ USD/nm$$

Cost of Navigation Fees: C_{nf}

According to Eqn.(5.52) and p.109, for domestic operations:

$$C_{nf} = 0$$

Cost of Registry Taxes: C_{rt}

This cost is estimated with Eqn.(5.53).

The factor f_{rt} is determined from Eqn.(5.54) as:

$$f_{rt} = 0.001 + 0.00123683 = 0.00224$$

Therefore, with Eqn.(5.53):

$$C_{rt} = 0.00224(DOC)$$

Total Direct Operating Cost of Landing Fees, Navigation and Registry Taxes: DOC_{lnr}

With Eqn.(5.47):

$$DOC_{lnr} = 0.165 + 0 + 0.00224(DOC) =$$

$$= 0.165 + 0.00224(DOC)\ USD/nm$$

5.4.2.5 Direct operating cost of financing: DOC_{fin}

Lacking better information, Eqn.(5.55) will be used:

$$DOC_{fin} = 0.07(DOC)$$

5.4.2.6 Total direct operating cost: DOC

The DOC cost components of sub-sub-sections 5.4.2.1 through 5.4.2.5 can be combined with Eqn.(5.19) to yield:

$$DOC = 1.808 + 0.02(DOC) + 2.474 + 2.180 + 0.165 +$$

$$+ 0.00224(DOC) + 0.07(DOC) =$$

$$= 6.627 + 0.09224(DOC) \text{ USD/nm}$$

or: $DOC = 7.300$ USD/nm

5.4.3 Calculation of Indirect Operating Cost: IOC

The Ourania indirect operating cost may be estimated with Eqn.(5.56) where the factor f_{ioc} is extrapolated from Figure 5.12 as:

$$f_{ioc} = 0.50$$

Therefore: $IOC = 0.50(DOC) = 0.5(7.300) =$

$$= 3.650 \text{ USD/nm}$$

5.4.4 Calculation of Program Operating Cost: C_{OPS}

The Ourania program operating cost may be estimated from Eqns (5.1) - (5.3).

It will be assumed that a total of 500 airplanes will be acquired and that all operators have the same route system:

$$N_{acq} = 500$$

The total annual block miles per airplane, $R_{bl_{ann}}$, may be found from Eqn.(5.4):

$$R_{bl_{ann}} = (390)(3,282) = 1,279,980 \text{ nm}$$

The number of years during which the airplane is being operated will be assumed to be equal to the airframe depreciation period, which is 10 years:

$$N_{yr} = 10 \text{ yrs}$$

Therefore, with Eqns (5.1) - (5.3):

$$C_{OPS} = (7.300)(1,279,980)(10)(500) +$$

$$+ (3.650)(1,279,980)(10)(500) =$$

$$= \text{USD } 70,078,905,000$$

5.4.5 Discussion of Results

The significance of the numerical results which were obtained in this section are discussed as follows:

5.4.5.1 Discussion of DOC results
5.4.5.2 Discussion of Program Cost results
5.4.5.3 Discussion of DOC trade studies

5.4.5.1 Discussion of DOC results

To determine the realism of the DOC predictions for the Ourania transport, several of its DOC components will be compared with DOC data of existing airplanes.

Table 5.9 provides the information. Aviation Week and Space Technology publishes DOC data regularly. The data in Table 5.9 were published in the August 28, 1989 issue, however: these data are for the year 1988.

Note that the data in Table 5.9 are given in USD per block hour. The Ourania DOC predictions of sub-section 5.4.2 are in USD per nautical mile. To convert it was necessary to multiply by the block speed: 390 kts.

The Ourania data are for the year 1990. The CEF-ratio from 1990 to 1988 is 1.003 according to Figure 2.7. This is so close to 1.0 that the Ourania data may be compared with the 1988 data for comparable airplanes.

The Ourania estimates in Table 5.9 appear to be very reasonable when compared to the data for comparable airplanes. However, the following question should be asked:

'Can a new design, such as the Ourania, be viable if its total DOC is not significantly better than that of existing airplanes?'

To address this question, the following point should be made: the DOC total for the Ourania is inflated by the selection of a 10 year depreciation period. Figure 5.13 presents the results of a trade study which shows the effect of the depreciation period on DOC. The depreciation period used by the airlines in conjunction with the data in Table 5.9 is not known. The author believes that this

Table 5.9 Direct Operating Cost Data in USD/blhr for Several Jet Transports

Source of airline data: Aviation Week and Space Technology, August 28, 1989*

Type No. of Airlines No. of Airplanes	Relative Cost	Crew	POL	Total Flight DOC	Airplane Maint.	Engine Maint.	Total Maint.	Total DOC
Boeing 727-200	High	568	834	1,633	266	195	539	2,306
11 airlines	Low	217	674	1,091	108	49	341	1,711
809 airplanes	Average	442	711	1,286	182	118	445	1,891
Boeing 737	High	449	549	1,707	269	178	484	2,250
13 airlines	Low	165	399	749	43	14	159	1,159
661 airplanes	Average	282	446	1,104	152	84	318	1,540
Boeing 757	High	547	574	1,546	152	241	415	2,136
4 airlines	Low	187	509	1,075	80	11	190	1,681
98 airplanes	Average	410	549	1,301	106	157	334	1,933
McDD MD-80	High	561	782	1,525	321	859	1,411	3,377
8 airlines	Low	196	455	1,028	44	32	124	1,380
305 airplanes	Average	405	530	1,245	127	203	416	1,884
McDD DC-9	High	485	781	1,953	311	337	773	2,749
8 airlines	Low	181	272	777	52	53	225	1,268
445 airplanes	Average	366	478	1,044	181	143	423	1,569
Ourania**	Estimated in							
500 airplanes	5.4.2***	227	479	766****	300	269	965	2,847

* 1988 Data ** 1990 Estimates

*** The data in 5.4.2 are in USD/nm. They must be multiplied by the block speed of 390 kts to obtain the data for this table.

**** Includes 0.02DOC for insurance

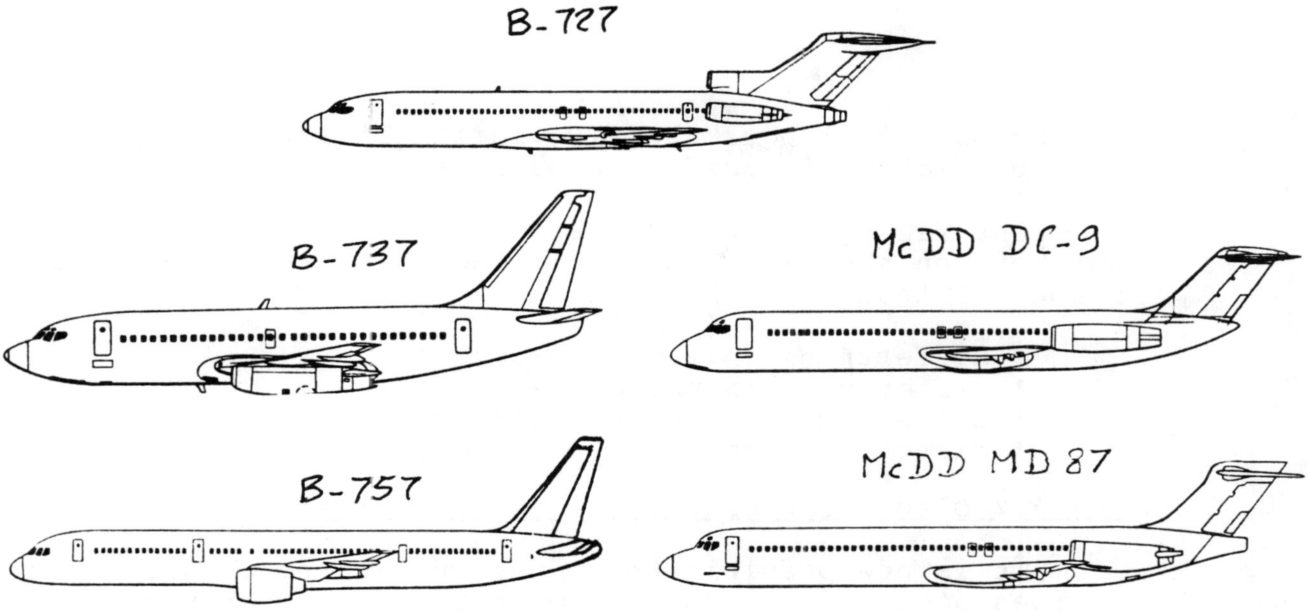

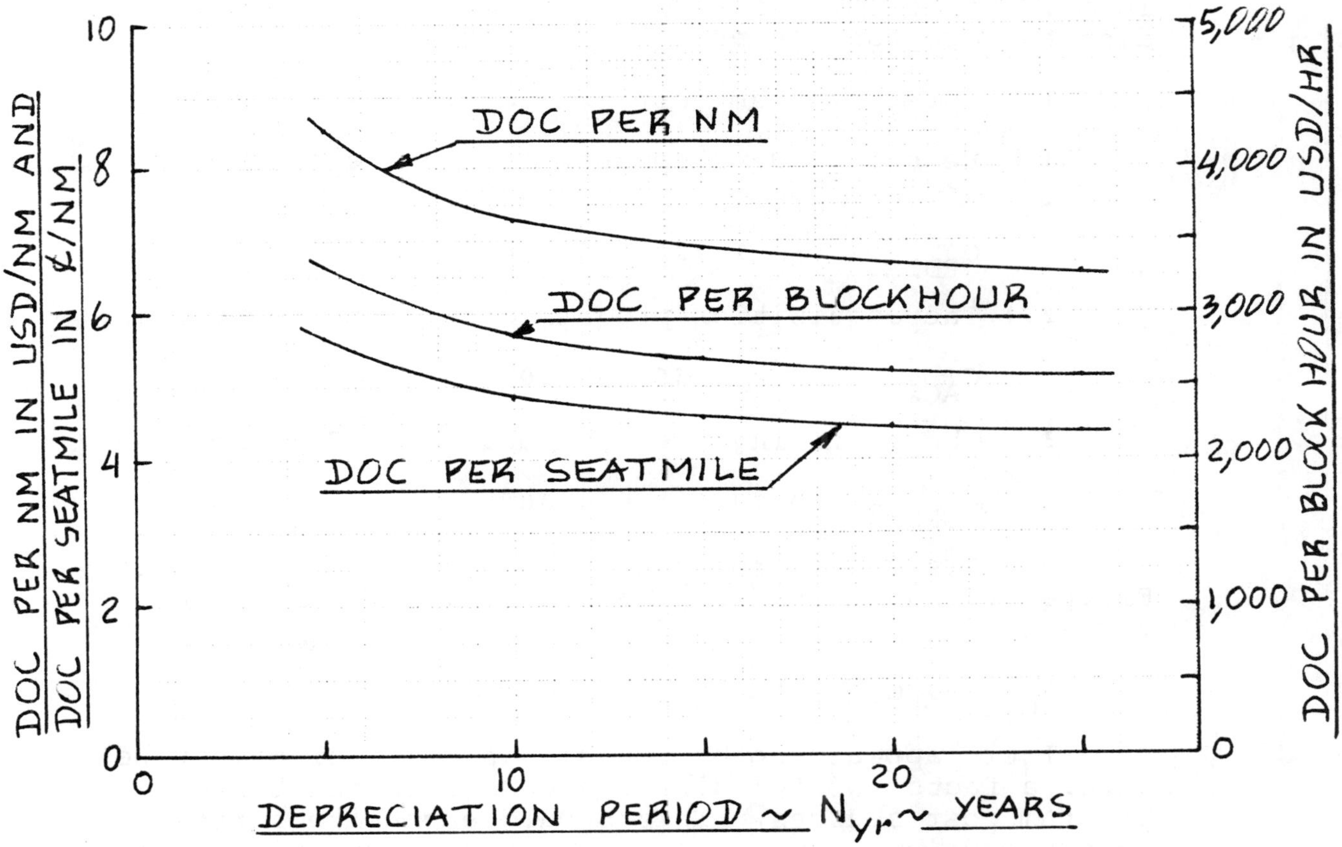

Figure 5.13 Effect of Depreciation Period on the Direct Operating Cost of the Ourania Transport

period is much longer than 10 years! If a 20 year period of depreciation is used, the Ourania data look better!

One final comment on the data in Table 5.9: the wide gap between the 'high' and the 'low' data is caused by a number of factors:

* differences in route structure
* differences in pay scales
* differences in depreciation strategies
* differences in age of airplanes

5.4.5.2 Discussion of program cost results

It is now possible to compare the three program cost sources of Eqn.(2.3) for the Ourania program with each other. The fourth cost source, C_{DISP}, was not considered because the expected life of the airplane is considerably longer than the 10 year depreciation period used in finding the program operating cost data.

The three program cost sources are as follows:

1.) Research, development, test and evaluation program cost, from Section 3.8:

$$C_{RDTE} = \text{USD } 1,232,000,000$$

2.) The acquisition program cost, from Section 4.6:

$$C_{ACQ} = \text{USD } 10,056,000,000$$

3.) The operating program cost, from Section 5.4:

$$C_{OPS} = \text{USD } 70,078,905,000$$

What is clear from all this is that the trends of Figure 2.5 and therefore inequality (2.4) are indeed correct:

$$C_{OPS} \gg C_{ACQ} \gg C_{RDTE}$$

There appears to be roughly an order of magnitude (i.e. a factor of 10) difference between the three major program cost sources. This should emphasize further the points made by Figure 2.3: early design decision making has a large impact on the ultimate cost of an airplane program!

5.4.5.3 Discussion of DOC trade studies

The purpose of this sub-sub-section is to present and discuss the results of several trade studies which show how DOC is affected by the following parameters:

1) Block distance, R_{bl}

2) Crew salary, SAL

3) Fuel Price, FP

4) Specific fuel consumption, c_j

5) Airframe and systems maintenance manhours per block hour, $MHR_{map_{bl}}$

6) Take-off weight, W_{TO}

The results of these trade studies are presented in Figures 5.14 through 5.19.

Figure 5.14 shows the strong effect of block distance on DOC: decreasing the block distance below 1,000 nm has a very inflationary effect on DOC.

The effect of crew salary on DOC is seen to be very weak in Figure 5.15: doubling the crew salary has only a marginal effect on DOC.

Figure 5.16 indicates the very strong effect of fuel price on DOC. Figure 5.17 shows that specific fuel consumption also has a strong effect on DOC. If fuel prices rise too much, there will be strong pressure on airplane designers to seek ways to:

1) reduce fuel consumption: lower drag!

and

2) decrease specific fuel consumption: more advanced engine technology!

It is seen from Figure 5.18 that the number of airplane maintenance manhours per block hour is a strong factor in determining airplane DOC. Anything that the designer can do to lower the need for maintenance will be benificial. However, read the caution notes below!

Finally, Figure 5.19 shows that an increase in take-off weight (as caused by an increase in empty weight) also drives up the DOC of an airplane. Thus, designing for low empty weight helps to keep the DOC down.

NOTES OF CAUTION:

I If a reduction in takeoff weight is achieved by use of more expensive materials or manufacturing techniques,

and:

If a reduction in drag or engine specifics is achieved by higher cost in manufacturing of airframe and/or engine,

then it will be necessary to evaluate the net benefit on DOC. The data presented here assume that ALL ELSE is kept constant. That is hardly ever the case!

II. Several cost contributions in this chapter were expressed as a fraction of DOC. One example is the IOC of sub-section 5.4.3. There is no reason why IOC should change if parameters such as takeoff weight, fuel price, engine specific fuel consumption, etc. are changed. The effect of these parameters on the total operating cost (TOC) should therefore be evaluated with the help of a baseline IOC. This baseline IOC must be kept constant while parameters which do not affect IOC are being varied.

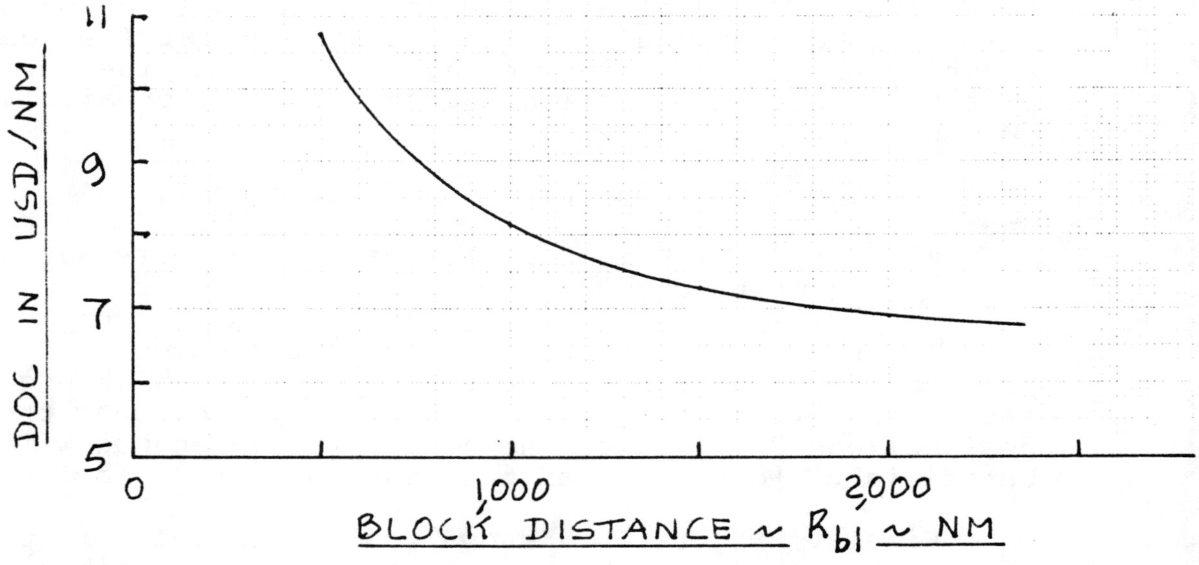

Figure 5.14 Effect of Block Distance on the Direct Operating Cost of the Ourania Transport

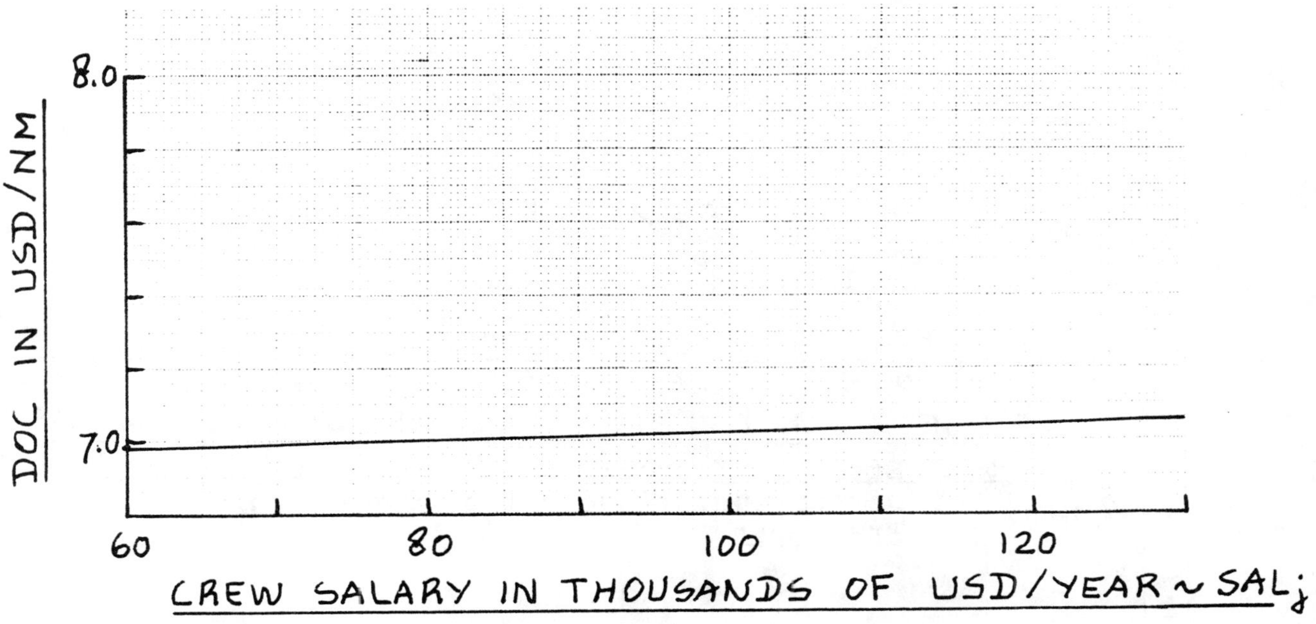

Figure 5.15 Effect of Crew Salary on the Direct Operating Cost of the Ourania Transport

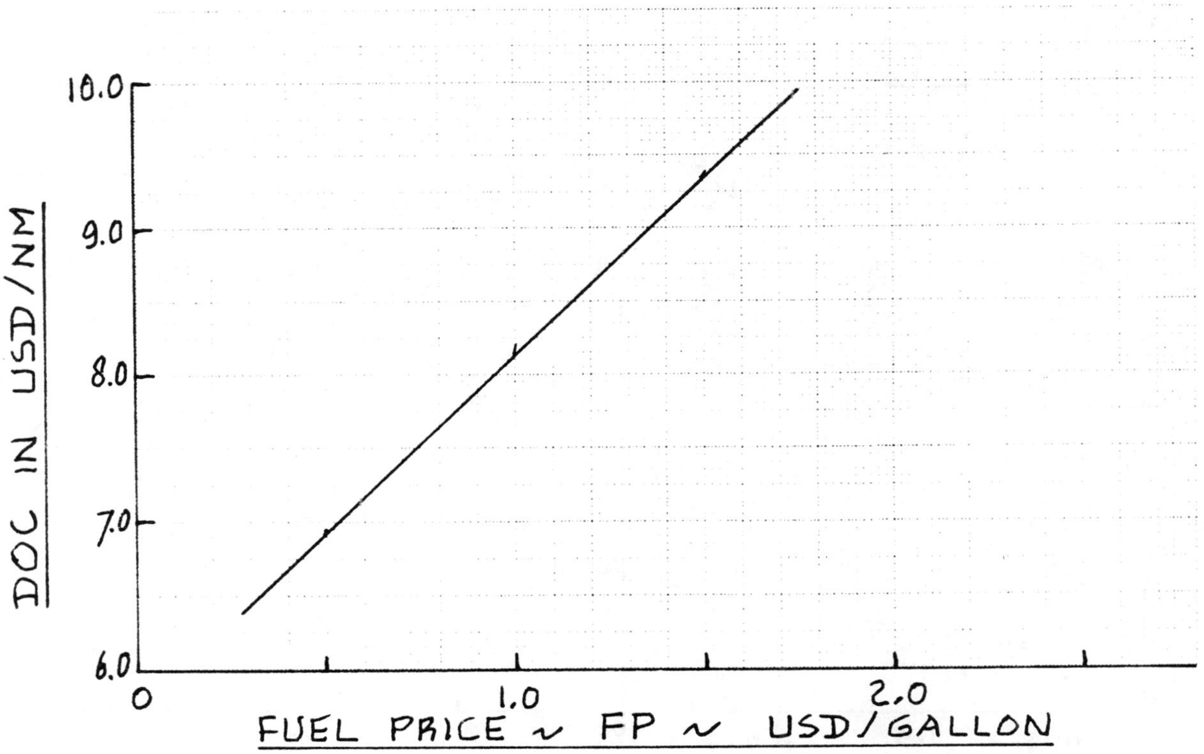

Figure 5.16 Effect of Fuel Price on the Direct Operating Cost of the Ourania Transport

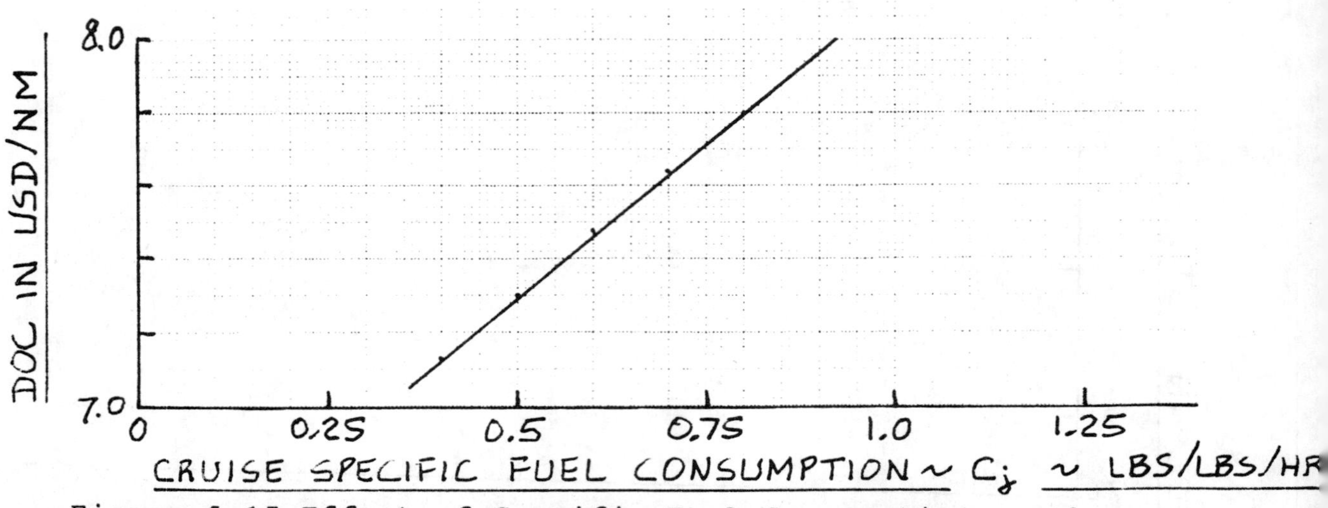

Figure 5.17 Effect of Specific Fuel Consumption on the Direct Operating Cost of the Ourania Transport

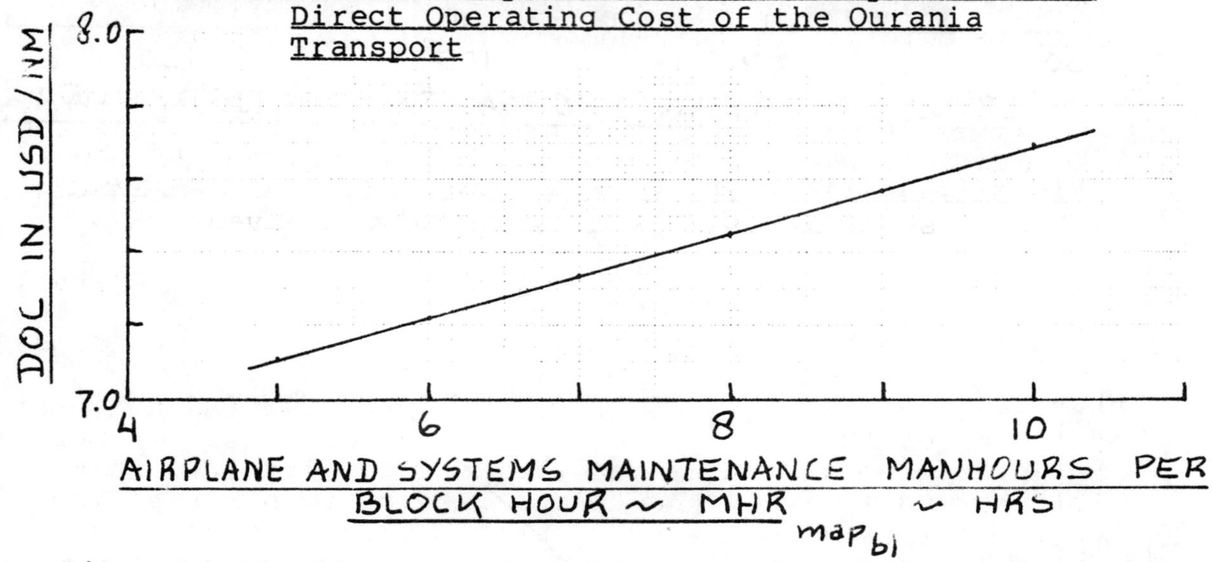

Figure 5.18 Effect of Maintenance Manhours per Block Hour on the Direct Operating Cost of the Ourania Transport

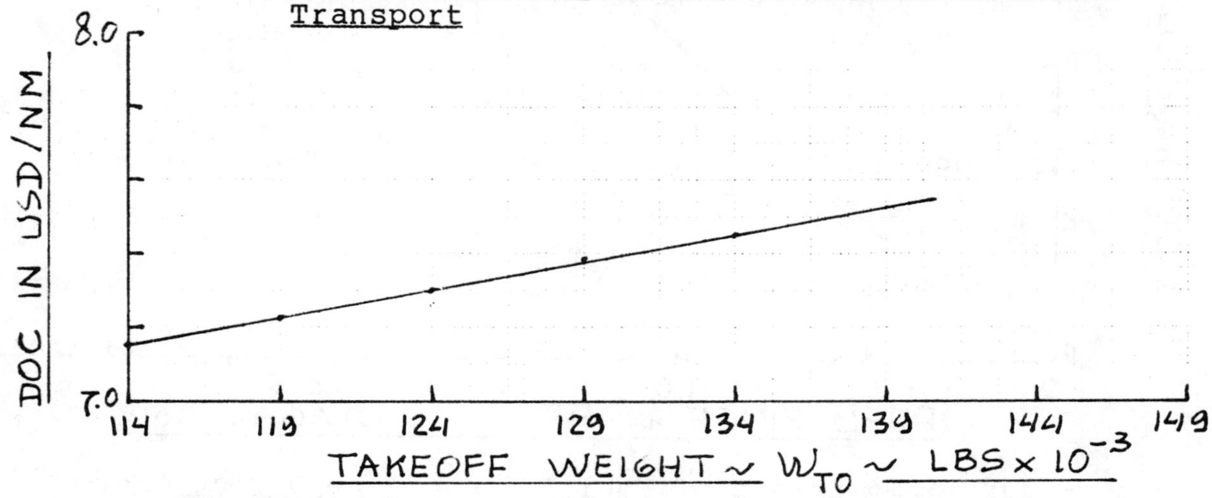

Figure 5.19 Effect of Takeoff Weight on the Direct Operating Cost of the Ourania Transport

5.5 SOME IMPORTANT OBSERVATIONS ABOUT COMMERCIAL AIRPLANE OPERATING COST

The purpose of this section is to discuss a number of important aspects of airplane operating costs: direct (DOC), indirect (IOC) and total (TOC). The total operating cost, TOC, is related to the direct (DOC) and indirect (IOC) operating costs by:

$$TOC = DOC + IOC \qquad (5.61)$$

Table 5.10 shows some historical trends of the operating cost components as defined by Equations (5.19) and (5.57). Note that the financing cost is not included because this is not normally disclosed by airlines.

Note that the DOC over the last decade has gradually decreased from 56 percent to 53 percent of TOC, while the IOC has increased from 44 percent to 47 percent of TOC. The reason for these changes is not known. Based on the data, it seems reasonable to assume for the year 1985:

$$DOC = 0.53(TOC) \qquad (5.62)$$

$$IOC = 0.47(TOC) \qquad (5.63)$$

Note also from Table 5.10 that the fuel and oil cost fraction of TOC has increased from 19 percent to 23 percent, despite a decrease in fuel prices (Fig.5.3) from 1981 to 1985! Also, the flight crew cost as a fraction of TOC has decreased from 9 percent to 7 percent. This may be because during this period:

1) inflation was brought under reasonable control
2) average crew salaries have decreased.

It is useful to keep the orders of magnitude of the various operating cost contributions in mind. The 1985 data suggest the following tabulation of TOC components:

$$DOC_{flt} = 0.33(TOC) \qquad (5.64)$$

where: $C_{crew} = 0.07(TOC) \qquad (5.64a)$

$$C_{pol} = 0.23(TOC) \qquad (5.64b)$$

$$C_{ins++} = 0.03(TOC) \qquad (5.64c)$$

$$DOC_{maint} = 0.10(TOC) \qquad (5.65)$$

Table 5.10 Trends in Airline Total (= DOC + IOC) Operating Cost Components

Cost Item	1976	1981	1982	1983	1984	1985
Direct Operating Cost (DOC)						
Flight crew, C_{crew}	0.086	0.072	0.072	0.072	0.069	0.068
Fuel, oil and lub., C_{pol}	0.193	0.292	0.272	0.245	0.233	0.232
Hull insurance, rentals and training, C_{ins++}	0.039	0.027	0.025	0.027	0.031	0.034
DOC_{flt}	0.318	0.391	0.370	0.344	0.333	0.333
DOC_{maint}	0.125	0.103	0.100	0.100	0.101	0.100
DOC_{depr}	0.080	0.064	0.068	0.072	0.072	0.071
DOC_{lnr}	0.037	0.035	0.033	0.033	0.030	0.030
DOC Total	0.560	0.593	0.571	0.549	0.536	0.534
Indirect Operating Cost (IOC)						
IOC_{pax}	0.094	0.086	0.091	0.092	0.092	0.092
IOC_{sta}	0.133	0.113	0.122	0.126	0.130	0.130
IOC_{pse}	0.149	0.147	0.155	0.164	0.166	0.165
IOC_{gaa}	0.064	0.061	0.061	0.069	0.076	0.079
IOC Total	0.440	0.407	0.429	0.451	0.464	0.466
TOC = DOC + IOC	1.000	1.000	1.000	1.000	1.000	1.000

Notes: 1) IOC_{frt} is included in IOC_{sta} 2) Source: Interavia, Feb.1987

$$DOC_{depr} = 0.07(TOC) \quad (5.66)$$

$$DOC_{lnr} = 0.03(TOC) \quad (5.67)$$

$$IOC_{pax} = 0.09(TOC) \quad (5.68)$$

$$IOC_{sta} = 0.13(TOC) \quad (5.69)$$

$$IOC_{pse} = 0.17(TOC) \quad (5.70)$$

$$IOC_{gaa} = 0.08(TOC) \quad (5.71)$$

It is of interest to consider the effect of block distance, R_{bl}, and of airplane type on the various contributions to TOC.

Figures 5.20 and 5.21 provide this information for four types of airplane for the four major DOC components. It is noted from Fig.5.20 that block distance has essentially no effect on the fractions of DOC_{flt} and DOC_{depr} of total DOC. However, airplane type has a significant effect on their relative contributions.

It is noted from Figure 5.21 that block distance and type of airplane both have a significant influence on the magnitudes of DOC_{maint} and DOC_{lnr} relative to total DOC.

Figures 5.22 - 5.24 provide similar information for the five major IOC components. It is noted from Fig.5.22 that neither the type of airplane nor block distance have much effect on the magnitudes of IOC_{pax} and IOC_{gaa} relative to total IOC.

However, as is seen in Figures 5.23 and 5.24, both airplane type and block distance have a considerable effect on IOC_{sta}, IOC_{pse} and IOC_{ascf} as a fraction of total IOC.

The author believes that even though the data in Figures 5.20 - 5.24 are from 1975, they can be expected to apply to 1990 (and perhaps beyond) because of the fact that (as Table 5.10 indicates) calendar time had only a minor effect on the distribution of TOC components.

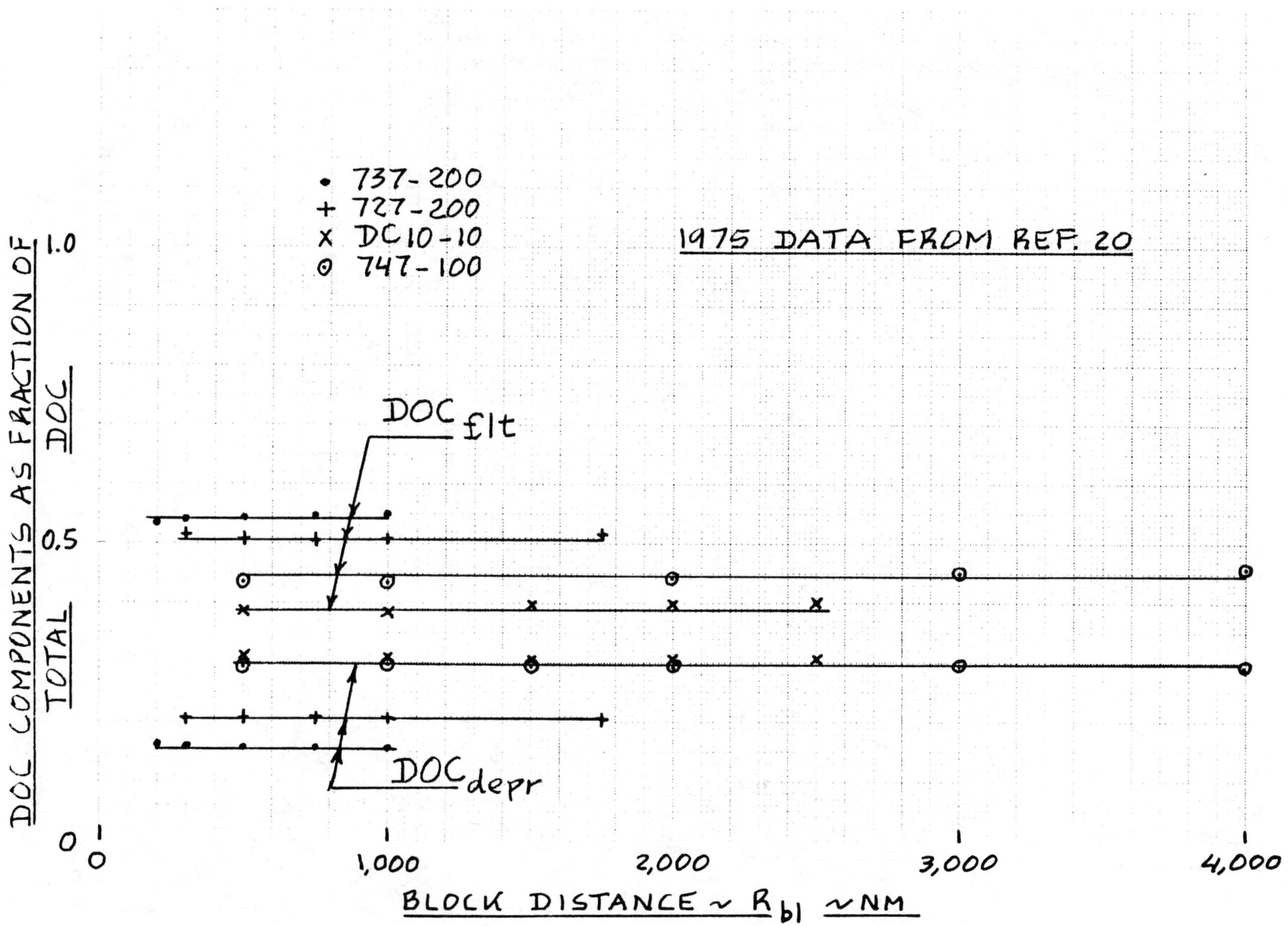

Figure 5.20 Effect of Block Distance and Airplane Type on the Direct Operating Cost for Flight and Depreciation as a Fraction of Total DOC

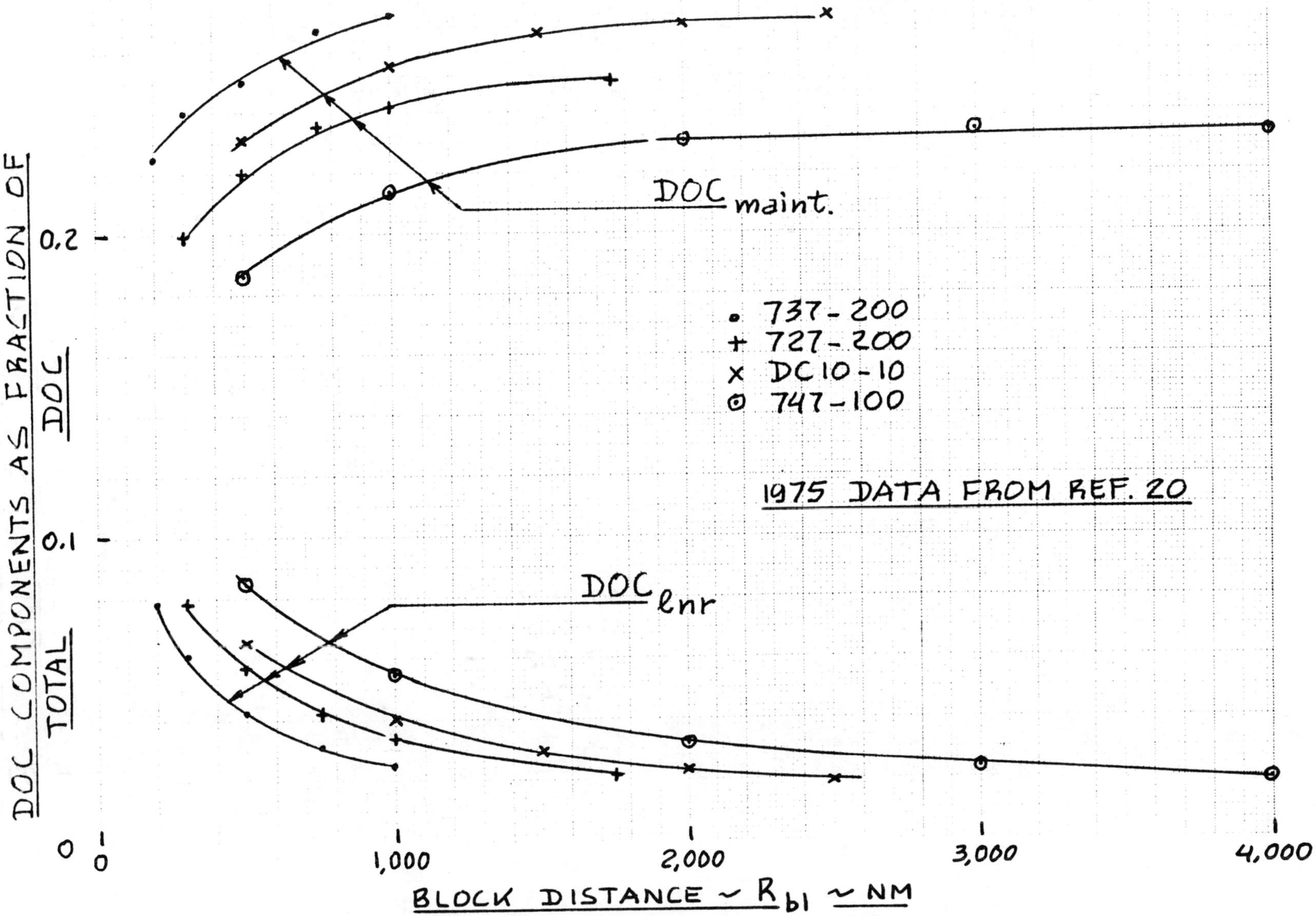

Figure 5.21 Effect of Block Distance and Airplane Type on the Direct Operating Cost for Maintenance and Landing, Navigation and Registration as a Fraction of Total DOC

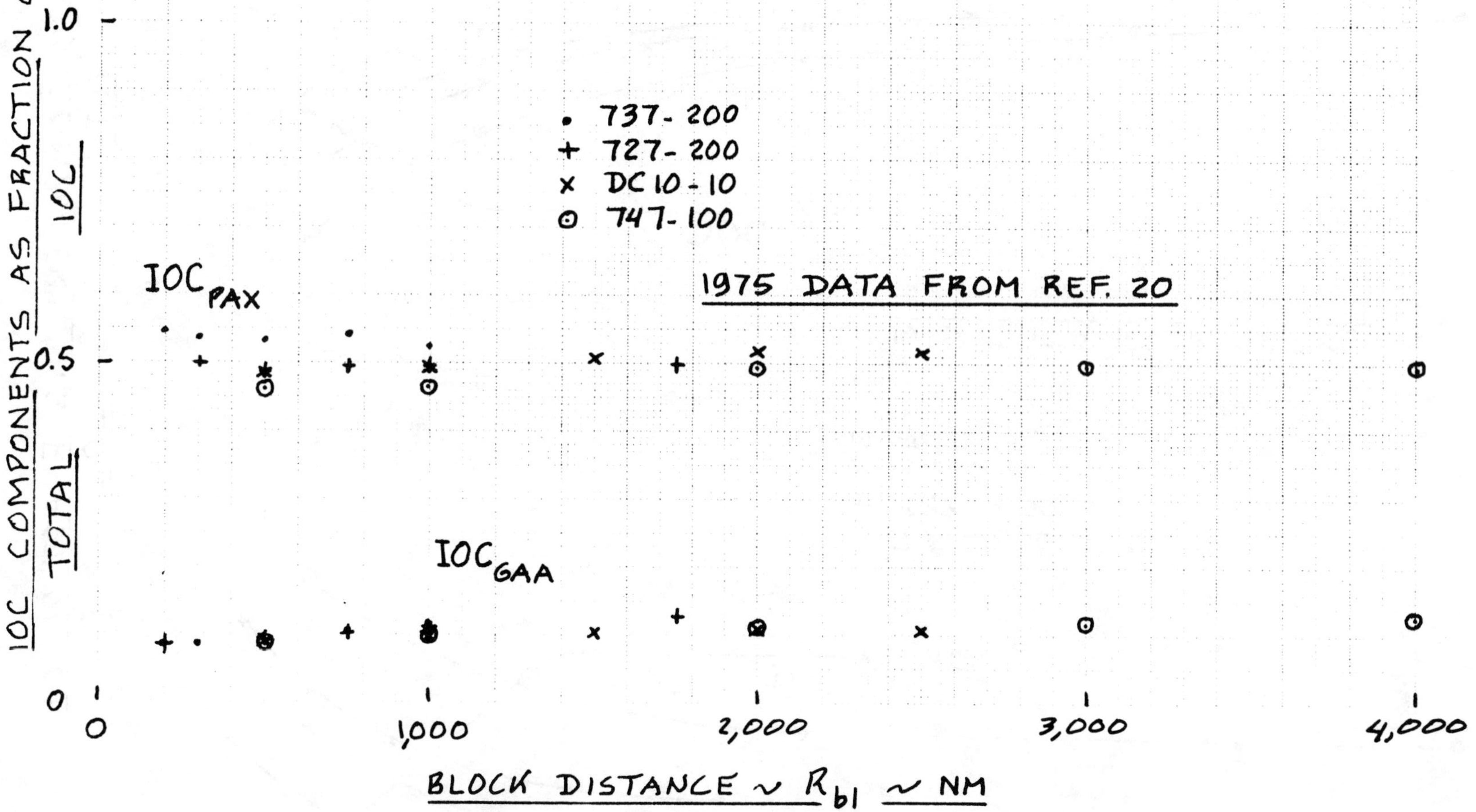

Figure 5.22 Effect of Block Distance and Airplane Type on the Indirect Operating Cost for Passengers and General Administrative Expenses as a Fraction of Total IOC

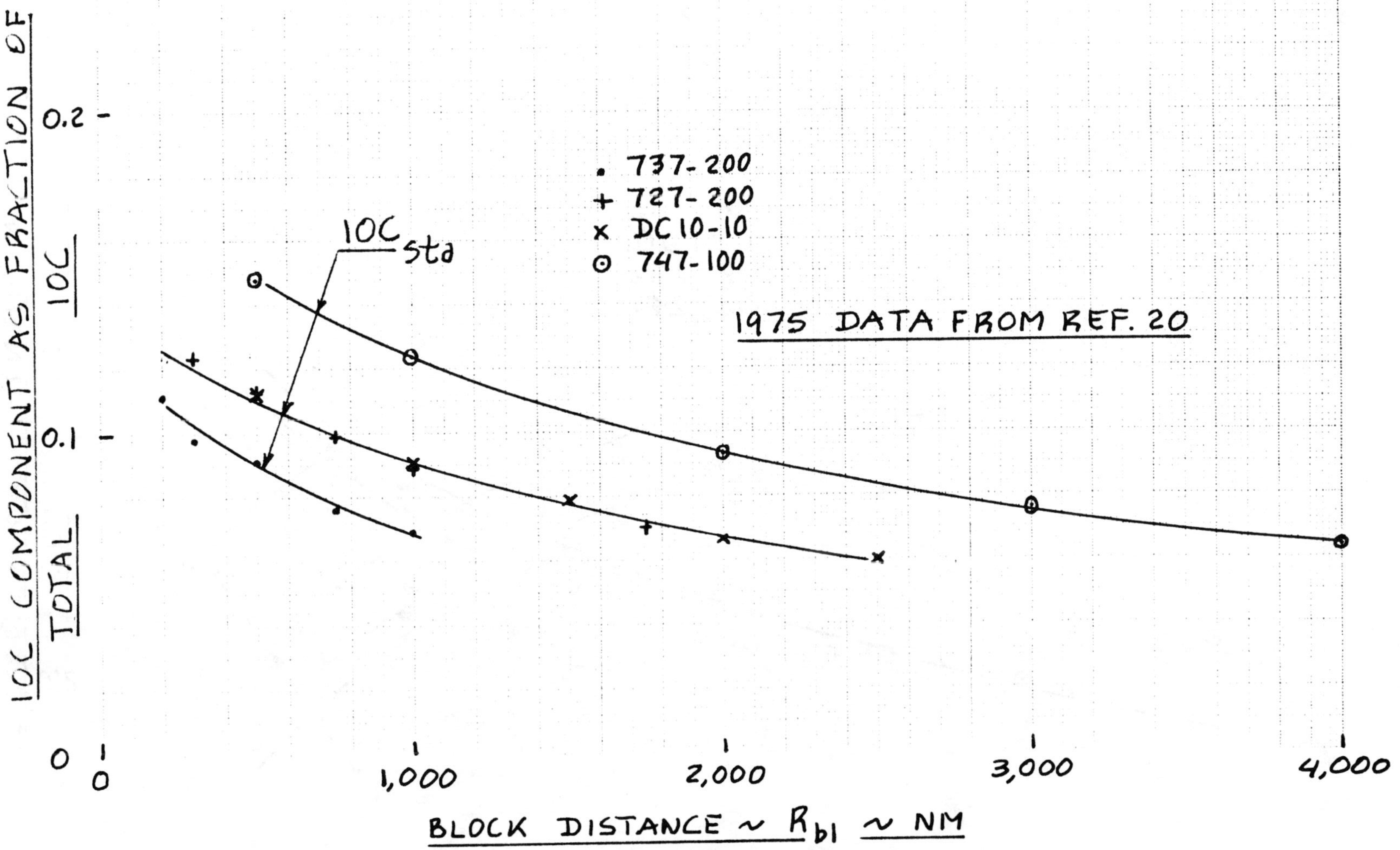

Figure 5.23 Effect of Block Distance and Airplane Type on the Indirect Operating Cost for Station Operation as a Fraction of Total IOC

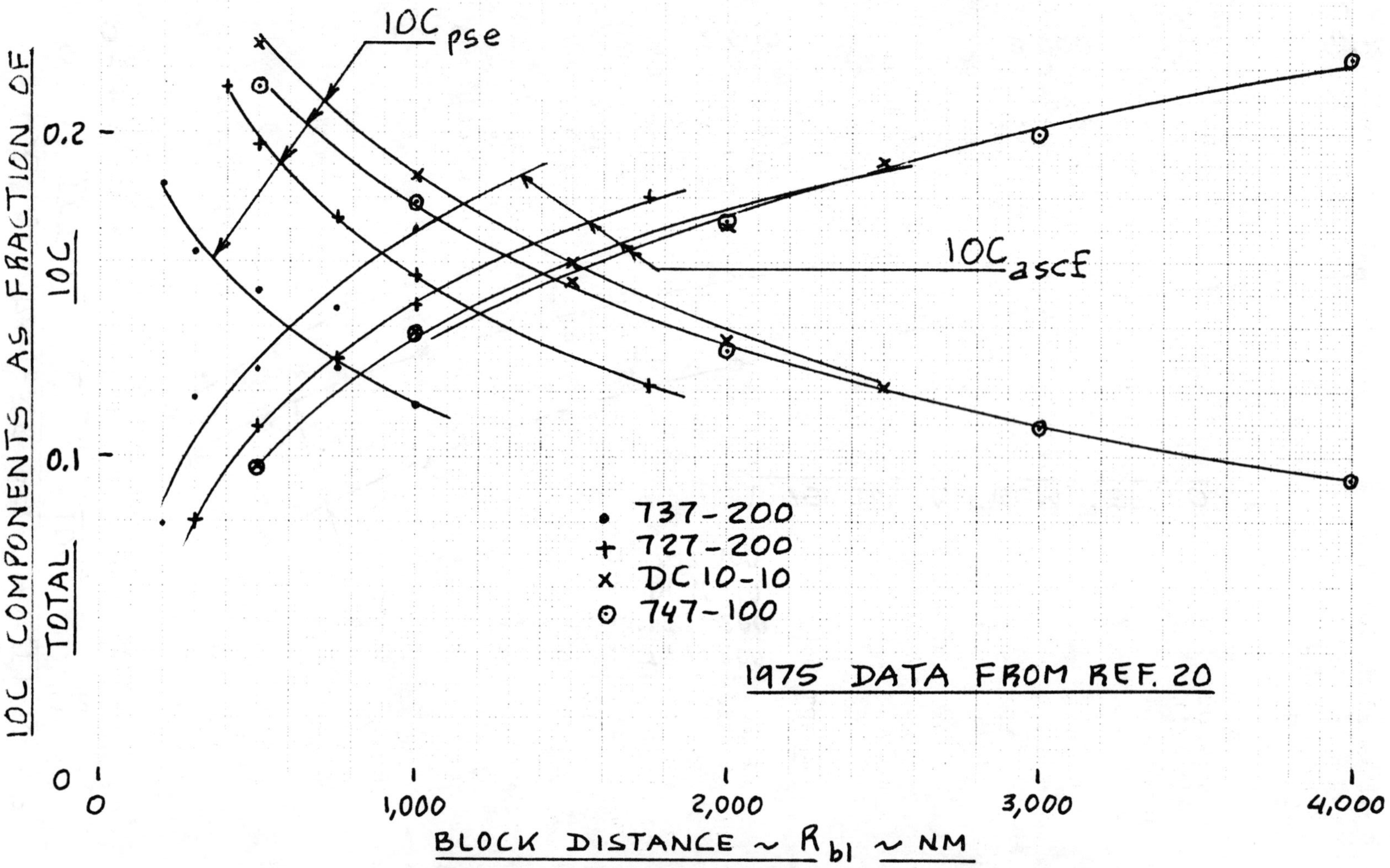

Figure 5.24 Effect of Block Distance and Airplane Type on the Indirect Operating Cost for Promotion, Sales and Entertainment and for Airplane Service, Control and Freight as a Fraction of Total IOC

6. METHOD FOR ESTIMATING OPERATING COST OF MILITARY AIRPLANES

The purpose of this chapter is to present a method for estimating the operating cost of military airplanes in peace time. A method for estimating the operating cost of commercial airplanes is presented in Chapter 5.

A detailed method for estimating operating cost for military airplanes is provided in Ref.10. Despite the level of detail given in Ref.10, it is not sufficient to give the preliminary designer specific insight into what can be done (from a design viewpoint) to lower operating cost of military airplanes. Therefore, the author has decided to combine the methods, ideas and data given in Refs 10 and 14 into a simpler method. This method should be used only for preliminary cost estimating purposes.

IMPORTANT ASSUMPTION:

Military airplanes are acquired in a manner depending considerably on budgetary and political constraints. It is not possible to translate such constraints into any rational format. Therefore, no attempt will be made to represent program operating cost as a summation over a number of calendar years. Instead, the assumption will be made that the number of airplanes to be acquired, N_{acq}, have indeed been acquired by the military service.

Furthermore, the assumption will be made that deliveries have been made to ONE military service only so that cost differences between services need not be accounted for.

The program operating cost of military airplanes can be broken down as follows:

$$C_{OPS} = C_{POL} + C_{PERSDIR} + C_{PERSIND} + C_{CONMAT} + C_{SPARES} + C_{DEPOT} + C_{MISC} \qquad (6.1)$$

where: C_{POL} is the program fuel, oil and lubricants cost: see Section 6.1

$C_{PERSDIR}$ is the program cost of direct personnel (that is aircrew and maintenance): see Section 6.2

$C_{PERSIND}$ is the program cost of indirect personnel: see Section 6.3

C_{CONMAT} is the program cost of consumable materials used in conjunction with maintenance: see Section 6.4

C_{SPARES} is the program cost of spares: see Section 6.5

C_{DEPOT} is the program cost associated with depots: see Section 6.6

C_{MISC} is the program miscellaneous cost: see Section 6.7

The total (or program) operating cost of military airplanes, C_{OPS}, is shown in relationship to other life cycle cost (LCC) sources in Eqn.(2.3) and in Figure 2.3.

A summary for the determination of the program operating cost for a military airplane, C_{OPS}, is presented in Section 6.8. An example application of the method is given in Section 6.9. Some important observations on the subject of military airplane operating cost are included in Section 6.10.

6.1 PROGRAM COST OF FUEL, OIL AND LUBRICANTS: C_{POL}

The program cost of fuel, oil and lubricants depends on the following:

* type of airplane
* mission of the airplane
* annual utilization
* number of airplanes in active service

The program cost for fuel, oil and lubricants can be estimated from:

$$C_{POL} = [(F_{OL})(W_{F_{used}})\{(FP)/(FD)\}* \\ *(N_{mission})(N_{serv})](N_{yr}) \quad (6.2)$$

where: F_{OL} is a factor which accounts for the cost of oil and lubricants. According to Ref.14 it is acceptable to use: $F_{OL} = 1.005$

$W_{F_{used}}$ is the mission fuel used in lbs. per mission. This can be found from Eqn.(2.14) in Part I or from the more detailed method of Section 5.4, Part VII.

FP is the price of fuel in USD per gallon. This is obtained from Figure 5.3. In most instances the price of JP-4 can be used.

FD is the fuel density in lbs per gallon. This may be obtained from p.88. In most instances the density of JP-4 can be used.

$N_{mission}$ is the number of missions flown per year. This can be estimated from:

$$N_{mission} = (U_{ann_{flt}})/(t_{mis}) \qquad (6.3)$$

where: $U_{ann_{flt}}$ is the annual utilization in flight hours. Table 6.1 provides typical data for military airplane utilization in peace time.

t_{mis} is the average mission time in hours. Figure 6.1 shows how this is defined.

NOTE: in the case of multi-mission airplanes, a scenario of typical 'mission mixes' per year must be predicted.

N_{serv} is the number of airplanes of the type in actual service. This number is averaged as:

$$N_{serv} = N_{acq} - N_{res} - 0.5(N_{loss}) \qquad (6.4)$$

where: N_{acq} is the number of airplanes which are acquired

N_{res} is the number of airplanes held in active reserve

N_{loss} is the number of airplanes lost through accidents over the service life, N_{yr}

How to determine these three airplane quantities is discussed next.

TABLE 6.1 MILITARY AIRPLANE UTILIZATION IN FLIGHT HOURS PER YEAR AND MILITARY CREW TO AIRPLANE RATIOS

Airplane Type	Flight Hours per airplane per year, $U_{ann_{flt}}$	Crew Ratio per airplane, R_{cr}	Number in crew N_{crew}
Trainers	600 - 1,000	-	2
Fighters	250 - 400	1.1	1 - 2
Bombers	300 - 600	1.5	2 - 5
Tankers	300 - 600	1.5	4 - 6
Transports	< 1,200	1.5	4 - 8
	1,200 - 2,400	2.5	4 - 8
	2,400 - 3,600	3.5	4 - 8

NOTE: average annual utilization for military airplanes is 350 hours

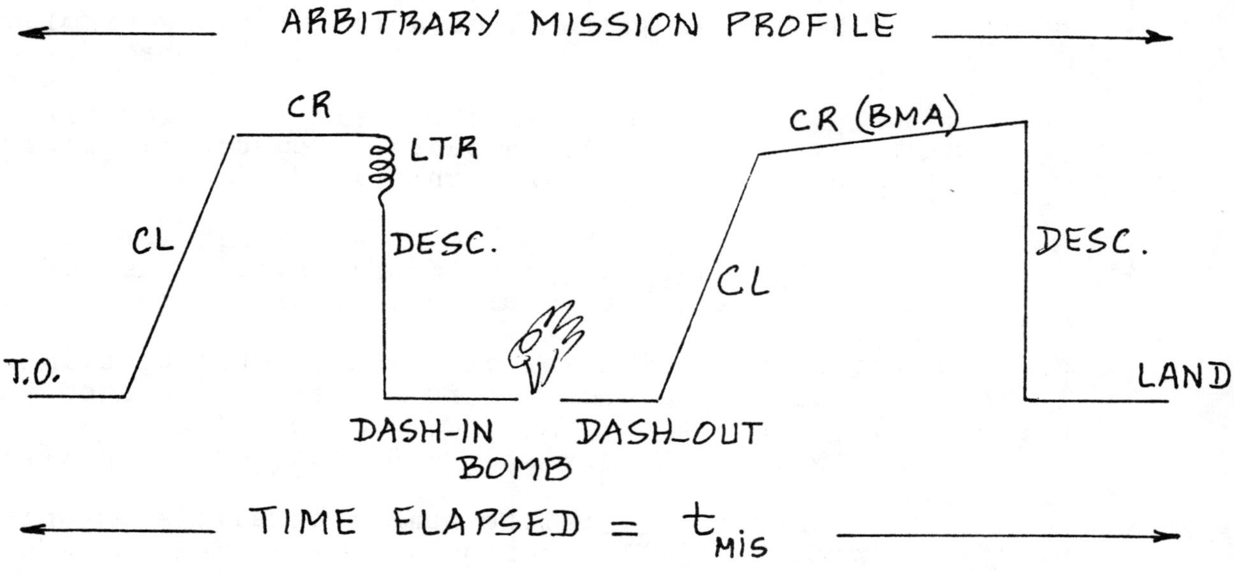

NOTE: FOR EXAMPLE MISSION PROFILES, SEE: P.61 IN PART I AND PAGES 244-248, AND 280-284 IN PART VII.

Figure 6.1 Definition of Mission Time, t_{mis}

The number of airplanes acquired, N_{acq}, is assumed to be equal to the number of airplanes built to production standards, N_m. This in turn is related to the total number of airplanes built, $N_{program}$ by:

$$N_{program} = N_m + N_{rdte} \quad (6.5)$$

where: N_{rdte} is the number of airplanes built during the research, development, technology and evaluation phase. For military programs this number is in the following range:

$$N_{rdte} = 6 \text{ to } 20 \quad (6.6)$$

NOTE: Table 4.1 gives examples of airplane program production runs for several military programs.

The number of airplanes held in reserve, N_{res}, depends on factors such as: politics and estimated loss rates in peace time. For preliminary cost estimating purposes it is suggested to use:

$$N_{res} = 0.10(N_{acq}) \quad (6.7)$$

The total number of airplanes lost due to accidents in peace time, N_{loss}, is estimated from:

$$N_{loss} = (L_R)(N_{serv})(U_{ann_{flt}})(N_{yr}) \quad (6.8)$$

where: L_R is the annual loss rate expressed in airplanes lost per 10^5 flying hours. Table 6.2 shows typical loss rates for USAF and for USN/USMC.

For the reader interested in loss rate data in times of military conflict, reference is made to Section 13.3 in Part IV. Methods to predict combat loss rates are given in Ref.22.

N_{yr} is the number of years the airplane is in active service. In peace time this number may be assumed to range from 20 to 40 years. Table 6.3 provides 1987 data for the average age of several U.S. military airplanes.

All other quantities were previously defined.

Table 6.2 Military Airplane Accident Summaries for 1984 and 1986

Type	1984 Number of Accidents	1984 Total Flight Hours (Fleet)	1984 Accident Rate, L_R	1986 Number of Accidents	1986 Total Flight Hours (Fleet)	1986 Accident Rate, L_R
USAF:						
T-37	1	333,000	0.3	1	333,300	0.3
T-38	3	375,000	0.8	4	363,600	1.1
F-4	11	275,000	4.0	12	245,000	4.9
RF-4	1	77,000	1.3	5	78,000	6.4
F-5	2	29,000	6.9	6	26,500	22.6
F-15	3	176,000	1.7	7	194,400	3.6
F-16	10	200,000	5.0	11	250,000	4.4
F-111	3	79,000	3.8			
A-7	6	86,000	7.0	1	83,300	1.2
A-10	6	222,000	2.7	3	214,300	1.4
C-130	3	375,000	0.8	2	400,000	0.5
C-135				1	250,000	0.4
C-141	1	333,000	0.3	1	333,300	0.3
B-52	2	105,000	1.9			
Totals:	52	2,665,000	2.0	54	2,771,700	1.9
USN and USMC:						
T-34C	1	179,597	0.6	2	180,900	1.1
T2-J	2	84,734	2.4	2	14,300	14.0
TA-4	9	102,739	8.8	6	118,300	5.1
F-4	4	86,331	4.6	5	54,400	9.2
F-14	4	110,727	3.6	10	120,600	8.3
F-18	1	47,256	2.1	2	107,500	1.9
A-4	4	49,472	8.1	3	59,200	5.1
A-6	7	106,637	6.6	6	104,900	5.7
EA-6	3	26,778	11.2	3	29,800	10.0
A-7	13	146,547	11.3	9	126,400	7.1
AV-8	1	18,297	5.5	3	25,400	13.1
Totals:	49	960,000	5.1	51	941,700	5.4

Data from: Aviation Week and Space Technology, March 11, 1985 and June 8, 1987

Note: Accident Rate = Loss Rate = L_R in airplanes per 100,000 flying hours

TABLE 6.3 SOME AIR FORCE AIRPLANES AND THEIR AGE (1987)

Airplane Type	USAF No. in Service	USAF Ave. Age (yrs)	ANG No. in Service	ANG Ave. Age (yrs)	AF Reserve No. in Service	AF Reserve Ave. Age (yrs)
T-33	54	29.4	21	31.7		
T-37	609	25.3				
T-38	810	21.5				
F-4	596	17.1	602	20.5	119	21.0
F-5	96	11.7				
F-15	732	6.9	66	11.4		
F-16	943	3.7	132	5.5	49	5.0
F-100	10	29.4				
F-106	5	28.1	33	27.9		
F-111	334	16.4				
A-7	31	15.9	342	14.1		
A-10	451	6.9	106	8.0	97	9.0
A-37	22	13.8	60	14.7		
C-5	76	12.1	5	16.9	15	17.0
C-9	23	16.5				
C-10	56	2.9				
C-12	75	6.2	6	1.9		
C-20	13	1.1				
C-21	79	2.7	3	0.1		
C-22	1	3.6	4	2.7		
C-23	18	2.4				
C-130	354	19.3	210	19.8	168	23.4
C-135	608	26.2	104	28.6	24	29.0
C-140	5	24.7				
C-141	255	21.0	8	21.3	8	22.0
E-3	33	7.9				
B-1	60	0.8				
B-52	263	27.0				
FB-111	62	17.1				

Data from: Air Force Magazine, May 1987

<u>NOTE:</u> This table may be used to estimate N_{yr} for a new military airplane program.

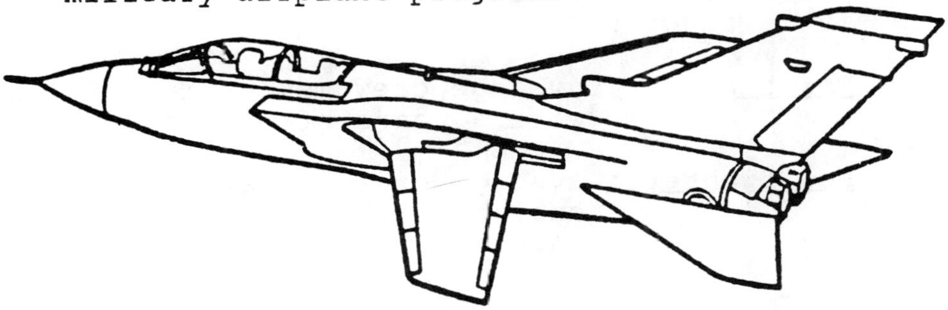

6.2 PROGRAM COST OF DIRECT PERSONNEL: $C_{PERSDIR}$

The program cost component of direct personnel can be split into two elements:

$$C_{PERSDIR} = C_{crewpr} + C_{mpersdir} \tag{6.9}$$

where: C_{crewpr} is the program cost for aircrews: see sub-section 6.2.1

$C_{mpersdir}$ is the program cost for direct maintenance personnel: see sub-section 6.2.2

6.2.1 Program Cost of Aircrews: C_{crewpr}

The program cost for aircrews may be estimated from:

$$C_{crewpr} = (N_{serv})(N_{crew})(R_{cr})(Pay_{crew})(OHR_{crew})(N_{yr}) \tag{6.10}$$

where: N_{serv} is found from Eqn.(6.4)

N_{crew} is the number of crew members required for the airplane: see Table 6.1.

R_{cr} is the crew ratio per airplane. This may be determined with the help of Table 6.1.

Pay_{crew} is the annual pay in USD per year. The annual pay depends on:

* pay grade
* years of service

Table 6.4 gives military pay data applicable to the calendar years 1989 and 1990. To determine Pay_{crew}, several assumptions must be made about

crew composition in terms of grade and years in service. For preliminary cost estimating purposes it is suggested to assume:

For trainers: average crew member is a 1st Lt. with 5 years of military service.

For fighters: average crew member has the rank of captain with 8 years of military service.

Table 6.4 1989 Military Pay Data for Aircrews*

Years in Service	Col/Capt	LtCol/Cmdr	Maj/LtCmdr	Capt/Lt	1stLt/LtJG	2ndLt/Ens
	Basic Pay in USD/year — Rank					
4	38,099	32,677	28,488	26,964	24,109	19,418
6	38,099	32,677	29,027	28,253	24,617	19,418
8	38,099	32,677	30,308	29,268	24,617	19,418
10	38,099	33,667	32,375	30,857	24,617	19,418
12	38,099	35,474	34,196	32,375	24,617	19,418
14	39,391	37,854	35,759	33,174	24,617	19,418
16	45,619	40,687	37,325	33,174	24,617	19,418
18	47,952	43,020	38,358	33,174	24,617	19,418
20	48,996	44,323	38,358	33,174	24,617	19,418
22	51,836	45,871	38,358	33,174	24,617	19,418

Aviation Career Incentive Pay (ACIP), excl.' Hazardous Duty Pay' in USD/month

Years of Aviation Officer Service	<or =2	>2	>3	>4	>6	>18	>20	>22
Monthly Rate	125	156	188	206	400	370	340	310

Aviation Continuation Pay (Re-up Bonus) in USD/year

Years of Active Federal Commissioned Service	6	7	8	9	10	11	12
1989 Bonus	9,000	9,000	8,250	8,250	7,125	6,000	4,875
Subsequent annual bonus	12,000	12,000	11,000	11,000	9,500	8,000	6,500

* Source: Professional Pilot Magazine, April, 1989

For bombers: average crew member has the rank of major with 12 years of military service.

For transports: average crew member has the rank of 1st Lt. with 5 years of military service. Note: transports often carry non-commissioned personnel!

IMPORTANT NOTE: the ranks given are for 'average' crew members in the USAF and the USMC only. For the USN use the 'equivalent' ranks of Table 6.4.

An example military pay calculation will now be given. For this example it is assumed that the crew pay needs to be found for a USAF fighter:

Page 152 suggests that for preliminary costing purposes it may be assumed that for a fighter the average crew member has the rank of captain with 8 years of military service. In that case, it is seen from Table 6.4 that:

$$Pay_{crew} = (29,268 + 12*400 + 12,000) =$$

$$= 46,068 \text{ USD/year}$$

The 12,000 pay factor assumes 1990 and beyond aviation continuation pay.

NOTE: Military pay scales are subject to congressional action. For years beyond 1990, the reader should check whether these pay scales are valid!

OHR_{crew} is an overhead rate factor, associated with the actual crew pay. Lacking actual military overhead data, it is suggested to use: $OHR_{crew} = 3.0$. This overhead rate factor is similar in magnitude to industry overhead rates.

N_{yr} is the number of years the airplane is in active service. This quantity may be estimated with the help of Table 6.3.

6.2.2 Program Cost of Direct Maintenance Personnel: $C_{mpersdir}$

The program cost of direct maintenance may be estimated from:

$$C_{mpersdir} = (N_{serv})(N_{yr})(U_{ann_{flt}})* \\ *(MHR_{flthr})(R_{m_{ml}}) \qquad (6.11)$$

where: N_{serv} is found from Eqn.(6.4)

N_{yr} is found from Table 6.3.

$U_{ann_{flt}}$ is found from Table 6.1

MHR_{flthr} is the number of maintenance manhours required per flight hour. This depends on the following factors:

* number of years in service
* equipment density (this affects accessibility)
* maturity of the technologies used in the airplane systems: propulsion, avionics, flight controls, brakes, wheels and tires.

Table 6.5 provides some historical data on the effect of 'years in service' on MHR_{flthr} for a number of military airplanes. Figure 6.2 shows some data based on actual USNavy experience for A4 type airplanes.

Figure 6.3 illustrates how equipment density affects the maintenance hours per flight hour.

The cost analyst should not attemp to arbitrarily 'lowball' a number for MHR_{flthr}. Instead, a list should be prepared of all factors which affect maintenance and which compare the new design with a comparable existing design for which MHR_{flt} is known. Only then can an educated estimate for MHR_{flthr} be made.

Nicolai (in Ref.14) implies that the number of maintenance manhours per flight hour is related

Table 6.5 Maintenance Man-Hours per Flight Hour for Military Airplanes

Type	1973	1974	1975	1976	1977	1978	1979	1980	1981	1982	1983
T-37	7.8										
T-38	10.0		8.8	11.6	8.7	9.0	12.2	9.0	10.3	6.8	10.9
F-100			19.8	17.1	17.1	16.1					
F-4D			55.6	41.9	54.2	56.2	47.7	44.8	28.4	26.9	42.0
F-4E	33.0										
RF-4C			39.6	31.9	33.8	34.2	32.8	31.1	22.9	18.2	36.1
F-5	17.0										
FB-111A			57.2	47.7	36.4	58.2	47.6	45.8	60.8	49.0	56.0
F-111D	40.0		19.8	17.1	17.1	16.1					
F-15		20.0	29.1	27.8	26.0	30.6	31.9	28.6	36.9	30.5	44.1
F-15C								13.6	24.3	22.3	32.8
F-16A	25.0						25.4	22.6	11.2	14.6	21.8
F-16B								18.4	13.9	15.3	30.6
A-7D	25.0										
A-10		13.0				7.2	11.8	12.0	11.2	8.9	13.9
C-130	20.0			25.1	37.1	22.4	19.6	19.8	21.6	19.7	24.7
C-141	21.0		17.5	18.6	18.2	17.3	15.0	15.0	20.7	19.3	25.2
C-5A	40.0		107.9	89.4	48.2	41.9	53.4	47.3	52.5	37.0	64.9
C-9	12.0										
KC-135	26.6		32.8	31.6	34.0	26.6	21.2	23.2	20.2	17.5	24.2
B-52D	37.0										
B-52G	49.0		46.5	46.9	44.9	45.5	42.1	40.6	39.6	34.0	50.7
B-52H			45.5	48.9	47.6	49.3	39.5	33.9	34.6	30.2	42.9

The following ranges are suggested for preliminary cost estimating purposes:

<u>For Trainers:</u> 6 - 10   <u>For Fighters:</u> 15 - 35
<u>For Bombers:</u> 25 - 50   <u>For Transports:</u> 25 - 60

Data from: Refs 14, 24 and Air Force Magazine, September 1983.

to the total area (in square feet) of all access panels in an airplane!

$R_{m_{ml}}$ is the military maintenance labor rate in USD/hr. This depends on many factors, including the ratio of civilian versus military personnel used in maintenance functions. For 1989, this labor rate, <u>including all overhead</u> contributions may be assumed to be:

$$(R_{m_{ml}})_{1989} = 45 \text{ USD/hr}$$

For future years it is suggested to use:

$$(R_{m_{ml}})_{\text{then year}} = (R_{m_{ml}})_{1989} * $$

$$*(CEF_{\text{then year}})/(CEF_{1989}) \quad (6.12)$$

where: the CEF-values are found from Figure 2.7.

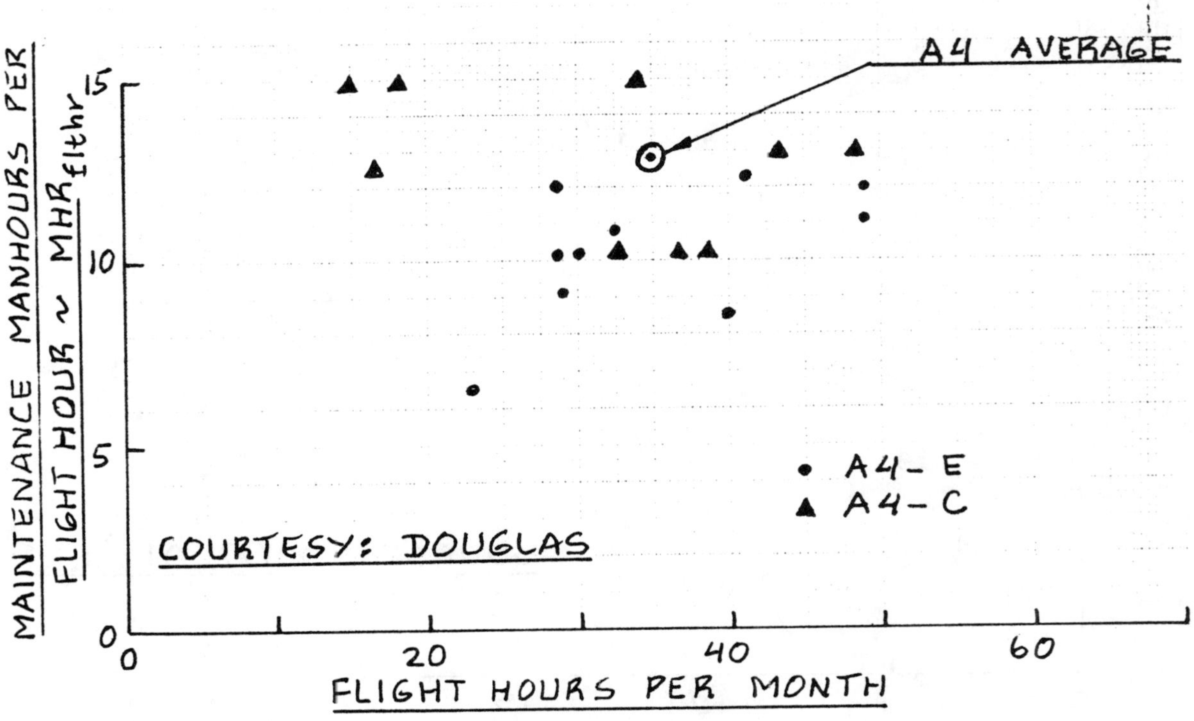

Figure 6.2 Effect of Flying Hours per Month on Maintenance Manhours per Flight Hour

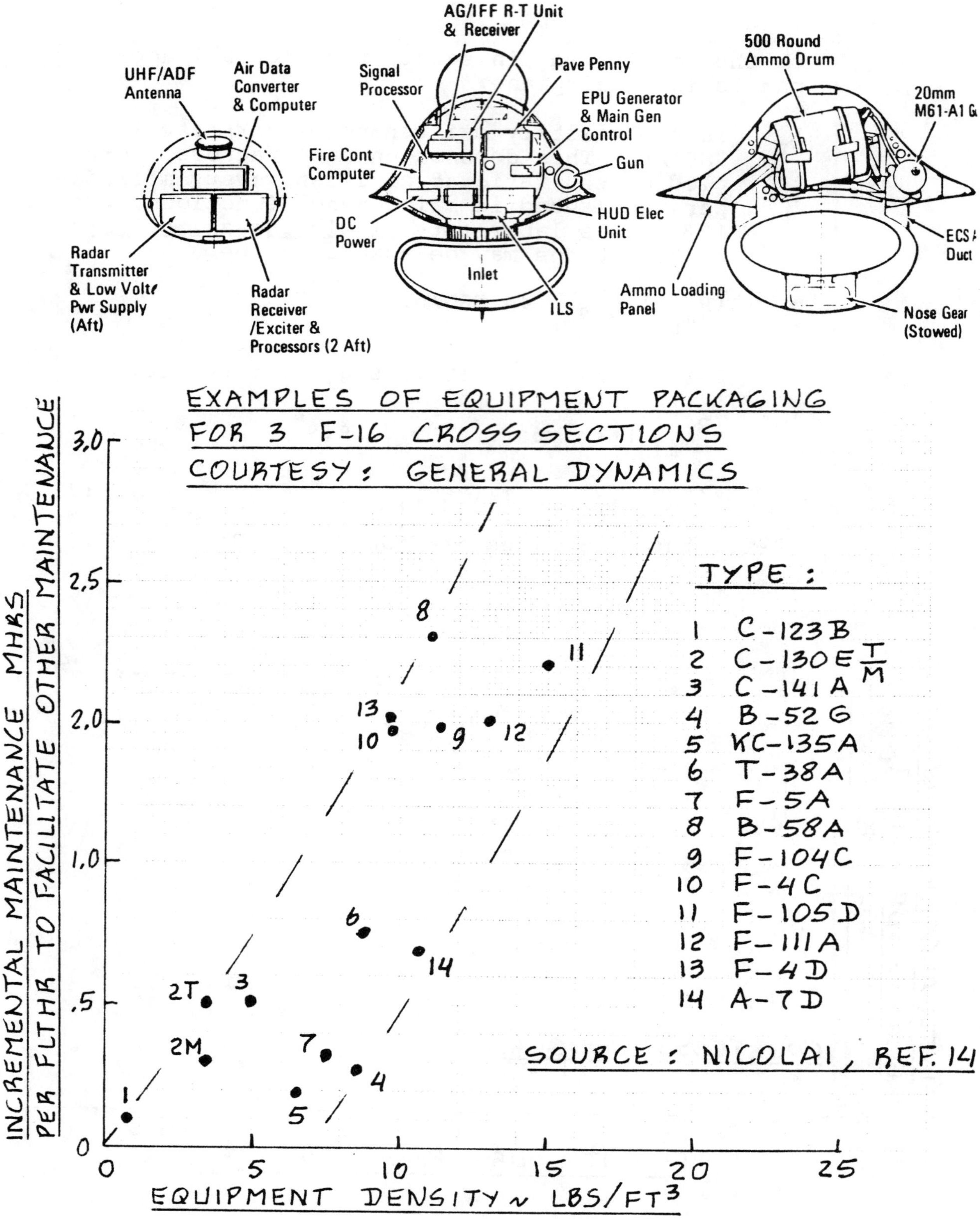

Figure 6.3 Effect of Equipment Density on Maintenance Manhours per Flight Hour

6.3 PROGRAM COST OF INDIRECT PERSONNEL: $C_{PERSIND}$

This cost component covers the cost of all squadron level personnel which are not directly involved in flight operations or airplane maintenance. This cost component may be estimated from:

$$C_{PERSIND} = (f_{persind})(C_{OPS}) \qquad (6.13)$$

where: C_{OPS} is given by Eqn. (6.1)

$f_{persind}$ may be estimated with the help of Table 6.6

6.4 PROGRAM COST OF CONSUMABLE MATERIALS: C_{CONMAT}

This cost component covers the cost of consumable materials used in conjunction with military airplane maintenance functions. This cost component may be found as follows:

$$C_{CONMAT} = (N_{serv})(N_{yr})(U_{ann_{flt}})*$$

$$*(MHR_{flthr})(R_{conmat}) \qquad (6.14)$$

where: R_{conmat} is the average cost in USD/hr (USD per maint. hour) for consumable materials. Lacking specific information it is suggested to use:

$$R_{conmat} = (6.50)(CEF_{then\ year})/(CEF_{1989}) \qquad (6.15)$$

where: the factor 6.50 is the 1989 cost of consumable materials in USD per maintenance hour.

All other terms were previously defined.

6.5 PROGRAM COST OF SPARES: C_{SPARES}

The program cost of spares may be estimated from:

$$C_{SPARES} = (f_{spares})(C_{OPS}) \qquad (6.16)$$

Table 6.6 Multipliers for Program Operating Cost of Military Airplanes

Operating Cost Component	Operating Cost Fraction (=Multiplier)	Fraction of Total Program Operating Cost For Each Airplane Type				
		B52G	C-141	F-111	F-4	A-7D
C_{POL}	f_{pol}	0.14	0.23	0.05	0.10	0.04
$C_{PERSDIR}$	$f_{persdir}$	0.34	0.32	0.28	0.38	0.38
$C_{PERSIND}$	$f_{persind}$	0.16	0.13	0.14	0.20	0.20
C_{SPARES}	f_{spares}	0.13	0.16	0.27	0.12	0.16
C_{DEPOT}	f_{depot}	0.20	0.15	0.22	0.13	0.16
C_{MISC}	f_{misc}	0.03	0.01	0.04	0.07	0.06
Totals:		1.00	1.00	1.00	1.00	1.00

Source: Ref.14, Figure 24.1

<u>NOTES:</u> 1) The fractional contribution of consumable materials cost, C_{CONMAT}, is included in the miscellaneous cost item in this table.

2) The following assumption may be made: $C_{CONMAT} = 0.25(C_{MISC})$

where: C_{OPS} is given by Eqn.(6.1)

f_{spares} may be estimated with the help of Table 6.6

6.6 PROGRAM COST OF DEPOT: C_{DEPOT}

Many overhaul and maintenance actions on military airplanes are not carried out at the squadron or wing level. Instead, these are carried out at so-called depots. The program depot cost associated with the operation of a given airplane type may be estimated from:

$$C_{DEPOT} = (f_{depot})(C_{OPS}) \tag{6.17}$$

where: C_{OPS} is given by Eqn.(6.1)

f_{depot} may be estimated with the help of Table 6.6

6.7 PROGRAM COST OF MISCELLANEOUS ITEMS: C_{MISC}

The following miscellaneous cost elements contribute to the operating cost of military airplanes:

* requirements for technical data to support maintenance functions
* requirements for training, training data and training equipment
* requirements for support equipment

Reference 10 provides detailed cost estimating relations for these items. For the purpose of preliminary cost estimating it is suggested to use:

$$C_{MISC} = (f_{misc})(C_{OPS}) \tag{6.18}$$

where: C_{OPS} is given by Eqn.(6.1)

f_{misc} may be estimated with the help of Table 6.6

Using the assumption made at the bottom of Table 6.6, it is possible to write C_{MISC} alternatively as:

$$C_{MISC} = 4C_{CONMAT} \tag{6.19}$$

6.8 SUMMARY FOR THE DETERMINATION OF THE PROGRAM OPERATING COST OF MILITARY AIRPLANES : C_{OPS}

The program operating cost of military airplanes, with the help of equations (6.1), (6.2), (6.9) and (6.13) through (6.18) may now be found from:

$$C_{OPS} = \frac{C_{POL} + C_{PERSDIR} + C_{CONMAT}}{(1 - f_{persind} - f_{spares} - f_{depot} - f_{misc})} \qquad (6.20)$$

By using Equation (6.19) it is also possible to cast Equation (6.20) in the following form:

$$C_{OPS} = \frac{C_{POL} + C_{PERSDIR} + C_{CONMAT} + C_{MISC}}{(1 - f_{persind} - f_{spares} - f_{depot})} \qquad (6.21)$$

or:

$$C_{OPS} = \frac{C_{POL} + C_{PERSDIR} + 5C_{CONMAT}}{(1 - f_{persind} - f_{spares} - f_{depot})} \qquad (6.22)$$

In comparing the program operating costs from one airplane to another it is often desirable to do this on the basis of their so-called program operating cost per hour: $C_{OPS/HR}$.

This program operating cost per flight hour may be obtained from:

$$C_{OPS/HR} = (C_{OPS})/\{(N_{serv})(N_{yr})(U_{ann_{flt}})\} \qquad (6.23)$$

where: C_{OPS} follows from Equations (6.20), (6.21) or (6.22).

All other terms were previously defined.

6.9 EXAMPLE APPLICATION

The purpose of this section is to present an example application of the method for determining the program operating cost of military airplanes. The material is organized in the following manner:

6.9.1 Program Operating Cost Components

6.9.2 Program Operating Cost Total

6.9.3 Program Operating Cost per Flight Hour

6.9.1 Program Operating Cost Components

The example will be constructed around the ground attack fighter airplane of Parts I and II (called Eris). The mission specification for this fighter is found in Table 2.19 of Part I.

The assumption will be made that all costs must be computed for the year 1990.

Program Fuel, Oil and Lubricants Cost: C_{POL}

These costs are computed with Eqn.(6.2).

The input requirements for Eqn.(6.2) are:

F_{OL} = 1.005, according to page 146.

$W_{F_{used}}$ = 18,500lbs, according to p.67 of Part I.

The assumption will be made that JP-4 is the fuel of choice for this ground attack fighter airplane.

FP = 0.75 USD/gallon (assumed from Fig.5.3, p.87).

FD = 6.55 lbs/gallon, from p.88.

$N_{mission}$ is computed from Eqn.(6.3) as:

$N_{mission}$ = 350/2.67 = 131 missions per year

This assumes an annual utilization of $U_{ann_{flt}}$ = 350 flight hours (see Table 6.1) and a mission duration of t_{mis} = 2.67 hours. The latter is pieced together with the help of Table 2.19 in Part I.

The following data are assumed for the number of airplanes involved in the Eris development, production and acquisition program:

$N_{program} = 750$ $N_{rdte} = 10$ $N_m = 740$

With the assumption made on page 145:

$N_{acq} = 740$ units

Next, the number of airplanes in active service will be determined from Eqn.(6.4).

It will be assumed that 10 percent of the Eris fleet will be held in active reserve: $N_{res} = 74$ units

The loss rate per 100,000 flight hours will be assumed to be: $L_R = 2$, which is an optimistic projection, based on Table 6.2.

The number of years during which the Eris will be in active service is assumed to be: $N_{yr} = 25$ years.

With Eqns (6.8) and (6.4):

$$N_{serv} = 740 - 74 - [\{(0.5)(2)(350)(25)\}/100,000]N_{serv} =$$

$$= 740 - 74 - 0.0875N_{serv}, \text{ so that:}$$

$N_{serv} = 666/1.0875 = 612$ units

Now with Eqn.(6.2):

$C_{POL} = \text{USD } 4,266,978,750$

Program Cost of Direct Personnel: $C_{PERSDIR}$

This cost may be determined with Eqn.(6.9).

The program cost for aircrews is found with the help of Equation (6.10). In this equation the following input data remain to be determined:

Since the Eris requires only one pilot:

$N_{crew} = 1.0$

The crew ratio, $R_{cr} = 1.1$ from Table 6.1.

The crew pay is assumed to be that of the example of page 154, so that:

$$Pay_{crew} = 46,068 \text{ USD/year}$$

The crew overhead factor, OHR_{crew} is assumed to be:

$$OHR_{crew} = 3.0 \text{ (p.154).}$$

With these data, the program cost for aircrews can be found from Eqn.(6.10) to be:

$$C_{crewpr} = (612)(1)(1.1)(46,068)(3)(25) =$$

$$= \text{USD } 2,325,973,320$$

The program cost for direct maintenance personnel is found with Eqn.(6.11). In this equation the following input data remain to be determined:

For maintenance manhours per flight hour, Table 6.5 has been used to project the following number:

$$MHR_{flthr} = 12$$

The maintenance labor rate per manhour for 1990 (including overhead) will be determined from Eqn.(6.12) as:

$$R_{m_{ml}} = 45(3.04)/(3.02) = \text{USD } 45.30$$

The program cost for direct maintenance personnel can now be computed:

$$C_{persdir} = (612)(25)(350)(12)(45.30) =$$

$$= \text{USD } 2,910,978,000$$

The program cost of direct personnel, aircrews and maintenance personnel is found with Eqn.(6.9) as:

$$C_{PERSDIR} = 2,325,973,320 + 2,910,978,000 =$$

$$= \text{USD } 5,236,951,320$$

<u>Program Cost of Indirect Personnel:</u> $C_{PERSIND}$

This cost component may be determined from Eqn.(6.13) and Table 6.6 as:

$$C_{PERSIND} = (0.20)(C_{OPS}), \text{ where } f_{persind} = 0.20$$

Program Cost of Consumable Materials: C_{CONMAT}

This cost is determined with Eqn.(6.14). The following input data remain to be determined:

With Eqn.(6.15):

R_{conmat} = 6.50(3.04)/(3.02) =

= 6.54 USD per maint.hour

With Eqn.(6.14):

C_{CONMAT} = (612)(25)(350)(12)(6.54) =

= USD 420,260,400

Program Cost of Spares: C_{SPARES}

This cost is found with Eqn.(6.16) and Table 6.6 as:

$C_{SPARES} = (0.14)(C_{OPS})$, where $f_{spares} = 0.14$

Program Cost of Depot: C_{DEPOT}

This cost is found from Eqn.(6.17) and Table 6.6 as:

$C_{DEPOT} = (0.15)(C_{OPS})$, where $f_{depot} = 0.15$

Program Cost of Miscellaneous Costs: C_{MISC}

This cost component is found from Eqn.(6.19) as:

$C_{MISC} = 4C_{CONMAT}$ = 4(420,260,400) =

= USD 1,681,041,600

6.9.2 Program Operating Cost Total

The program operating cost total may now be computed with Equation (6.21) as follows:

$$C_{OPS} = \frac{(4,266,978,750 + 5,236,951,320 + 420,260,400 + 1,681,041,600)}{(1 - 0.20 - 0.14 - 0.15)} =$$

$$= 11,605,232,070/0.51 = \text{USD } 22,755,000,000,$$

which has been rounded off to the nearest million.

6.9.3 Program Operating Cost per Flight Hour

The program operating cost per flight hour may be determined with Equation (6.23) as:

$$C_{ops/hr} = 22,755,000,000/\{(612)(25)(350)\} =$$

$$= 4,249 \text{ USD/hour}$$

The reader will have observed the fact that neither cost of depreciation nor cost of financing is included in the military program operating cost per flight hour.

Since military airplanes 'belong' to the tax payers, they are not depreciated in any commercial sense of the word. Also, the cost of financing is not included, because military operating costs are deferred out of annual budgets alotted by the congress.

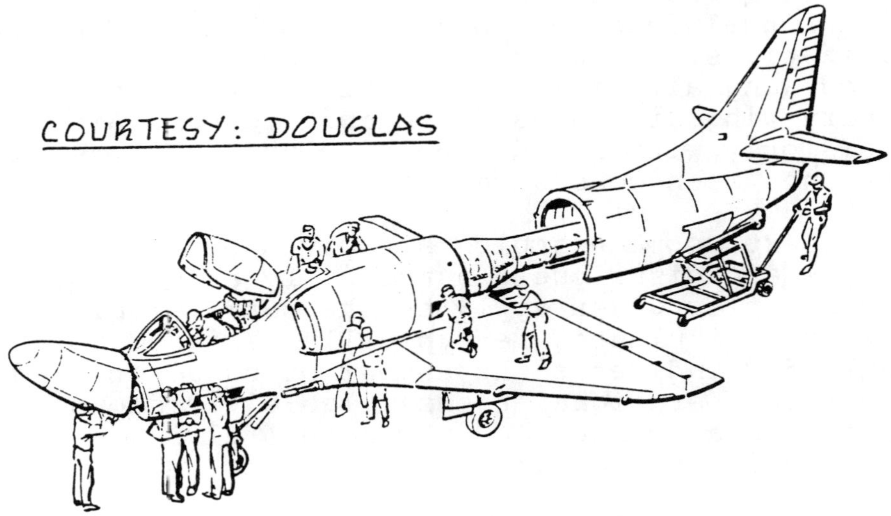

6.10 SOME OBSERVATIONS ON OPERATING COST OF MILITARY AIRPLANES

The purpose of this section is to provide the reader with additional insight and with additional references relative to the operating cost of military airplanes. The material is organized as follows:

6.10.1 Military Service Organization, Typical Airplane Numbers in Service and Reasons for High Operating Cost

6.10.2 Discussion of Operating Cost Distribution

6.10.3 Examples of System Maintainability and Reliability

6.10.1 Military Service Organization, Typical Airplane Numbers in Service and Reasons for High Operating Cost

Most military airplane operations are conducted at the squadron or wing level. The number of airplanes per squadron, the number of squadrons per wing and the number of wings per Air Force or per Navy Group depends on many tactical, strategic and political considerations which are beyond the scope of this text. For the interested reader, Air Force Magazine and the United States Naval Institute Proceedings (both are monthly magazines) publish annual details about USAF, USN and USMC service organizations. Although the US Army does operate several fixed wing airplane types, this service concentrates on rotary wing airplanes for combat roles. Since this text deals only with fixed wing airplanes, USArmy airplanes have not been considered.

Next, some examples are given of how military airplanes are distributed numerically at the squadron level. The examples given apply to the USAF. Table 6.7 shows how many airplanes are typically assigned to a squadron. Table 6.8 indicates the number of squadrons dedicated to specific operational tasks. Table 6.9 shows the number of active airplanes and the annual flying hours produced by each for recent years.

Because military forces have to be able to operate relatively independently of large fixed bases during war times, a certain amount of duplication in support and infrastructure is inevitable. This in turn implies that the operation of military airplanes cannot be conducted

TABLE 6.7 NUMBER OF AIRPLANES PER SQUADRON

Airplane Type	Number per Squadron	Airplane Type	Number per Squadron
A-10	18 or 24	B-1	16
F-4	12 or 24	B-52	13 - 19
RF-4	18	C-5	15 or 16
F-5	11, 18 or 20	C-9	3 or 11
F-15	15, 18 or 24	C-130	16
F-16	18 or 24	AC-130	10
F-106	15	KC-10	19
F-111	12, 18 or 24	KC-135	13 - 25
FB-111	8 or 11	C-141	13 - 17
		E-3	2, 4 or 16

Source: Air Force Magazine, May 1988

TABLE 6.8 USAF FLYING SQUADRONS BY MISSION TYPE

Active Forces	FY'85	FY'87	FY'89*
Strategic Bomber	22	22	25
Air Refueling	35	36	35
Strategic Comm. and Ctrl	6	6	6
Intelligence	3	3	3
Strategic Reconnaissance	1	1	1
Strategic Interceptor	4	3	2
Fighter	78	81	79
Tactical Reconnaissance	8	7	5
Tactical Electr. Warfare	3	4	4
Special Operations Forces	5	5	5
Tact. Air Comm. Ctrl Syst.	3	3	3
Tactical Air Ctrl Syst.	7	7	7
Weather	2	2	1
Rescue	8	9	7
Tactical Airlift	14	13	12
Strategic Airlift	17	17	18
Special Mission	1	1	1
Aeromedical Airlift	3	3	3
Total Active Forces	220	223	217
Reserve Forces:			
ANG Selected Reserve	91	91	91
Air Force Reserve	56	57	58
Total Squadrons	367	371	366

Source: Air Force Magazine, May 1988 * Estimated

TABLE 6.9 NUMBER OF ACTIVE AIRPLANES AND FLYING HOURS

Airplane Type	Number of Active Airplanes in: FY'85	FY'87	FY'89*
Strategic Bomber	330	393	420
Tanker	559	576	566
Fighter/Interceptor/Attack	3,057	3,033	2,998
Recon./Electronic Warfare	418	432	438
Cargo/Transport	859	848	852
Search and Rescue (fixed wing only)	37	35	31
Helicopter (incl. rescue)	234	191	171
Trainer	1,613	1,595	1,515
Utility/Observation/Other	180	110	112
Total USAF	7,287	7,213	7,103
Air National Guard Total	1,688	1,732	1,736
Air Force Reserve Total	468	502	514
Total Active USAF, ANG and USAF Reserve	9,443	9,447	9,353
	Flying Hours in Thousands:		
USAF	2,914	2,837	2,770
ANG	423	435	447
AF Reserve	140	153	156
Total Fl. Hrs in Thousands	3,477	3,425	3,373
Average Flying Hours per year, per airplane:	368	363	361

Source: Air Force Magazine, May 1988 * Estimated

COURTESY: SAAB

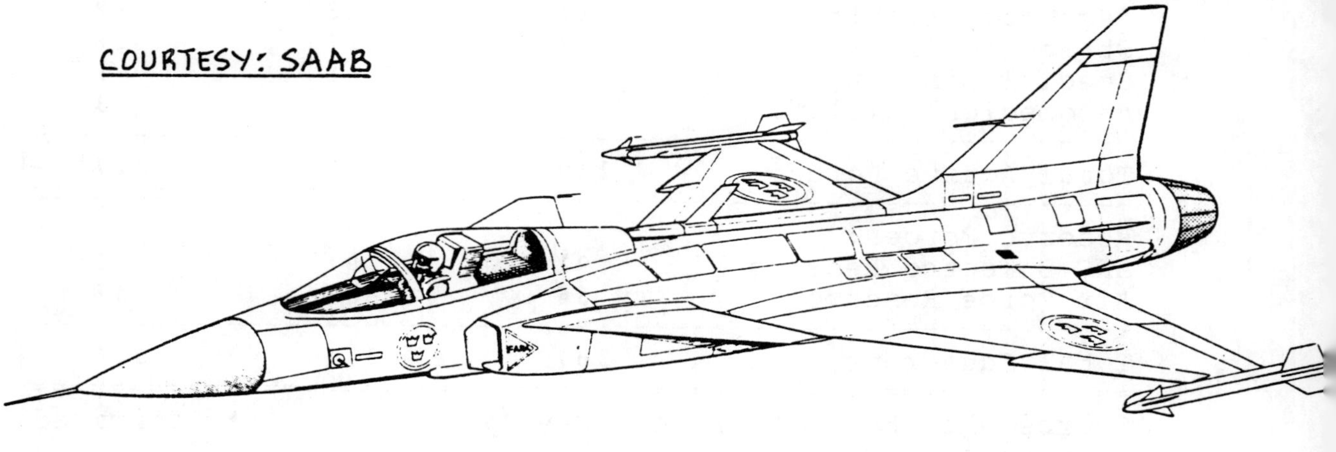

at levels of efficiency comparable to commercial airplane operations. This is one reason why the operating cost per hour for airplanes such as the Eris ground attack fighter is predicted to be rather high: USD 4,249 per flight hour as determined on page 167.

The challenge to the airplane designer is to design military airplanes and their systems layout in such a way that their mission can be accomplished at an affordable cost to the taxpayers.

The following subsections present observations on several aspects of operating cost distribution as well as on system reliability and maintainability.

6.10.2 Discussion of Operating Cost Distribution

Figure 6.4 shows the contribution to total operating cost of each cost component considered in Equation (6.1) for the Eris fighter. The cost data were calculated in Section 6.9. The percentage contribution of the seven operating cost components can be summarized as follows:

1)	C_{POL} (Fuel, oil and lubricants):		19
2)	$C_{PERSDIR}$ (Direct personnel):		23
	aircrew	10	
	maintenance	13	
3)	C_{SPARES} (Spares):		14
4)	C_{CONMAT} (Consumable materials):		2
5)	$C_{PERSIND}$ (Indirect personnel):		20
6)	C_{DEPOT} (Depot):		15
7)	C_{MISC} (Miscellaneous):		7
	C_{OPS} (Total operating cost percentage)		100

The following operating cost items are directly affected by the airplane designer and the design decisions he makes, mostly those in the early phases of preliminary design:

* Fuel, oil and lubricants (see Parts I and VII, specifically for what governs W_F and W_{fto})

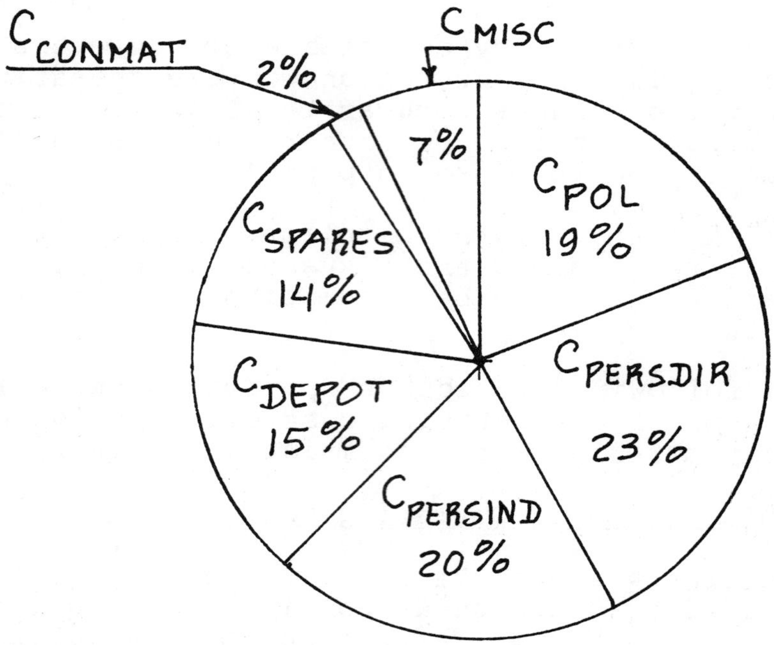

Figure 6.4 Distribution of Program Operating Cost Components for the Eris Fighter Program

Table 6.10 CLASSICAL EXAMPLES OF POOR DESIGN PRACTICES

===

EXAMPLE 1: In the F-4 fighter, the radio was a high failure rate item. It was installed underneath the rear ejection seat. To remove and to replace this radio this ejection seat had to be removed first. This made radio removal and replacement very expensive.

EXAMPLE 2: In the AV-8B V/STOL fighter, it is required to remove the wing before the engine can be removed. This adds significantly to the cost and time of engine removal.

EXAMPLE 3: In the B-1B bomber, a 'quick access' port was installed to allow for rapid checking and adding of engine oil levels. However, the oil gauge itself was hidden behind the very panel which included the 'quick access' panel. Maintenance personnel had to remove that entire panel before they could access the oil gage. That defeated the entire purpose of the 'quick access' port.

===

* Aircrew and maintenance (see Parts II, III and IV, for details on configuration, cockpit, structure and system layout design practices)

From the tabulation on page 171 it is clear that the designer has direct leverage over 58 percent of the operating cost of a typical ground attack fighter. Two examples illustrate this:

EXAMPLE 1: Maintenance Manhours

In Section 6.9 it was shown that the program operating cost for the Eris fighter is USD 22,755,000,000 over a projected 25 year life. The maintenance manhours per flight hour were assumed to be 12.

Through careless design it is conceivable that this number could increase to 24 maintenance manhours per flight hour. Table 6.5 shows that this can indeed occur. In such a case, the program operating cost increases by:

USD 1.2x2x2,910,978,000 (see p.165) = USD 6,986,347,200.

This means an increase of 31 percent of the total program operating cost!

If, on the other hand, by careful design the maintenance manhours per flight hour could be decreased from 12 to 6, a program operating cost savings of:

USD 1.2x0.5x2,910,978,000 (see p.165) = USD 1,746,586,800

would be the result. That is a savings of 8 percent!

EXAMPLE 2: Mission Fuel Used

It was shown on p.164 that the Eris program operating cost for fuel, oil and lubricants is approximately: USD 4,300,000,000. The designer should be able to reduce this cost by paying careful attention to details of:

* engine inlet design
* aerodynamic design
* nozzle and boattail design
* c.g. excursion diagram

A ten percent reduction in this cost item should be within reach: this shaves USD 430,000,000 off the program operating cost!

A thoughtless designer can needlessly increase operating cost. Table 6.10 gives three classic examples of poor design decisions from a maintenance viewpoint.

Chapter 8 contains a tabulation of specific design guidelines aimed at reducing cost.

NOTE OF CAUTION:

On p.171 all seven of the operating cost contributions are expressed as a fraction of the total operating cost. Whenever fundamental airplane design parameters are changed, there is no intrinsic reason why the cost items 4-7 should change as the fractions would dictate.

Therefore, whenever a cost trade study is conducted, a baseline should be established for the cost items 4-7. These baseline costs should then be kept constant while the effect of certain design parameters which affect cost items 1-3 is being investigated.

6.10.3 Observations on System Maintainability and Reliability

Chapter 8 and References 25-28 are recommended for further reading about ways to improve airplane maintainability and reliability.

Many military missions require the use of very complex avionics systems. Such systems are required to allow the airplane to function as a weapon system rather than merely as a gunnery or bombing platform.

Also, the nature of military missions, particularly in terms of speed, altitude and weather capability often dictates the use of very advanced technology. This is inherently expensive, risky and to some extent can cause frequent breakdowns.

As a result of poor system and component reliability as well as poor accessibility of those components which require frequent attention, the operational readiness of military airplanes has tended to be poor: see Table 6.11! Table 6.12 indicates how various avionics components affect the time between maintenance. Recollecting that, on the average, military airplanes fly about 350 hours per year, the need for maintenance occasions per year is very high, as shown in Table 6.13. Airplane designers should be able to straighten out their act! Again, Chapter 8 contains some tabulated guidelines which can be helpful in reducing cost.

Table 6.11 EXAMPLES OF FIGHTER READINESS

Airplane Type	Airplane Complexity	Not Mission Capable in percentage	Mean Flight Hours between Failure	Maintenance Events per Sortie	Maintenance Manhours per Sortie
USAF					
A-10	Low	32.6	1.2	1.6	18.4
A-7D	Medium	38.6	0.9	1.9	23.8
F-4E	Medium	34.1	0.4	3.6	38.0
F-15	High	44.3	0.5	2.8	33.6
F-111F	High	36.9	0.3	9.2	74.7
F-111D	High	65.6	0.2	10.2	98.4
USN					
A-4M	Low	27.7	0.7	2.4	28.5
AV-8A	Low	39.7	0.4	3.0	43.5
A-7E	Medium	36.7	0.4	3.7	53.0
F-4J	Medium	34.2	0.3	5.9	82.7
A-6E	High	39.3	0.3	4.8	71.3
F-14A	High	47.1	0.3	6.0	97.8

Source: Aviation Week and Space Technology, October 6, 1980

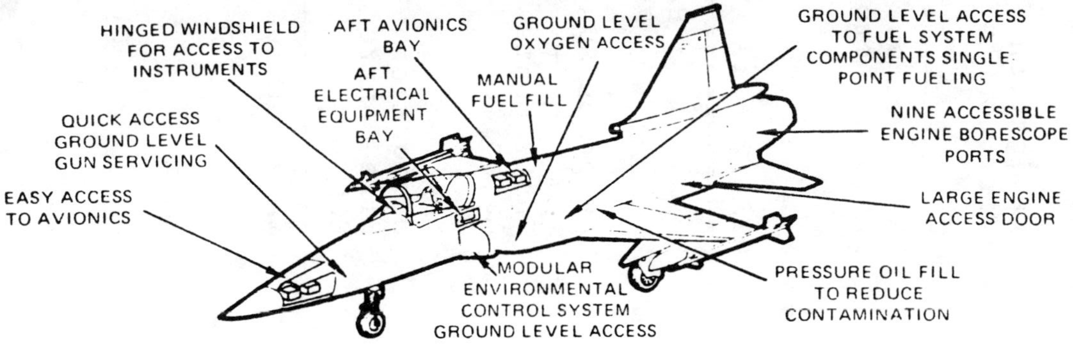

COURTESY: NORTHROP

ACCESSIBILITY FEATURES

TABLE 6.12 AVIONICS SYSTEM OPERATIONAL TIME BETWEEN MAINTENANCE, IN FLIGHT HOURS

Avionics System	F-111	F-4	F-16	A-7	F-15
Instruments	15	17	60	45	32
Autopilot	13	43	NA	NA	61
UHF	42	22	57	64	24
IFF	35	NA	44	89	15
Navigation	66	4	172	59	18
Fire Control	71	3	9	26	6
ECM	7	7	25	20	22

NA = not available
Source: Aerospace America, August 1985

TABLE 6.13 AVIONICS SYSTEM DOWN OCCURRENCES PER YEAR BASED ON 350 FLIGHT HOURS PER YEAR

Avionics System	F-111	F-4	F-16	A-7	F-15
Instruments	23	21	6	8	11
Autopilot	27	8	NA	NA	6
UHF	8	16	6	5	15
IFF	10	NA	8	8	23
Navigation	5	88	2	6	19
Fire Control	5	117	39	13	58
ECM	50	50	14	18	16

NA = not available
Source: Table 6.12

COURTESY: SAAB

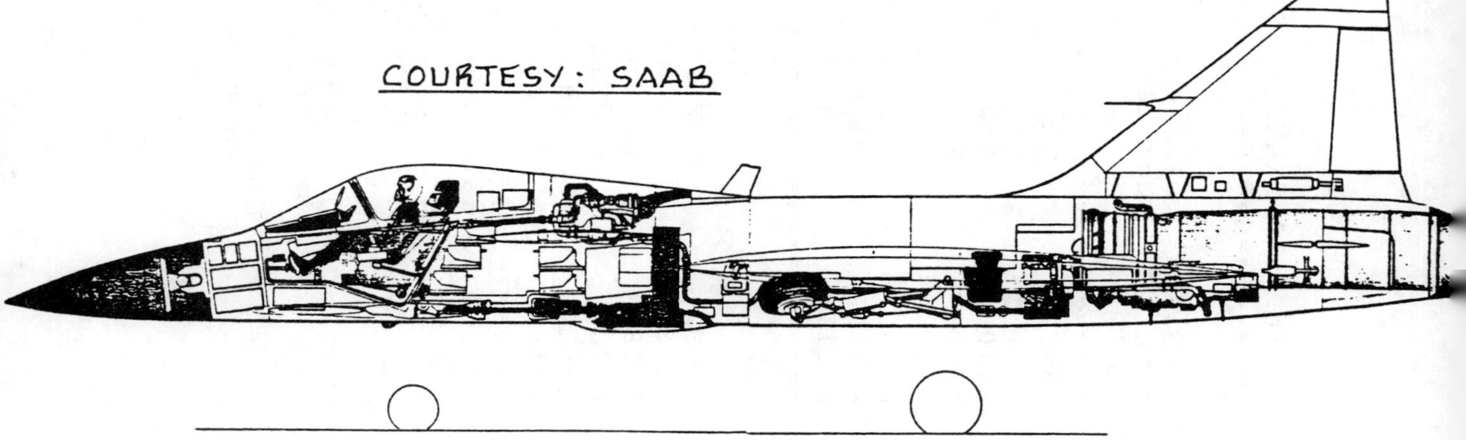

7. EXAMPLE OF LIFE CYCLE COST CALCULATION FOR A MILITARY AIRPLANE

The purpose of this chapter is to present an example of the calculation of life cycle cost (LCC) of a military airplane. The example has been constructed for the Eris ground attack fighter of Parts I and II. An equation for life cycle cost (LCC) was given in Ch.2, as Eqn.(2.3):

$$LCC = C_{RDTE} + C_{ACQ} + C_{OPS} + C_{DISP} \qquad (7.1)$$

where: C_{RDTE} is the total cost of research, development, test and evaluation

C_{ACQ} is the program acquisition cost

C_{OPS} is the program operating cost

C_{DISP} is the program disposal cost

More detailed definitions of these cost sources are provided in Chapter 2, pages 9 - 11. Methods for determining the first three sources of LCC are developed in Chapters 3 - 6. Applications of these methods to the Eris fighter program are given in Sections 7.1 - 7.3. An approach to the determination of disposal cost (C_{DISP}) is given in Section 7.4. The life cycle cost of the Eris fighter is computed in Section 7.5. Definitions for airplane unit cost are given and illustrated in Section 7.6. The material is organized as follows:

7.1 RESEARCH, DEVELOPMENT, TEST AND EVALUATION COST OF THE ERIS GROUND ATTACK FIGHTER

In this section the research, development, test and evaluation cost, C_{RDTE}, will be estimated for the Eris ground attack fighter. The mission specification for this fighter is stated in Table 2.19 of Part I. The assumption will be made that the cost estimates are needed for calendar year 1990. The method used is that of Ch.3. To use the method the following input data are needed:

W_{TO} = 64,905 lbs from p.14, Part V.

W_E = 33,500 lbs from p.14, Part V.

V_{max} = 450 kts from p.45, Part V.

N_{rdte} = 10 assumed from p.26.

N_{st} = 2. However, it will be assumed that even the two static test airplanes will be fully equipped with engines and avionics.

F_{diff} = 2.0 assumed from p.26 as well as the design decision to use 'de-facto' stability, which was made in sub-section 11.4.3 of Part II.

F_{cad} = 0.8 assumed from p.27.

From Eqn.(3.5):

$$W_{ampr} = \text{invlog}\{0.1936 + 0.8645(\log 64{,}905)\} =$$

$$= 22{,}585 \text{ lbs}$$

Airframe Engineering and Design Cost: C_{aed_r}

From Equation(3.2):

$$MHR_{aed_r} = 0.0396(22{,}585)^{0.791}(450)^{1.526}(10)^{0.183} *$$

$$*(2.0)(0.8) = 3{,}002{,}542 \text{ hrs}$$

R_{e_r} = 62.00 USD/hr from Figure 3.3, which assumes that the Eris program does not require a high level of security.

From Eqn.(3.3):

C_{aed_r} = 3,002,542(62.00) = USD 186,157,604

<u>Development Support and Test Cost:</u> C_{dst_r}

CEF for 1990 is 3.03 as determined from Figure 2.7.

From Eqn.(3.7):

$C_{dst_r} = 0.008325(22,585)^{0.873}(450)^{1.890}(10)^{0.346}*$

$*(3.03)(2.0)$ = USD 73,166,699

<u>Flight Test Airplanes Cost:</u> C_{fta_r}

This follows from Eqn.(3.8). First, all five cost components in this equation will be determined.

<u>Cost of engine and avionics:</u> $C_{(e + a)_r}$

The required takeoff thrust per engine, T_{TO_e}, follows from p.191, Part I:

T_{TO_e} = 29,670/2 = 14,835 lbs

From Eqn.(B9) in Appendix B, the 1990 engine price may be estimated as follows:

EP_{1990} = [invlog{2.3044 + 0.8858log(14,835)}]*

*(3.03/3.02) = USD 1,001,757

The avionics cost, $C_{avionics}$, will be assumed to be 25 percent of the Eris unit flyaway cost, AMP_{1990}. The latter is estimated from Eqn.(A14) in Appendix A:

AMP_{1990} = (3.03/3.02)*

*[invlog{2.3341 + 1.0586(log64,905)}] =

= USD 26,903,782

Thus, $C_{avionics}$ = 26,903,782/4 = USD 6,725,946

From Eqn.(3.9):

$$C_{(e + a)_r} = \{2(1,001,757) + 6,725,946\}(10) =$$

$$= \text{USD } 87,294,600$$

where it has been assumed that all 10 prototypes, including the 2 used for static testing, will be fully equipped with engines and avionics.

Manufacturing cost of flight test airplanes: C_{man_r}

From Eqn.(3.11):

$$MHR_{man_r} = 28.984(22,585)^{0.740}(450)^{0.543}(10)^{0.524})*$$

$$*(2) = \text{USD } 8,906,551 \text{ hrs}$$

R_{m_r} = 35 USD/hr from Figure 3.4

From Eqn.(3.10):

$$C_{man_r} = 35(8,906,551) = \text{USD } 311,729,285$$

Materials cost for the flight test airplanes: C_{mat_r}

From Eqn.(3.12):

$$C_{mat_r} = 37.632(1.0)(22,585)^{0.689}(450)^{0.624}*$$

$$*(10)^{0.792}(3.03) = \text{USD } 31,943,800$$

In this application it is assumed that the Eris is manufactured with primarily conventional materials (so that F_{mat} = 1.0) and that the 1990 CEF value is 3.03.

Tooling cost for the flight test airplanes: C_{tool_r}

From Eqn.(3.14):

$$MHR_{tool_r} = 4.0127(22,585)^{0.764}(450)^{0.899} *$$

$$*(10)^{0.178}(0.33)^{0.066}(2.0) =$$

$$= 5,784,225 \text{ hrs}$$

where the factor 0.33 represents the assumed manufacturing rate per month for the prototype phase, see p.33.

The tooling hourly rate is assumed to be:

R_{t_r} = 44 USD/hr from Figure 3.5.

From Eqn.(3.13):

$$C_{tool_r} = 44(5,784,225) = \text{USD } 254,505,900$$

<u>Quality control cost for the flight test airplanes:</u> C_{qc_r}

From Eqn.(3.15):

$$C_{qc_r} = 0.13(311,729,285) = \text{USD } 40,524,807$$

Finally, from Eqn.(3.8):

$$C_{fta_r} = 87,294,600 + 311,729,285 + 31,943,800 +$$

$$+ 254,505,900 + 40,524,807 = \text{USD } 725,998,392$$

<u>Flight Test Operations Cost:</u> C_{fto_r}

From Eqn.(3.16):

$$C_{fto_r} = 0.001244(22,585)^{1.160}(450)^{1.371}(8)^{1.281} *$$

$$*(3.03)(2) = \text{USD } 52,739,614$$

<u>Test and Simulation Facilities Cost:</u> C_{tsf_r}

In sub-section 11.4.3 of Part II the decision was made to use 'de-facto' stability in the Eris design. Therefore, from Eqn.(3.17):

$C_{tsf_r} = 0.20(C_{RDTE})$ as suggested on p.35.

<u>Profit over Flight Test Airplanes:</u> C_{pro_r}

From Eqn.(3.18):

$$C_{pro_r} = 0.1(C_{RDTE})$$

<u>Cost to Finance the Flight Test Airplanes:</u> C_{fin_r}

It will be assumed that the manufacturer will have to arrange for the financing of the Eris fighter program. Consistent with Section 3.7, the financing rate will be assumed to be 0.10 (p.36) so that, with Eqn.(3.19):

$$C_{fin_r} = 0.1(C_{RDTE})$$

<u>Total Research, Development, Test and Evaluation Cost:</u> C_{RDTE}

From Eqn.(3.1):

$$C_{RDTE} = (186{,}157{,}604 + 73{,}166{,}699 + 725{,}998{,}392 + + 52{,}739{,}614)/(1 - 0.2 - 0.1 - 0.1) =$$

$$= 1{,}038{,}062{,}309/0.6 = \text{USD } 1{,}730{,}103{,}848$$

Rounding this off:

$$C_{RDTE} = 1{,}730{,}000{,}000$$

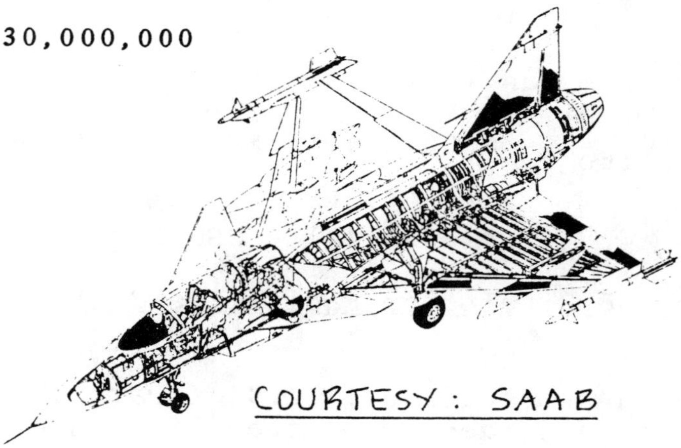

COURTESY: SAAB

7.2 MANUFACTURING AND ACQUISITION COST OF THE ERIS GROUND ATTACK FIGHTER

In this sub-section the manufacturing and acquisition costs, C_{MAN} and C_{ACQ}, will be estimated for the Eris ground attack fighter. The mission specification for this fighter is stated in Table 2.19 of Part I. The assumption will be made that the cost estimates are needed for calendar year 1990. The method used is that of Ch.4. To use the method the input data for Section 7.1 as well as the following input data are needed:

$N_{program}$ = 750 airplanes will be assumed.

N_{rdte} = 10 from page 178.

This yields with Eqn.(4.1):

N_m = 740

The Eris will be assumed to be manufactured with rather conventional materials, thus: F_{mat} = 1.0 as was also assumed on page 180.

Airframe Engineering and Design Cost of Production Airplanes: C_{aed_m}

This will be determined from Eqn.(4.5b). First, the engineering manhours for the Eris program are found from Eqn.(4.6):

$$MHR_{aed_{program}} = 0.0396(22,585)^{0.791}(450)^{1.526}*$$

$$*(750)^{0.183}(2.0)(0.8) =$$

$$= 6,616,455 \text{ hrs}$$

The engineering hourly rate will be assumed to be 62.00 USD/hr, as on p.178. Next, with Eqn.(4.5b) and with the C_{aed_r} of p.179:

$$C_{aed_m} = (6,616,455)(62.00) - 186,157,604 =$$

$$= 410,220,210 - 186,157,604 = \text{USD } 224,062,606$$

Airplane Program Production Cost: C_{apc_m}

This cost will be computed with Eqn.(4.7). First, all six cost components in this equation will be found.

Cost of engine and avionics of production airplanes: $C_{(e+a)_m}$

This follows from Eqn.(4.8). It will be assumed that the engine and avionics cost for the production version of the Eris fighter is the same as that for the RDTE version. With the cost data from pages 179 and 180:

$$C_{(e+a)_m} = \{(2)(1{,}001{,}757 + 6{,}725{,}946)\}(740) =$$

$$= (8{,}729{,}460)(740) = \text{USD } 6{,}459{,}800{,}400$$

Cost of the interiors of production airplanes: C_{int_m}

This cost is determined from Eqn.(4.9) with $F_{int} = 0$ for military airplanes:

$$C_{int_m} = 0$$

Manufacturing cost of production airplanes: C_{man_m}

Equation (4.10b) will be used to estimate this cost item. First, the program manufacturing manhours are computed from Eqn.(4.11):

$$MHR_{man_{program}} = 28.984(22{,}585)^{0.740}(450)^{0.543} *$$

$$*(750)^{0.524}(2) = 85{,}554{,}269 \text{ hrs}$$

The manufacturing labor rate will be assumed to be the same as that used in the RDTE phase on p.180:

$R_{m_m} = 35$ USD/hr. Thus, with Eqn.(4.10b) and with the data of p.180:

$$C_{man_m} = (85{,}554{,}269)(35.00) - 311{,}729{,}285 =$$

$$= 2{,}994{,}399{,}415 - 311{,}729{,}285 = \text{USD } 2{,}682{,}670{,}130$$

Materials cost for production airplanes: C_{mat_m}

The materials cost of the production program is determined with Equation (4.12). First, the materials cost of the entire program is found from Eqn.(4.13):

$$C_{mat_{program}} = 37.632(1.0)(22{,}585)^{0.689}(450)^{0.624} *$$

$$*(750)^{0.792}(3.03) = \text{USD } 975{,}966{,}932$$

where $F_{mat} = 1.0$ and $CEF_{1990} = 3.03$ were obtained from page 180.

Next, the cost of materials for the production program follows from Eqn.(4.12) as:

$$C_{mat_m} = 975{,}966{,}932 - 31{,}943{,}800 = \text{USD } 944{,}023{,}132$$

Tooling cost for production airplanes: C_{tool_m}

This cost item will be calculated with Eqn.(4.14b). First, the program tooling manhours are determined with Eqn.(4.15):

$$MHR_{tool_{program}} = 4.0127(22{,}585)^{0.764}(450)^{0.899} *$$

$$*(750)^{0.178}(10)^{0.066}(2) =$$

$$= 15{,}623{,}711 \text{ hrs}$$

In this calculation it was assumed that the Eris will be manufactured at a rate of $N_{r_m} = 10$ airplanes per month. For a production program of 740 units, this represents a 6 year program. That is quite reasonable for this type of airplane.

The tooling hourly rate will be assumed to be equal to that used in the RDTE phase on p.181:

$$R_{t_m} = 44.00 \text{ USD/hr}$$

Now, with Eqn.(4.14b) and the data from p.181:

C_{tool_m} = (15,623,711)(44.00) - 254,505,900 =

= 687,443,284 - 254,505,900 = USD 432,937,384

Quality control cost for production airplanes: C_{qc_m}

This cost will be found with Eqn.(4.16):

C_{qc_m} = 0.13(2,682,670,130) = USD 348,747,117

It is now possible to determine the entire airplane production cost from Eqn.(4.7) as follows:

C_{apc_m} = 6,459,800,400 + 0 + 2,682,670,130 +

+ 944,023,132 + 432,937,384 + 348,747,117 =

= USD 10,868,178,163

Cost of Flight Test Operations for Production Airplanes: C_{fto_m}

This cost is estimated with Eqn.(4.17). On p.167 it was found that for the Eris: $C_{ops/hr}$ = 4,249 USD/hr.

However, this operating cost includes the effects of military overhead, depot, spares and other factors. It is necessary to eliminate these factors from the p.167 operating cost before Eqn.(4.17) can be applied. Instead of doing this, it is assumed here that the operating cost per hour to the manufacturer is roughly 2,000 USD/hr.

Production airplane flight test hours per airplane are assumed to be:

t_{pft} = 20 hrs/airplane as stated on p.55.

The manufacturers flight test overhead factor is assumed to be: F_{ftoh} = 4.0 (on p.55). So, with Eqn.(4.17):

C_{fto_m} = (740)(2,000)(20)(4.0) = USD 118,400,000

Manufacturing Program Financing Cost: C_{fin_m}

The assumption will be made that the financing cost

for the manufacturing phase of the Eris program will be carried by the manufacturer and/or his lenders. A rate of 10 percent will be assumed. Thus, with Eqn.(4.18):

$$C_{fin_m} = 0.10(C_{MAN})$$

Program Manufacturing Cost: C_{MAN}

This cost can now be found from Eqn.(4.4) and the data determined in this section:

$$C_{MAN} = (224,062,606 + 10,868,178,163 +$$

$$+ 118,400,000)/(1 - 0.1) = 11,210,640,769/0.9 =$$

$$= \text{USD } 12,456,000,000 \text{ after some rounding off.}$$

Profit over the manufacturing phase: C_{pro_m}

According to the suggestion made on p.56, a 10 percent profit for the manufacturing phase is reasonable. Therefore, with Eqn.(4.19):

$$C_{PRO} = (0.10)(C_{MAN}) = \text{USD } 1,245,600,000$$

Program Acquisition Cost: C_{ACQ}

The total Eris program acquisition cost is estimated from Eqn.(4.2) as:

$$C_{ACQ} = 12,456,000,000 + 1,245,600,000 =$$

$$= \text{USD } 13,701,600,000$$

or, after rounding off:

$$C_{ACQ} = \text{USD } 13,702,000,000$$

7.3 OPERATING COST OF THE ERIS GROUND ATTACK FIGHTER

The program operating cost of the Eris fighter was determined with the method of Chapter 6 in Section 6.9. The result from p.167 is:

$$C_{OPS} = \text{USD } 22,755,000,000$$

7.4 DISPOSAL COST OF THE ERIS GROUND ATTACK FIGHTER

There comes a point in time when any airplane no longer has commercial or military value. That point is normally reached when:

1. The airplane has reached the end of its safe structural life and structural repairs are judged to be not economical.

2. The airplane has reached the end of its economic life: it can no longer compete effectively in the face of more modern airplanes.

3. The airplane ceases to have military value. This can occur if either its mission is no longer a viable one, or improvements in technology have rendered it obsolete.

4. It has been damaged beyond repair (such as in certain crashes or by damage caused by weather).

Whenever an airplane ceases to have commercial or military value it needs to be disposed of. The cost associated with such disposal is referred to as the disposal cost. The cost of disposing of airplanes depends on the types of materials used in construction and/or in operation. Disposal usually consists of:

1. Temporary storage

2. Draining of liquids and disposal thereof

3. Disassembly of engines and certain systems (such as computers and instruments)

4. Cutting up of the airframe and disposal of the resulting materials

There are costs associated with each of these actions. However, it is quite possible that there is a salvage value associated with actions 2-3. Examples are: resale value of engines, instruments, scrap metals (for re-cycling) and other materials. Any disposal costs are therefore partially offset by these resale values.

In some airplanes, materials and liquids are used which pose a problem for the environment if they are not disposed of with care. Examples are:

* Beryllium alloys
* Most composites (because they are not inherently degradable)
* Many oils and liquids (such as hydraulic fluids)

The author believes that it should be the responsibility of designers to include in their design decision making some serious thinking (and actions) about disposal problems. It is not ethical to assume that future generations will be able to 'take care' of problems created by design decisions of today.

To responsibly account for future disposal problems will require that designers enlist the help of chemical and environmental engineers in the design decision making process. Only that way can the cost of disposal be estimated with credibility.

One of the reasons that the earth environment is in serious trouble today is the fact that disposal problems are not included in design thinking and design decision making at most levels of industrial activity today.

Examples of potential problems which were caused by the lack of disposal considerations in design thinking are given in Table 7.1.

The author is not aware of any reliable methods for the estimation of airplane disposal costs. Reference 8 mentions disposal cost but provides no guidelines for its estimation.

In view of this sad state of affairs, it is suggested to use the following 'fudge' factor to account for the disposal of airplanes:

$$C_{DISP} = 0.01(LCC) \tag{7.2}$$

Whether or not this accounts fairly for the actual balance between resale values and disposal costs is very much an open question.

7.5 LIFE CYCLE COST OF THE ERIS GROUND ATTACK FIGHTER

The life cycle cost of the Eris fighter program will now be determined with Equation (7.1). The cost sources in this equation have the following magnitudes for the Eris fighter program:

TABLE 7.1 EXAMPLES OF POTENTIAL ENVIRONMENTAL PROBLEMS

Item	Problem
Nuclear waste	Insufficiently considered in the original decision to move ahead with fission power generation.
Lithium batteries	Used in thousands of personal computers. What will happen to this hazardous waste when personal computers are discarded?
Americium	This hazardous material is used in many smoke detectors. What will happen to this material when smoke detectors are discarded?
Non-biodegradables	These materials are used in nearly all 'fast food' containers today. Their disposal is creating massive waste piles in the environment.
Composites in airplanes and cars	The use of various types of composites is rapidly increasing. What will happen to these materials when disposal time comes around?

Source: This table is based on ideas by Mr. Ben Williams of the Air Force Institute of Technology.

C_{RDTE} = USD 1,730,000,000 (from p.182)

C_{ACQ} = USD 13,702,000,000 (from p.187)

C_{OPS} = USD 22,755,000,000 (from p.187)

C_{DISP} = USD 386,000,000 (with Eqn.(7.2))

With Equation (7.1) this yields:

$LCC_{Eris\ program}$ = USD 38,573,000,000

Observe, that for the Eris program:

$$C_{OPS} >> C_{ACQ} >> C_{RDTE}$$

This was also noted on p.11 in Chapter 2. It was further substantiated by the Ourania jet transport cost calculations which are summarized on p.132 in Ch.5.

The reason why the operating cost of the Eris program is 'only' twice that of the acquisition cost is that Eris is a relatively small airplane. The Eris has a crew of only one and consumes much less fuel per mission than bomber or transport airplanes. Also, the annual utilization of the Eris is much less than that of typical civil transports.

The relative magnitudes of the first three cost sources in the LCC equation do suggest one more time:

THE PRELIMINARY DESIGN PHASE (PART OF RDTE) LOCKS IN MOST OF THE LIFE CYCLE COST OF AN AIRPLANE!

This lesson should not be lost on designers nor on design managers.

7.6 UNIT COST OF THE ERIS GROUND ATTACK FIGHTER

The problem with determining the unit cost of military airplanes is largely one of definition. The author has decided to use the following definitions:

1) The manufacturing unit cost is defined as:

$$C_{muc} = (C_{MAN})/(N_m) \tag{7.3}$$

2) The acquisition unit cost is defined as:

$$C_{auc} = (C_{ACQ})/(N_m) \tag{7.4}$$

3) The program unit cost is defined as:

$$C_{puc} = (C_{ACQ} + C_{RDTE})/(N_m) \tag{7.5}$$

4) The life cycle unit cost is defined as:

$$C_{lcuc} = (LCC)/(N_m) \tag{7.6}$$

The reader will observe that the third definition corresponds to what for commercial airplanes was called the airplane estimated price, AEP, in Equation (4.3).

In the popular press the concept 'unit flyaway cost' is often used. However, the definition of this 'unit flyaway cost' is usually omitted, leaving the reader to guess how it is defined. The author believes that the definitions 1), 2) and 3) are useful in comparing unit costs of airplanes with similar missions. Airplane designers should keep in mind the fact that only the fourth definition reflects the actual unit cost to the taxpayer!

It is instructive to use the Eris fighter as an example of what these unit costs actually are.

With Eqn.(7.3) and p.187, the manufacturing unit cost is estimated as:

$$C_{muc} = 12,456,000,000/740 = \text{USD } 16,832,432$$

or, rounded off:

$$C_{muc} = \text{USD } 17,000,000$$

With Eqn.(7.4) and p.187, the acquisition unit cost is estimated as:

$$C_{auc} = 13,702,000,000/740 = \text{USD } 18,516,216$$

or, rounded off:

$$C_{auc} = \text{USD } 19,000,999$$

With Eqn.(7.5), p.187 and p.182, the program unit cost is estimated as:

$$C_{puc} = (13,702,000,000 + 1,730,000,000)/740 =$$

$$= \text{USD } 20,854,054$$

or, rounded off:

$$C_{puc} = \text{USD } 21,000,000$$

With Eqn.(7.6) and p.191, the life cycle unit cost is estimated as:

$$C_{lcuc} = (38,573,000,000)/740 = \text{USD } 52,125,676$$

or, rounded off:

$$C_{lcuc} = 52,000,000$$

Note that the airplane market price estimate for 1990 as obtained from Equation (A14) in Appendix A (see p.179) represents an over-estimate! Since the avionics price estimate of p.180 is based on a percentage of AMP the actual Eris cost estimates should be revised downward accordingly. This was not done in this text.

8. AIRPLANE DESIGN OPTIMIZATION AND DESIGN-TO-COST

A major question which every airplane designer asks himself at one time or another is: 'How can airplanes be designed so that they are the best possible?'

The problem with the use of the word 'best' is that it is not immediately clear what is meant by it. Therefore, it is 'better' to ask the following questions:

A) How can airplanes be designed so that their performance is maximized?

B) How can airplanes be designed so that their empty weight is minimized?

C) How can airplanes be designed so that their cost is minimized?

D) How can airplanes be designed to a given cost?

In mathematics, the words MAXIMUM and MINIMUM are both related to the concept: OPTIMUM. Maxima or minima arise by attempts to optimize certain functions of independent variables. Such functions are given the generic name: cost functions. The mathematical theory of optimization of cost functions is well developed and documented: References 29 and 30 are typical examples.

In airplane design, examples of cost functions are:

* Range
* Endurance
* Fuel consumption
* Empty weight
* Operating cost
* Return on investment

Before these cost functions can be optimized it will be necessary to express them functionally in terms of a finite set of design variables. After reviewing the material in Parts I through VII it will be agreed that the number of airplane design variables is very large indeed. Also, many of these variables are not truly independent, AND their interrelationship is not always definable.

That is why the airplane design process in Parts I and II was referred to as non-unique and iterative. It is useful to review these terms before returning to the idea of design optimization.

The airplane design process is non-unique because:

1. It is possible that more than one configuration approach results in an acceptable airplane. This is clearly illustrated by Figure 1.1 in Part II!

2. It is possible that more than one approach to the interior layout design and systems design leads to an acceptable airplane. Parts II, III and IV contain many examples of this.

The airplane design process is iterative because:

1. Most design decisions made early in the airplane design process turn out to have major (and often not foreseen) impact on design decisions which are made later. If that impact is unfavorable, many earlier decisions have to be reconsidered: i.e. the design has to be iterated*.

2. Except for a relatively small number of design variables (to be discussed later), it has not yet been found possible to relate design variables to each other in a mathematically unique manner. Besides, many airplane design variables are not even independent in the mathematical sense.

To arrive at an acceptable airplane design solution, many design decisions are therefore made on the basis of experience or intuition. This in turn makes it necessary to iterate through previous design decisions. The need for this was amply illustrated in Parts I, II and V. The iterative nature of the design decision-making process is 'formalized' through the step-by-step design procedures called: Design Sequences I and II as outlined in Part II.

It should be clear from Parts I - VII and also from Chapters 3 - 7 in this volume that it is nearly always desirable to keep the takeoff weight of an airplane (for a given mission) as low as possible. It is therefore essential that the designer understand the sensitivity of his design to those mission parameters and those design variables which have been found by experience to have a major impact on the takeoff weight of a new airplane. The problem of takeoff weight sensitivity was dealt with in some detail in Part I, Chapter 2. Applications and interpretations of the takeoff weight sensitivity of a new airplane are presented in Section 8.1.

---

* The process of arriving at a satisfactory weight and balance situation (Part II, Chapter 10) is one example!

Note the use of the word 'acceptable' (not optimum) in previous paragraphs. To find a true optimum design solution, the following steps must be completed:

1) One or more cost functions must be defined. These cost functions must be consistent with the mission objectives of the airplane AND they must not conflict with each other.

2) Mathematical relationships between each cost function and certain design variables must be formulated.

3) A rule must be formulated which makes clear how the cost function(s) is (are) to be evaluated: maximum or minimum AND whether or not design constraints are to be accounted for.

4) Design constraints and their relationship to certain design variables must be identified.

5) A theory (mathematical procedure) by which the optimum will be found must be agreed upon.

It depends on the specifics of a particular design problem how these steps are to be taken. Sections 8.2 through 8.6 contain suggestions and guidelines to help airplane design students find their way through the maze called: design optimization. Several examples of design optimization studies are discussed in Section 8.7.

Decisions to go ahead with an airplane program are rarely made without economic considerations. Minimizing economic cost is therefore a 'driver' in many airplane design and airplane program decisions.

More and more, cost itself is being viewed as a constraint. In the design and development of military airplanes, budget limitations are playing an increasingly important role. Such budget limitations must be viewed by the designer as a constraint.

Similarly, in the design and development of commercial airplanes, the acquisition cost (cost of ownership) as well as the direct operating cost are being viewed as design constraints.

This raises the question of how to design airplanes within certain cost constraints: design-to-cost. An introduction to this subject is given in Section 8.8.

Assume that the 'best' of all available design optimization tools have been employed in the design of a new airplane. There still are many aspects of the design which were untouched by the process of optimization, if only because the number of variables accounted for in any formal design optimization procedure is relatively small.

Examples of design variables left untouched by most design optimization processes are found in areas such as:

* interior layout design
* systems design and integration
* materials selection
* application of new, maturing technologies

It is in these areas where the experience and the intuition of the designer are the only method to keep cost down. Therefore, it is useful for the designer to have available a list of 'detail design guidelines' which have been found helpful in lowering airplane cost. Such guidelines are given in Section 8.9.

The material in this chapter is therefore organized as follows:

8.1 Takeoff Weight Sensitivity
8.2 Example Cost Functions in Airplane Design
8.3 Relation Between Cost Functions and Airplane Design Variables
8.4 Design Constraints in Airplane Design
8.5 Rules for Evaluating Cost Functions
8.6 Methods for Design Optimization
8.7 Examples of Design Optimization Studies
8.8 An Introduction to the Design-to-Cost Problem
8.9 Design Guidelines for Lowering Cost

8.1 TAKEOFF WEIGHT SENSITIVITY

It was shown in Chapters 3 - 7 that airplane takeoff weight plays an almost dominant role in determining cost. For that reason it is essential that airplane designers understand what 'drives' the takeoff weight. The purpose of this section is to review and discuss the sensitivity of airplane takeoff weight to mission performance parameters and to airframe design parameters.

A mathematical derivation of the sensitivity of airplane takeoff weight to such parameters can be found in Chapter 2 of Part I. In that chapter, the takeoff weight sensitivity to certain design parameters is expressed as the partial derivative: $\partial W_{TO}/\partial y$, where: y is any one of the following parameters:

* Empty weight, W_E
* Payload weight, W_{PL}

and, for any given mission phase i*:

* Range, R_i
* Endurance, E_i
* Lift-to-drag ratio, $(L/D)_i$
* Specific fuel consumption, c_{j_i} or c_{p_i}
* Propeller efficiency, η_{p_i}

The validity of these takeoff weight sensitivities depends on the validity of a statistical relationship between airplane takeoff weight and airplane empty weight which can be written as:

$$\log W_{TO} = A + B\log W_E \tag{8.1}$$

where: W_{TO} is the takeoff weight in lbs

W_E is the empty weight in lbs

A and B are regression coefficients determined from weight data on existing airplanes

It was shown in Chapter 2 of Part I that this type of relationship exists for airplanes with similar mission orientation. The scatter of actual takeoff weight and empty weight values relative to Eqn.(8.1) was shown to be

---

*For a definition of what is meant by mission phases, see Chapter 2 in Part I.

small. Numerical values for the regression coefficients A and B are given in Table 2.15 of Part I for twelve types of airplanes with similar mission orientation.

Whenever new, lighter structural materials are used in a new design, the validity of Eqn.(8.1) is in doubt. A method to account for the use of advanced materials is given in Appendix B of Part I.

Anytime a statistical method is employed it might be expected that the method becomes highly suspect whenever new, untried configurations are proposed. A discussion of circumstances which can 'drive' a designer toward new and untried configurations is presented in Reference 31.

The significance of takeoff weight sensitivities to an airplane designer is best illustrated with an example. For this example an airplane designed to fill the market void identified as Void I in Figure 9.4 (next chapter!) will be used. A preliminary mission specification for such an airplane is given in Table 8.1. For lack of a better name the airplane will be referred to as the SLRT: Small Long Range Transport.

The takeoff weight sensitivities for the SLRT are determined with the help of a user-friendly, preliminary design computer program called AAA: Advanced Aircraft Analysis program. The main features of this program are described in Reference 32. The takeoff weight sensitivities of the SLRT are determined in three steps:

<u>Step 1:</u> Calculation of the regression coefficients A and B in Eqn.(8.1).

Figure 8.1 shows a 'screendump' of the AAA program after entering takeoff weight and empty weight data for the 23 jet transports identified in Table 8.2. Note the following values for A and B:

$$A = 0.15 \qquad\qquad B = 1.02$$

Figure 8.2 shows a logarithmic plot of the weight data from Table 8.2. Note the small amount of scatter, thus supporting the validity of using Equation 8.1.

The mission specification of Table 8.1 is a very aggressive one: there is no such airplane in existence in 1990! It can be expected that to design a workable SLRT, use of more advanced structural materials (compared to those used in the 23 airplanes of Table 8.2) will be required. For that reason, Equation (B2) of Appendix B

Table 8.1 Mission Specification for the Small, Long Range Transport (SLRT)

Passengers:	125 (175 lbs each plus 40 lbs baggage)
Crew:	2 + 4 = 6 (175 lbs each plus 40 lbs baggage)
Range:	5,000 nm still air range
Average cruise altitude:	45,000 ft
Design cruise speed:	450 kts
Fuel reserves:	10 percent of mission fuel used

Table 8.2 Transport Jet Data Used in Obtaining the Regression Coefficients for Eqn.(8.1)

No.	Type	Takeoff Weight (lbs)	Empty Weight (lbs)
1.	Canadair 600	41,100	22,650
2.	601	43,100	23,940
3.	Airbus A300-600	363,765	188,291
4.	600R	375,885	187,251
5.	Airbus A310-200	305,560	166,742
6.	300	330,695	167,477
7.	Airbus A320-100	145,505	83,062
8.	-200	158,730	84,586
9.	Fokker F-100-A	95,000	51,720
10.	Yakovlev 42	124,560	76,180
11.	Tupolev 204	206,130	124,780
12.	British Aerospace 146-100	84,000	48,150
13.	-200	93,000	49,505
14.	Boeing 737-200	124,500	59,738
15.		115,500	59,500
16.	737-400	138,500	71,578
17.		150,000	72,520
18.	757-LR	250,000	124,380
19.	767	300,000	176,470
20.		400,000	195,770
21.	Gulfstream GIV	71,700	40,312
22.	McDonnell Douglas MD-81-A	140,000	77,291
23.	-C	160,000	79,333

Source: Jane's All the Worlds Aircraft, 1989

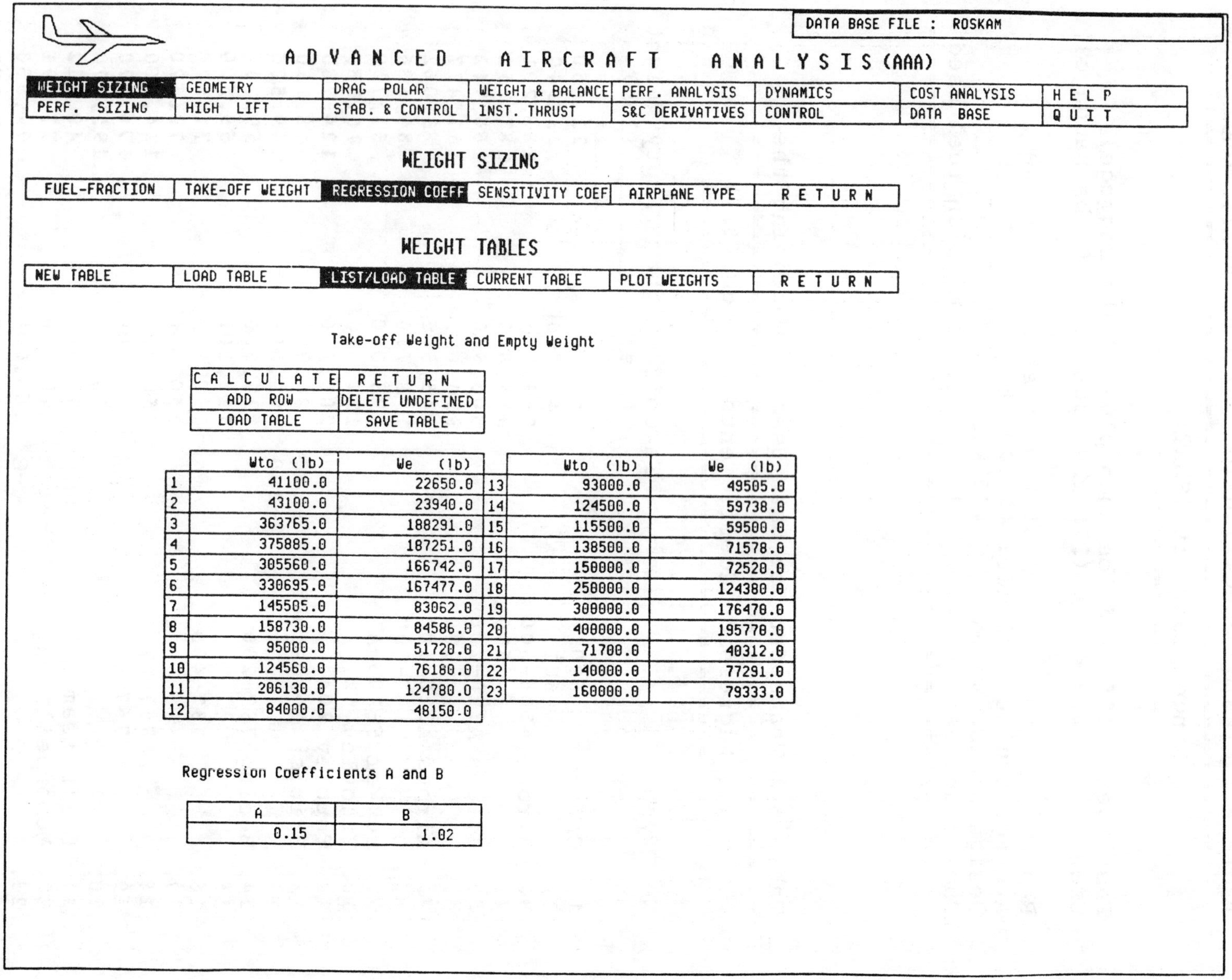

	Wto (lb)	We (lb)		Wto (lb)	We (lb)
1	41100.0	22650.0	13	93000.0	49505.0
2	43100.0	23940.0	14	124500.0	59738.0
3	363765.0	188291.0	15	115500.0	59500.0
4	375885.0	187251.0	16	138500.0	71578.0
5	305560.0	166742.0	17	150000.0	72520.0
6	330695.0	167477.0	18	250000.0	124380.0
7	145505.0	83062.0	19	300000.0	176470.0
8	158730.0	84586.0	20	400000.0	195770.0
9	95000.0	51720.0	21	71700.0	40312.0
10	124560.0	76180.0	22	140000.0	77291.0
11	206130.0	124780.0	23	160000.0	79333.0
12	84000.0	48150.0			

A	B
0.15	1.02

Figure 8.1 AAA Screendump of Takeoff Versus Empty Weight Regression Analysis

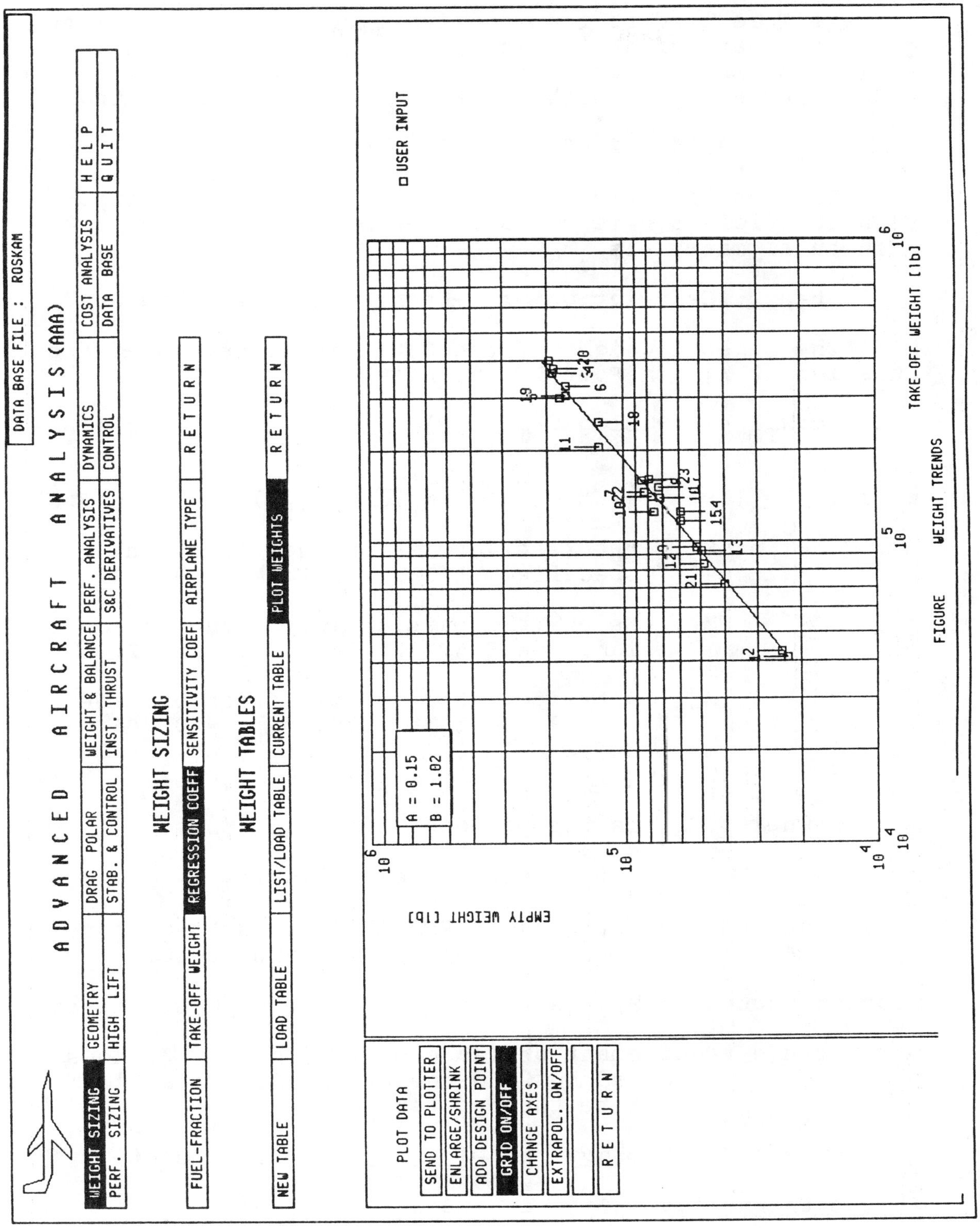

Figure 8.2 AAA Screendump of Logarithmic Plot of Takeoff Weight Versus Empty Weight

in Part I is used to find an 'improved' value for the regression coefficient A. The assumption is made that by judicious use of composites and Li-Al alloys a 5 percent reduction in empty weight is achievable: $\eta = 0.95$, thus:

$$A_{new} = 0.15 - 1.02\log(0.95) = 0.173$$

The assumption has been made here that the regression coefficient B is NOT affected by the choice of new structural materials.

Step 2: Calculation of takeoff weight for the SLRT.

The takeoff weight of the SLRT can be estimated from Equation (2.24) in Part I, which is:

$$\log W_{TO} = A + B\log(CW_{TO} - D) \quad (8.2)$$

where: $C = \{1 - (1 + M_{res})(1 - M_{ff}) - M_{tfo}\}$ (8.3)

with: M_{res} the reserve mission fuel fraction, specified in Table 8.1: $M_{res} = 0.10$.

M_{ff} the overall mission (used) fuel fraction, see Eqn.(2.13), p.16, Part I.

M_{tfo} = trapped fuel and oil fraction, see p.69, Part I. Use: $M_{tfo} = 0.005$.

$$D = W_{PL} + W_{crew} \quad (8.4)$$

where: W_{PL} is the payload weight in lbs

W_{crew} is the crew weight in lbs

Next, the AAA program of Ref.32 is used to compute the overall mission fuel fraction, M_{ff}. The critical mission phase for M_{ff} is the one corresponding to the extreme range requirement of Table 8.1. Figure 8.3 shows the input data needed to find the cruise fuel fraction: $(W_{i+1}/W_i)_{cruise}$. Note the aggressive use of:

aerodynamics technology: $(L/D)_{average} = 21.0$ and of

engine technology: $c_j = 0.5$ lbs/lbs/hr

Figure 8.4 shows the result of the takeoff weight calculation: $W_{TO} = 152,548$ lbs.

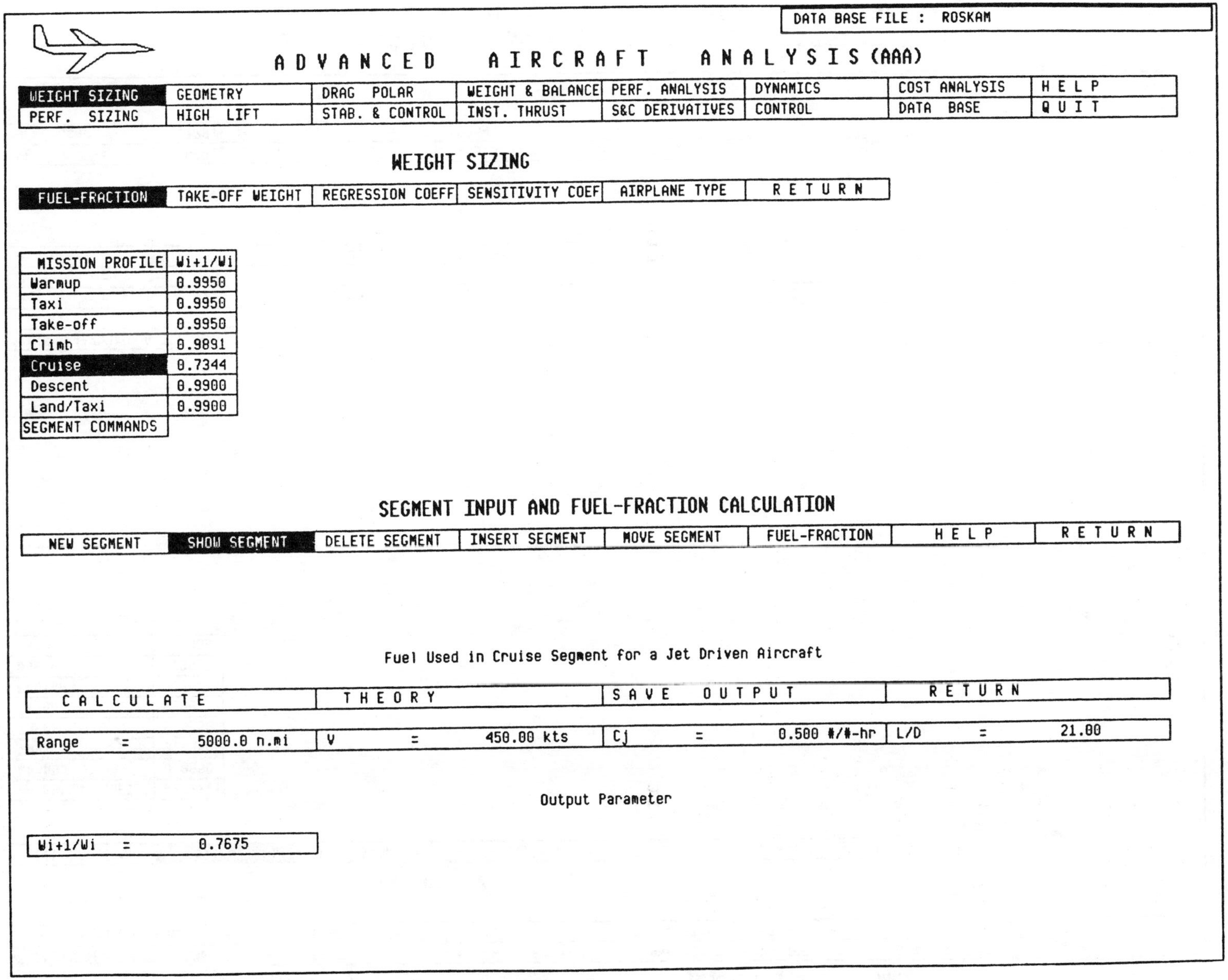

Figure 8.3 AAA Screendump of Fuel Fraction Calculation for a Cruise Mission Phase

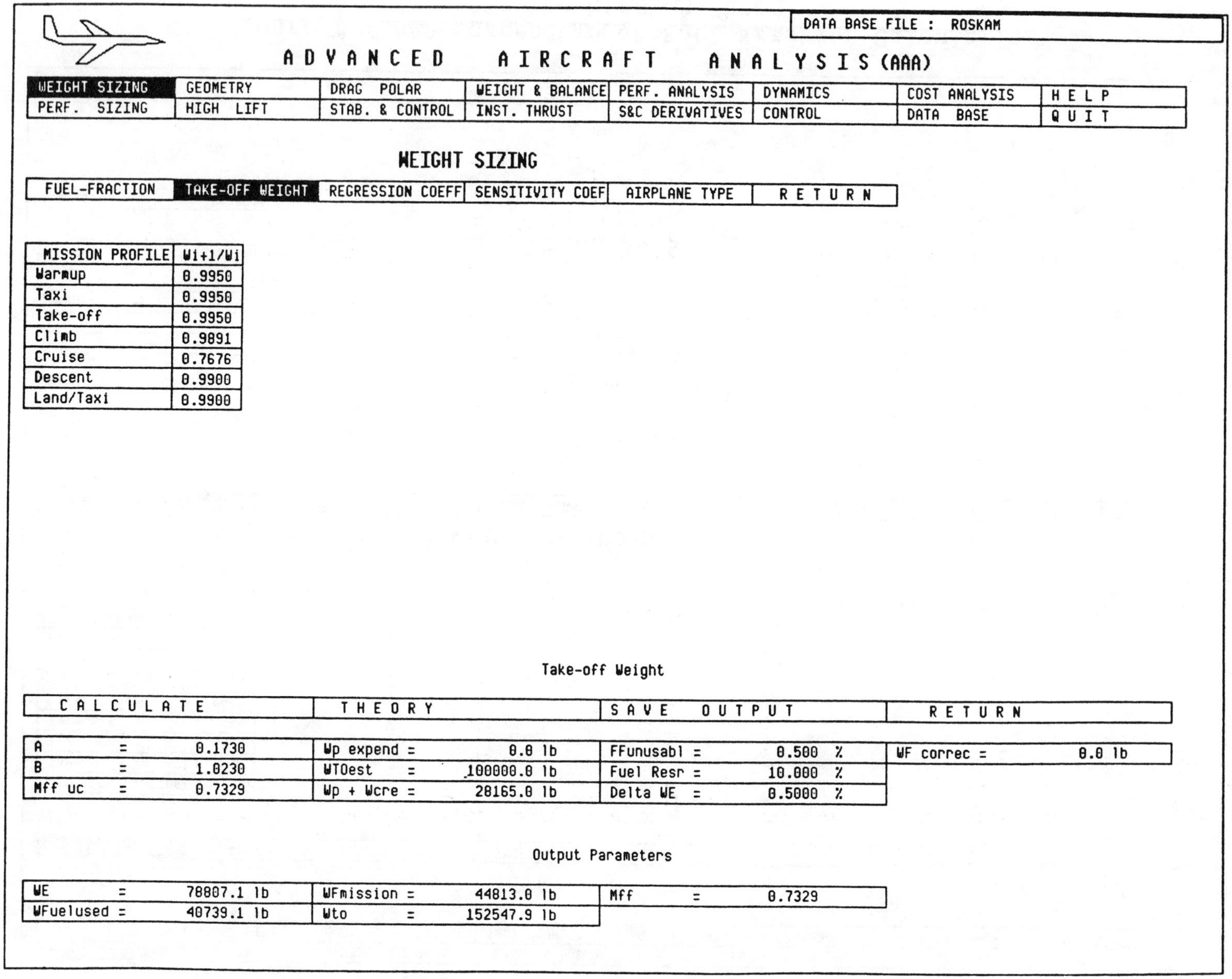

Figure 8.4 AAA Screendump of Overall Mission Fuel Fraction and Takeoff Weight Calculation

Step 3: Calculation of takeoff weight sensitivities.

Formulas for $\partial W_{TO}/\partial y$ were derived in Section 2.7 of Part I. Because of the proposed aggressive use of the following technologies:

* materials: A * aerodynamics: L/D * engine: c_j

it is particularly important that the sensitivity of takeoff weight to these three parameters be understood.

Figure 8.5 shows a AAA screendump of the relevant sensitivity calculations. An interpretation of these results is now given:

1) Sensitivity of takeoff weight to empty weight: W_E

From Figure 8.5:

$\partial W_{TO}/\partial W_E$ = 1.98 (2.0 rounded off)

This signifies that for each extra pound of empty weight, the takeoff weight grows by 2 lbs. Such growth in weight might occur as a result of not being able to achieve the assumed 5 percent empty weight reduction and/or by careless structural design. Management should exercise extra careful weight control during the detail design phase!

2) Sensitivity of takeoff weight to cruise specific fuel consumption: c_j

From Figure 8.5:

$\partial W_{TO}/\partial c_j$ = 331,591 (332,000 lbs/lbs/hr/lbs, rounded off)

This signifies that if the engine s.f.c is not 0.5, but say 0.55, the takeoff weight would increase by ΔW_{TO} = 0.05x332,000 = 16,600 lbs!

Management has to watch out that the engine manufacturer can indeed achieve the promised levels of cruise specific fuel consumption!

3) Sensitivity of takeoff weight to cruise lift-to-drag ratio: L/D

From Figure 8.5:

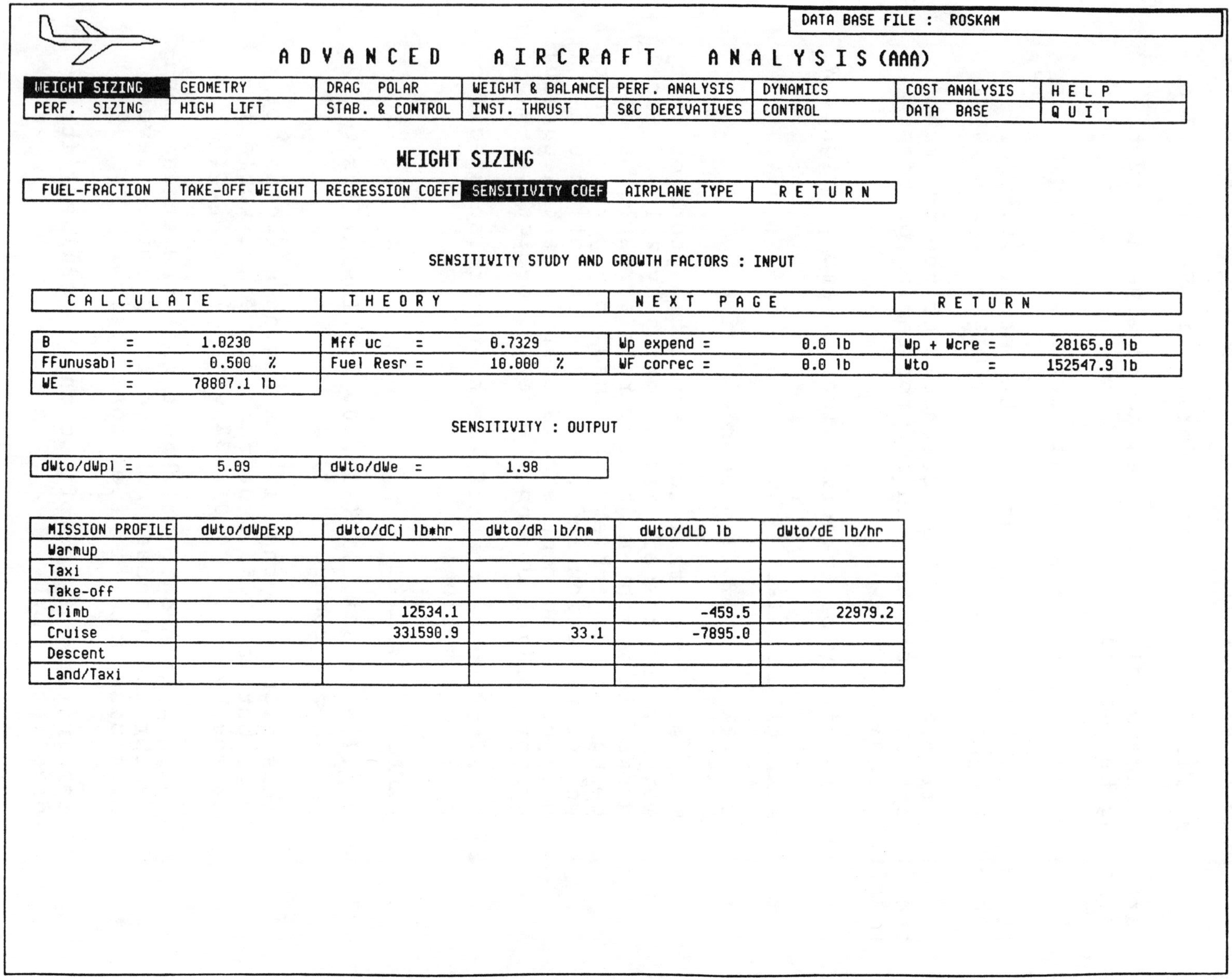

Figure 8.5 AAA Screendump of Calculation of Takeoff Weight Sensitivity to Mission and Technology Parameters

$\partial W_{TO}/\partial(L/D)$ = -7895.0 (-7,900 lbs per unit L/D, rounded off)

This signifies that if the average cruise L/D is not 21.0, but say 19.0, the takeoff weight would increase by ΔW_{TO} = 2x7,900 = 15,800 lbs!

Management has to exercise careful control over the aerodynamics department to ensure that the projected high cruise L/D can indeed be achieved!

4) Sensitivity of takeoff weight to a change in the required minimum fuel reserves of 10 percent as stipulated in Table 8.1.

Suppose that the airworthiness authorities would stipulate an increase in minimum fuel reserves from 10 percent to 15 percent. This increase can be interpreted as a change in the requirement for cruise range from 5,000 nm to 5,250 nm while keeping the old ten percent reserve number.

From Figure 8.5:

$\partial W_{TO}/\partial R$ = 33.1 (33.0 lbs/nm rounded off)

This signifies that an increase of 250 nm range due to an increase in fuel reserve minimums would force an increase in takeoff weight of:

ΔW_{TO} = 250x33.0 = 8,250 lbs

Before launching a project such as the SLRT it is essential that the designer assures himself that critical fuel reserve rules are accepted by all parties concerned!

It should be clear from this discussion that takeoff weight is critically dependent on several design parameters. The prelimininary designer should determine the sensitivity of takeoff weight to these parameters early in the design process. In this manner, management can be guided to take the appropriate action with regard to the establishment of controls on critical design parameters.

Having thusly established the sensitivity of a new design to critical design parameters the designer should initiate studies of airplane design optimization. The first step on the road to optimization studies is the identification of cost functions. Section 8.2 presents a discussion of potential airplane design cost functions.

8.2 EXAMPLE COST FUNCTIONS IN AIRPLANE DESIGN

Airplane designers are confronted with a wide range of cost functions which are potential candidates for optimization. Cost functions in airplane design can be generically classified as follows:

1) Performance Cost Functions
2) Weight Cost Functions
3) Economic Cost Functions

Sub-sections 8.2.1 through 8.2.3 provide several examples of each. A major problem in the conduct of airplane design optimization studies is the very selection of cost functions. Cost functions, to be useful, must be consistent with the mission objective of an airplane, AND they must not conflict with each other. A brief discussion of this problem is given in sub-section 8.2.4.

8.2.1 Examples of Performance Cost Functions

The reader should review the example mission specifications of Tables 2.17 through 2.19 in Part I. It is seen that in these mission specifications, several performance characteristics appear repeatedly:

1. Cruise speed, V_{cr} 2. Time-to-climb to altitude, t_{cl}

3. Cruise range, R_{cr} 4. Loiter time (or endurance), E

As long as everything else is the same, performance cost functions 1, 3 and 4 beg to be maximized while cost function 2 begs to be minimized, particularly in the case of fighter interceptors.

Before these cost functions can be optimized it is necessary to relate them to a set of design parameters. How to relate typical cost functions to airplane design parameters is discussed in Section 8.3.

8.2.2 Examples of Weight Cost Functions

The reader is referred again to the mission specifications of Tables 2.17 through 2.19 in Part I. Payload is seen to be a prominent requirement in all. Furthermore, from the performance sizing procedures outlined in Part I, it is obvious that takeoff weight, empty weight and fuel weight all play an important role. Therefore, the following weights suggest themselves as potential candidates for weight cost functions:

1. Payload weight, W_{PL}

2. Takeoff weight, W_{TO}

3. Empty weight, W_E

4. Fuel weight, W_F

It is clear that weight cost functions 2 and 3 beg to be minimized. In any given airplane, the sum of payload weight and fuel weight is referred to as the useful load:

$$W_{US} = W_{PL} + W_F \tag{8.5}$$

It is the useful load (sum of the cost functions 1 and 4) which beg to be maximized.

In Part V, page 26, Eqn.(4.2), the empty weight, W_E, was expressed as:

$$W_E = W_{struct} + W_{pwr} + W_{feq} \tag{8.6}$$

where: W_{struct} is the airplane structural weight

W_{pwr} is the powerplant weight

W_{feq} is the fixed equipment weight

Each of these three empty weight components can be viewed as a weight cost function: their minimization is certainly desirable.

Before any optimization can take place, a mathematical relation between the weight cost functions and a set of design variables will have to be developed. That is discussed in Section 8.3.

8.2.3 Examples of Economic Cost Functions

For airplane operators the economic aspects of airplanes is of prime importance. This is certainly so for commercial operators. However it is true to a large extent even for military operators. Therefore, cost must be made as low as possible. From discussions in Chapters 2 through 7 it follows that candidates for economic cost functions could be:

1. Manufacturing cost, C_{MAN}

2. RDTE cost, C_{RDTE}

3. Direct operating cost, DOC

4. Life cycle cost, LCC

In addition, the following economic cost functions suggest themselves:

5. Return on investment, ROI 6. Profit potential, P_{pot}

Before any of these cost functions can be used in an optimization study it is necessary to relate them to airplane design variables. Examples of how some of these economic cost functions are related to design variables are discussed in Section 8.3.

8.2.4 Consistency of, and Conflict Between Cost Functions

For any airplane design cost runction to be useful, it must satisfy two criteria:

1. It must be consistent with the mission objectives of the airplane.

2. It must not conflict with another cost function that has already been selected.

An example of a cost function which is not consistent with a mission objective would be:

Minimizing wave drag in an airplane that spends only a small amount of time in supersonic flight conditions.

Examples of cost functions which conflict with each other are:

A) Optimize lift-to-drag ratio, L/D with wing aspect ratio, A, as a variable, while also minimizing wing structural weight with A as a variable.

B) Minimizing structural weight with properties of structural materials as a variable, while also minimizing manufacturing cost with only materials weight as a variable. It may be, that the cost of advanced materials 'outweighs' their weight advantage!

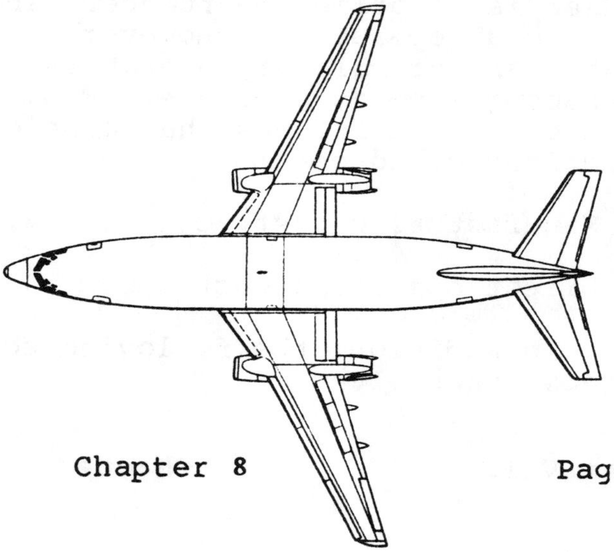

8.3 RELATION BETWEEN COST FUNCTIONS AND AIRPLANE DESIGN VARIABLES

In this section, some of the example cost functions of Section 8.2 will be expressed in terms of appropriate airplane design variables. Finding such expressions for other example cost functions is left as an exercise for the reader. In addition, a discussion is presented on how some of these design variables themselves depend on or relate to each other. The material is organized as follows:

8.3.1 Performance Cost Functions
8.3.2 Weight Cost Functions
8.3.3 Economic Cost Functions

8.3.1 Performance Cost Functions

The following two performance cost functions of sub-section 8.2.1 will be discussed:

* Cruise speed, V_{cr}
* Cruise Range, R_{cr}

8.3.1.1 Cruise speed as a cost function

It will be assumed that the classical parabolic drag polar format is applicable:

$$C_D = C_{D_o} + (C_L)^2/(\pi Ae) \tag{8.7}$$

Below the absolute ceiling, each airplane can fly level at two speeds: a low speed and a high speed. The high speed is referred to as the cruise speed. All this applies at cruise thrust settings only! It may be seen from Equation (3.60) in Part I that airplane cruise speed, V_{cr}, can be written as follows:

$$V_{cr} = \tag{8.8}$$

$$= [(W/S)_{cr} \frac{(T/W)_{cr} + \{\{(T/W)_{cr}\}^2 - (4C_{D_o})/(\pi Ae)\}^{1/2}}{(\rho C_{D_o})}]^{1/2}$$

where: $(W/S)_{cr}$ is the wing loading for the cruise flight condition being considered

$(T/W)_{cr}$ is the thrust-to-weight ratio for the cruise flight condition being considered

C_{D_o} is the zero lift drag coefficient

A is the wing aspect ratio

e is Oswald's efficiency factor

If these five parameters are viewed as independent design variables (which, as will be shown, they are NOT), and cruise speed is seen as a cost function which should be maximized, four conclusions can be drawn. V_{cr}:

1) increases with increasing wing loading,
2) increases with increasing thrust-to-weight ratio,
3) increases with decreasing zero lift drag coefficient,
4) increases with increasing product Ae

In other words, a maximum cruise speed value does not exist. There are several flaws in this reasoning:

Flaw 1: Wing loading cannot be increased independently.

NOTES: 1. The cruise wing loading differs from the takeoff wing loading because fuel has been used between takeoff and cruise.

2. The takeoff wing loading is constrained by a number of performance constraints as was shown in Chapter 3 of Part I.

Therefore: The cruise wing loading, $(W/S)_{cr}$, cannot be varied independently.

Flaw 2: Thrust-to-weight ratio cannot be increased independently.

NOTES: 1. The cruise thrust-to-weight ratio differs from takeoff thrust-to-weight ratio because of differences in thrust (throttle) setting and also because of altitude effects on engine output.

2. The takeoff thrust-to-weight ratio is constrained on its low side by performance considerations as shown in Chapter 3 of Part I.

3. Selecting a high thrust-to-weight ratio to attain a high cruise speed is probably self defeating: engine weight and cost will increase, as will engine fuel specifics.

Thus: the cruise thrust-to-weight ratio, $(T/W)_{cr}$, cannot be varied independently.

<u>Flaw 3:</u> The zero lift drag coefficient cannot be independently decreased. See Chapter 4 in Part VI for methods to compute C_{D_o} for airplanes.

NOTES: 1. At low subsonic speeds, C_{D_o} depends on:

a) airplane wetted area, S_{wet}. Wetted area in turn depends on the overall layout of the configuration and on the wing area: see Chapter 12 in Part II.

The reader is reminded of the fact that wetted area tends to correlate in a logarithmic manner with airplane takeoff weight for airplanes with similar mission orientation. See Part I, Figures 3.22.

b) skin friction coefficient, c_f. This in turn depends on Reynolds Number, Mach Number and on the smoothness of the exterior surfaces.

c) airplane interference drag and separation drag. These factors depend on the detailed layout of the external surfaces.

2. At high subsonic speeds, C_{D_o} depends on all previous factors AND on the dragrise behavior of an airplane. The dragrise depends on such detail design variables as: thickness ratio, sweep angle and aspect ratio of wing and empennage as well as on the overall progression of crosssectional areas: area ruling!

3. At supersonic speeds, C_{D_o} depends on all previous factors AND on lift induced wave drag.

<u>Therefore:</u> the zero-lift drag coefficient, C_{D_o}, can not be varied independently.

<u>Flaw 4:</u> The product Ae cannot be increased independently.

NOTES: 1. Increasing wing aspect ratio has the following consequences:

a) wing weight increases (This is seen from the wing weight equations in Ch. 5 of Part V).

b) aeroelastic effects may become unacceptable. Examples of aeroelastic effects are:

* aileron reversal
* flutter
* aeroelastically induced shift of the wing aerodynamic center

Aeroelastic effects depend on:

* structural arrangement
* properties of structural materials
* aerodynamic load distribution
* mass distribution
* dynamic pressure of the flight condition

Reference 33 is an excellent introduction to the theory of aeroelasticity.

c) wing span may become too large because of operational considerations.

2. Oswald's efficiency factor, e, depends on the following variables:

* wing aspect ratio
* wing sweep angle
* airfoil leading edge radius
* wing lift curve slope (the latter also is a function of aeroelastic effects)

Chapter 4 of Part VI contains a method for estimating 'e'.

<u>Therefore:</u> the product Ae cannot be varied independently.

The cruise speed of the airplane therefore depends not on four variables (which are not independent), but on many more. Most of these additional design variables or design factors are not easily related to each other, certainly not during the early design phases.

8.3.1.2 Cruise range as a cost function

It is shown in Chapter 5 of Part VII that the cruise range depends on how the cruising flight is carried out: cruise at constant altitude, cruise at constant Mach number or some combination of these. Eqns (5.33) through (5.36) in Part VII provide simplified analytical formulas for cruise range under either scenario. Only the case of constant speed cruise of a jet airplane will be taken up here. From Equation (5.36) in Part VII:

$$R_{cr} = (U_1/c_j)_{cr}(L/D)_{cr}\ln(W_{initial}/W_{end}) \quad (8.9)$$

where: U_1 is the true airspeed in kts

c_j is the engine specific fuel consumption in (lbs/hr)/lbs

L/D is the lift-to-drag ratio

NOTE: all of these quantities are for the cruise flight condition

$W_{initial}$ is the airplane weight at the start of cruise in lbs

W_{end} is the airplane weight at the end of cruise in lbs

Observe that:

$$U_1 = M_1 a \quad (8.10)$$

where: M_1 is the steady state (cruise) Mach number

a is the speed of sound at the cruise altitude in kts

Substituting Eqn.(8.10) into Eqn.(8.9) yields:

$$R_{cr} = (a/c_j)_{cr}\{M(L/D)\}_{cr}\ln(W_{initial}/W_{end}) \quad (8.11)$$

It is clear from this equation that for:

* a given cruise altitude (which defines a)
* a given engine cruise thrust setting (which, with the given altitude and known engine inlet conditions, defines c_j)

* a given weight ratio, $W_{initial}/W_{end}$

the cruise range is maximized for maximum $\{M(L/D)\}$.

Figure 8.6 illustrates how L/D of a typical transport jet varies with Mach number. Note that $\{M(L/D)\}$ becomes maximum at a Mach number different from that at which (L/D) becomes a maximum. All of this must be tempered by the fact that the airplane also must be flying at a lift coefficient which is consistent with the L/D at $\{M(L/D)\}_{max}$. From Table 3.1 in Part VI:

$$C_{L_{cr}} = W/\bar{q}S = f\{C_{D_o}(\pi Ae)\}^{1/2} \quad (8.12)$$

where: f is a factor, such that $0<f<1.0$, which accounts for the fact that $C_{L_{cr}}$ is not the same as the value C_L takes on at maximum L/D.

For a discussion of the design variables which are determinant for C_{D_o}, A and e, the reader should read sub-sub-section 8.3.1.1. The reader should also appreciate the fact that the weight ratio, $W_{initial}/W_{end}$, in Eqn.(8.11) is itself not an independent variable. Values for this ratio, typical for transport jets, are shown in Table 8.3. The actually available weight ratio depends on the overall configuration of the design, in particular on the general structural arrangement and on materials used in the structure.

TABLE 8.3 TYPICAL CRUISE WEIGHT RATIOS, $W_{initial}/W_{end}$ FOR JET TRANSPORTS

Transport Type	W_{TO}(lbs)	W_F(max, lbs)	$W_{initial}/W_{end}$
Boeing 747-400	800,000	387,000	1.70
Boeing 767-200ER	300,000	113,000	1.47
Boeing 757	250,000	75,000	1.34
McD.D. MD81A	140,000	39,000	1.31
McD.D. DC-10-30	572,000	246,000	1.58
Airbus A320-100	150,000	28,000	1.19
Airbus A320-200	162,000	42,000	1.28

Note: data assume that 15 percent of the fuel is used for reserves and for conditions other than cruise.

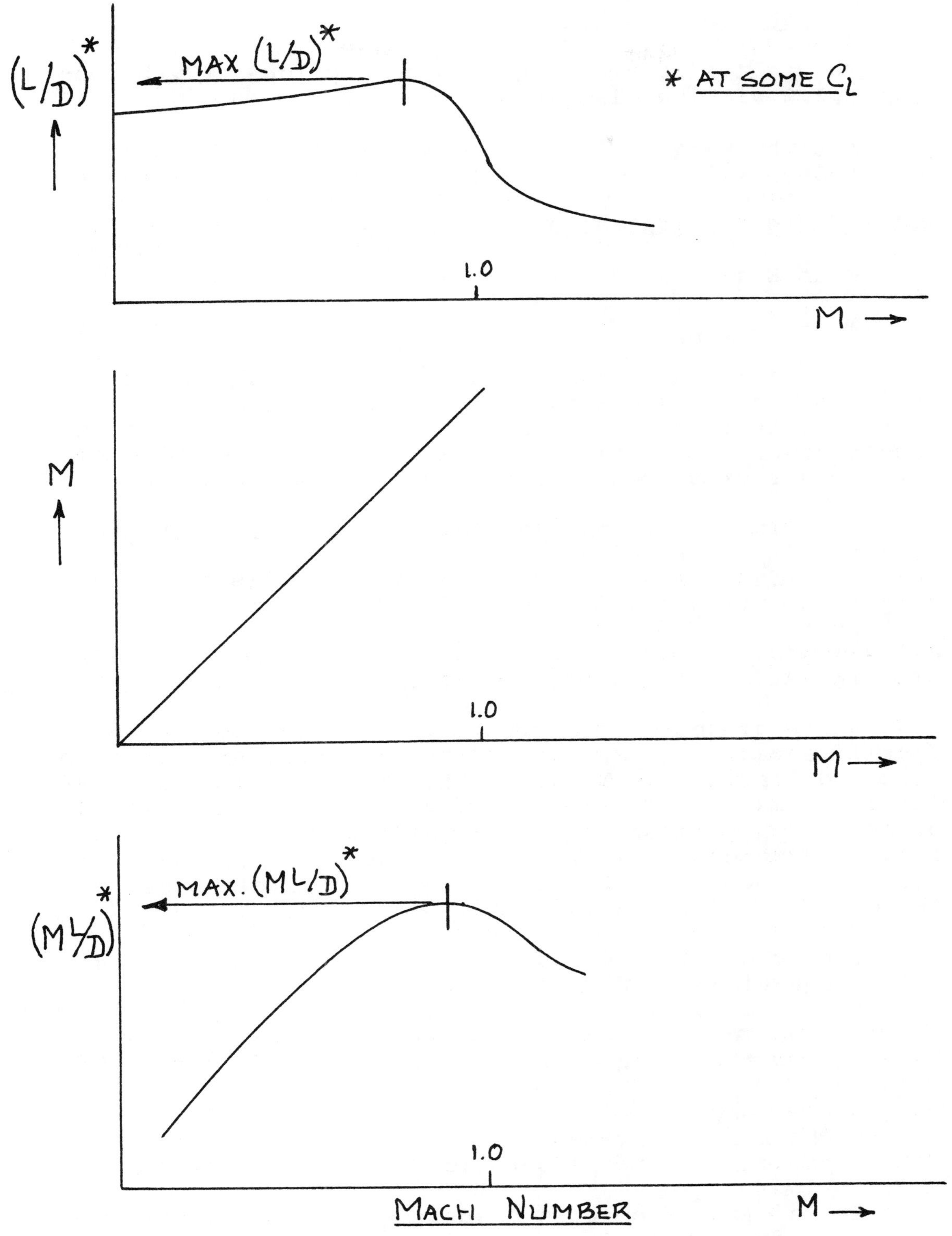

Figure 8.6 Effect of Mach Number on L/D and on M(L/D)

8.3.2 Weight Cost Functions

The following two weight cost functions of sub-section 8.2.2 will be discussed:

* Empty weight
* Useful load

8.3.2.1 Empty weight as a cost function

From Equation (8.6):

$$W_E = W_{struct} + W_{pwr} + W_{feq} \tag{8.13}$$

Minimizing any one of these empty weight contributions is clearly desirable. A problem with this statement is that in many instances attempts to reduce weight can be costly (in terms of dollars!). Thus, a balance must be struck between weight decrease and cost increase.

The Class II weight estimation formulas of Chapters 5, 6 and 7 of Part V can be used to relate airplane empty weight to fundamental design variables. The Class II weight equations are derived from a data base which consists primarily of airplane structures made from aluminum alloy types of materials. When more advanced materials are planned, downward adjustments have to be made.

There are no simple methods to account for the effect of advanced materials and/or manufacturing proceses on dollar cost. To do justice to the effect of dollar cost, much more detailed knowledge about the design of the airplane must be available than is normally the case in advanced design. An interesting approach to the problem of dollar-cost-weight trades in airplane design is found in Reference 34.

The reader should observe that many of the component weight equations in Chapter 5 of Part V require the limit load factor as an input. The limit load factor itself is not an independent variable! The V-n diagrams of Ch.4 of Part V show the design load factor to be a function of:

* the takeoff weight
* the flight envelope
* the gust response behavior

It is clear from this discussion that the relationship between empty weight and design variables is a very complicated one.

8.3.2.2 Useful load as a cost function

The useful load of an airplane is defined in Eqn.(8.5) as:

$$W_{US} = W_{PL} + W_F \tag{8.14}$$

where: W_{PL} is the payload weight

W_F is the fuel weight

Payload weight and payload type are normally defined in the mission specification of an airplane. It turns out that payload weight and payload type both have a significant effect on airplane empty weight. Examples which illustrate this effect are:

1) In a passenger jet transport, the number of passengers carried, the class of service offered and the requirements for on-board meals and other services have a major impact on the weight of cabin interior amenities.

2) In a cargo jet transport, the type of cargo has a major effect on the structural design and layout of the cargo hold:

* if containers are to be carried, roller-rails and tie-down provisions are required

* if concentrated loads are to be carried (any type of roadable or off-highway type vehicle), major floor reinforcements are required

* if front, rear and/or side loading of the payload is/are required, major structural provisions for these features will be required

* if weapons have to be carried and launched, major provisions in the structure to take the various weapons loads will be required

The reader is referred to Parts III and IV for more detailed discussions of these effects.

The point to be made here is that payload itself can be a complex function of design variables involved in the empty weight of an airplane, mostly in the structural and fixed equipment part of this weight.

8.3.3 Economic Cost Functions

The following three economic cost functions of sub-section 8.2.3 will be discussed:

* Manufacturing cost (C_{MAN}
* Direct operating cost (DOC)
* Return on investment (ROI)

8.3.3.1 Manufacturing cost as a cost function

A method for determining airplane manufacturing cost was presented in Chapter 4. The method, as reflected by Equations (4.4), (4.5), (4.7), (4.17) and (4.18), shows that airplane manufacturing cost depends on the following design parameters:

1) Takeoff weight, W_{TO} (through the AMPR weight of Equation (3.4) or (3.5))
2) Maximum design speed, V_{max}, as defined on p.26
3) Number of airplanes produced to manufacturing standard, N_m
4) Labor rates for engineering, manufacturing and tooling, R_{e_m}, R_{m_m} and R_{t_m}
5) Number of passengers, N_{pax} (only for commercial airplanes
6) Manufacturing rate, N_{r_m}
7) Various judgement factors, F_{diff}, F_{cad}, F_{int}, F_{mat}, F_{ftoh} and F_{fin_m}
8) Cost escalation factor, CEF
9) Operating cost per hour, $C_{ops/hr}$, and number of production flight test hours, t_{pft}
10) Cost of engines and avionics

Design variables 1) and 2) are the most important ones. It turns out that the engine cost is related to the takeoff weight.

8.3.3.2 Direct operating cost (DOC) as a cost function

A method for determining airplane direct operating cost is presented in Chapters 5 and 6. The DOC of an airplane depends (according to Equations (5.19) and (6.1) and their components) on the following design variables:

1) Takeoff weight, W_{TO}, because many of the DOC component costs are expressed in terms of the takeoff weight

2) Cockpit crew requirements: this is determined by the mission of the airplane

3) Fuel used: this depends on lift-to-drag ratios and on engine efficiencies

4) Accessibility and ease of maintenance

For a discussion of the relative significance of DOC component contributions, see Chapters 5 and 6.

It turns out that takeoff weight, aerodynamic and fuel efficiency are normally the 'driving' factors behind the DOC of commercial and military airplanes.

8.3.3.3 Return-on-investment (ROI) as a cost function

The following definition of return-on-investment (ROI) will be used:

$$ROI = \qquad (8.15)$$

$$= \left\{\frac{(REV - DOC - IOC)V_{bl}}{(AEP)(1 - tx_{inv})}\right\}(1 - tx_{rev})U_{ann_{bl}}(100)$$

where: REV is the revenue in USD per nm

DOC is the direct operating cost in USD per nm

IOC is the indirect operating cost in USD per nm

V_{bl} is the block speed in nm/hr

AEP is the airplane estimated price in USD (See page 62)

tx_{inv} is the investment tax credit rate extended by the government to the operator. For preliminary design purposes it is suggested to use: $tx_{inv} = 0.10$ (ten percent).

tx_{rev} is the 'income' or 'revenue' tax rate paid to the government. For preliminary design studies, use: $tx_{rev} = 0.20$.

<u>Note:</u> tax rates are determined by the congress and therefore are a variable over which the designer has little control!

$U_{ann_{bl}}$ is the annual airplane utilization in hours

Note that ROI as defined in Eqn.(8.15) comes out as a percentage.

It is obviously desirable to maximize ROI. However, there are a number of problems:

* The ticket price charged per nautical mile, REV, must be competitive or no passengers will be willing to pay the price

* The ticket price will influence market demand to some extent: the demand for passenger miles is elastic

* DOC and IOC should be kept as low as possible. Methods for computing DOC and IOC are given in Chapters 5 and 6 of this volume. Note the very large number of airplane design and operating variables which enter the picture!

* Annual utilization should be kept as high as possible. There are several limitations to this:

  a) if annual utilization goes above some number, maintenance and repairs become a problem

  b) the route structure of the airline may not make it feasible to have high annual utilization

8.4 DESIGN CONSTRAINTS IN AIRPLANE DESIGN

Aeronautics is one of the most heavily regulated activities of mankind. The reason for this is: SAFETY. To ensure minimum standards of safety in airplane design and in airplane operation, governments adopt and enforce minimum standards of airworthiness and operating safety.

These standards have been referred to as the 'regulations' in this series of books. Many aspects of these regulations were covered in Parts I through VII. Copies of those regulations which directly impact airplane performance, stability and control (and thereby the configuration design) are included as appendices to Part VII.

From a design optimization viewpoint, these regulations should be interpreted as: design constraints. Design constraints limit the numerical ranges of design parameters which are available to the designer.

In addition to regulatory design constraints, there are many other examples of design constraints. Airplane design constraints are generically classified as follows:

1. Design constraints related to airworthiness

2. Design constraints related to operational safety

3. Design constraints related to operational factors other than safety

4. Design constraints related to economics

5. Design constraints related to the infra-structure

6. Design constraints related to typical military requirements

7. Design constraints related to the environment

There is no implied priority in the listing of these design constraints. They can all play important roles in arriving at key design decisions. Examples of airplane design constraints and the role played in design decision making are provided in sub-sections 8.4.1 through 8.4.7.

8.4.1 Design Constraints Related to Airworthiness

Civilian and military airworthiness is ensured by specifying minimum standards of:

1. Airplane performance with and without all engines operating

2. Airplane handling qualities with and without all engines operating

3. Structural integrity in the presence of extreme atmospheric and/or other operational loads

4. Reliability and/or redundancy of flight crucial systems

Parts I through VII contain many examples of the impact of airworthiness regulations on airplane design. In the following, one example of each of these airworthiness constraints will be given.

8.4.1.1 Example of a performance constraint

It was shown in Chapter 3 of Part I that takeoff and landing distances are regulated by certain definitions of fieldlength: ground distance, air distance and obstacle height. Also, factors such as air density, temperature and humidity must be acounted for. If an airplane is to safely operate from a field with a given minimum fieldlength, then the result of these fieldlength constraints is to limit the available choices of wing loading and of thrust-to-weight ratio. This was shown in Chapter 3 of Part I in the form of a plot of takeoff thrust-to-weight ratio, $(T/W)_{TO}$, versus takeoff wing loading, $(W/S)_{TO}$.

Figure 8.7 is an example of this. The range of design parameter values which are denied by the fieldlength constraints is clearly indicated. A complicating factor is that the size of the 'denied' ranges is a function of the flap capability of the airplane, as expressed through the maximum lift coefficients: $C_{L_{max_{TO}}}$ and $C_{L_{max_L}}$

Table 3.1 in Part I gives ranges of numerical magnitudes for these maximum lift coefficients for mechanical flaps. There are various forms of augmentation (blowing is one example) which can be used to attain significantly higher values of maximum lift coefficient. In any case, there are limits to attainable maximum lift coefficients. These limits can have a number of causes: aerodynamic, flap weight, flap system complexity and flap system cost are only a few of these.

It is possible to formulate these maximum lift coef-

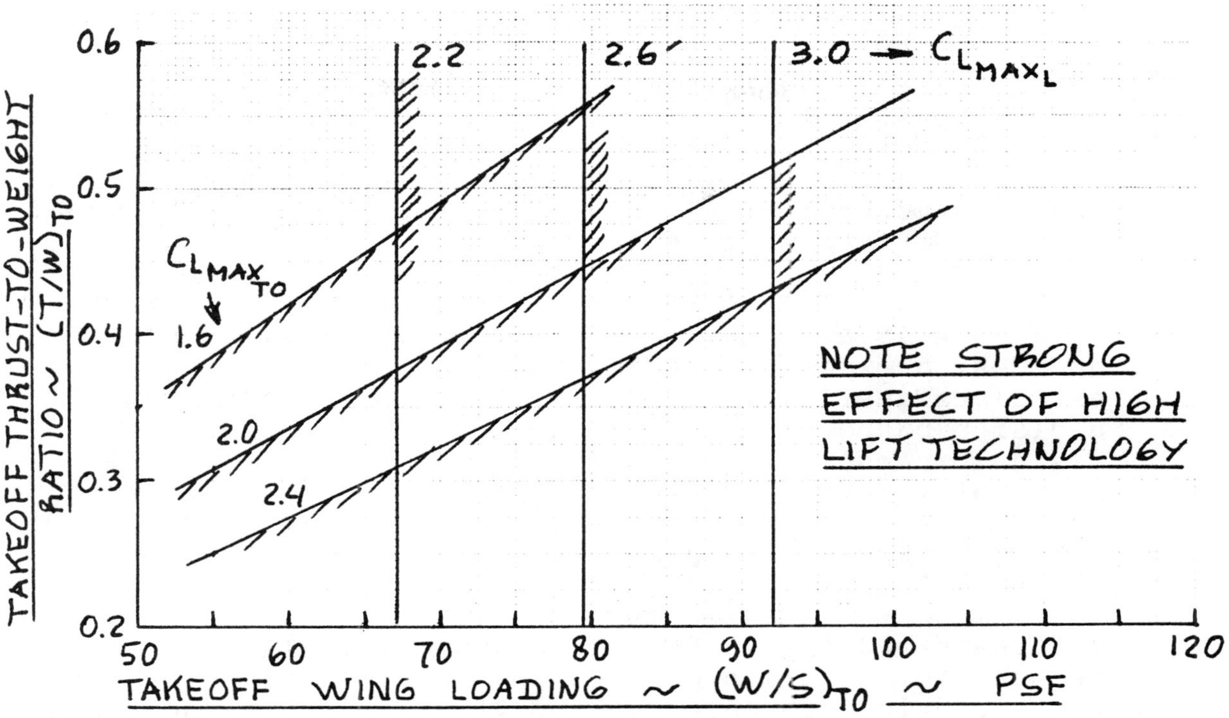

Figure 8.7 Example of Fieldlength Constraints on Wing Loading and Thrust-to-Weight Ratio

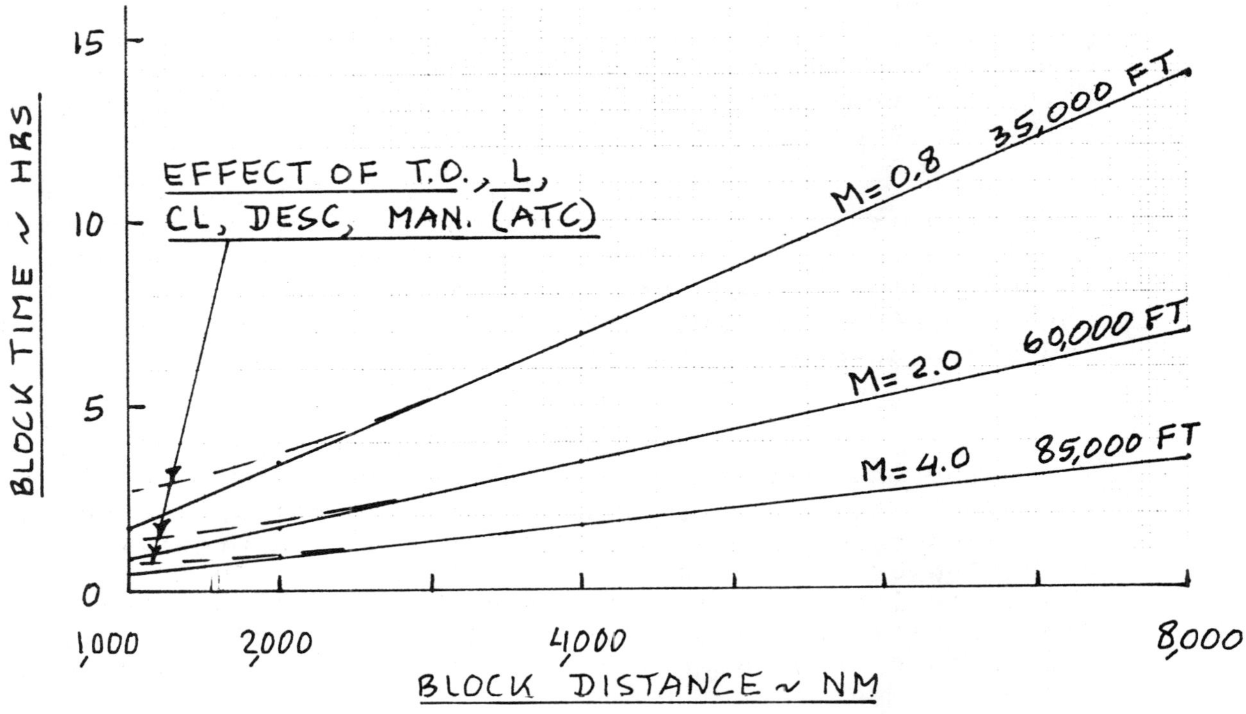

Figure 8.8 Effect of Cruise Mach Number and Block Distance on Block Time

ficient limits in a suitable mathematical format. This is most easily done for the case of mechanical flaps. The lift prediction methods of Part VI are examples of how this can be done. Considerations involving calendar time and money restrictions make it difficult to arrive at similar formulations for items such as: flap weight, flap system complexity and flap system dollar cost.

Here is an example of a case where design intuition and design experience will enter the picture. An experienced designer will not hesitate to make a flap design decision based on intuition. However, this makes the design process no longer optimizable in the rational sense of the word.

8.4.1.2 Example of a handling quality constraint

Minimum requirements for safe handling qualities are discussed briefly in Part II and in Part VII. Analytical methods for determining handling quality parameters are presented in References 35 and 36. It can be shown that each requirement for handling qualities translates into some form of design constraint. One example will be discussed: the minimum control speed problem.

In multi-engine airplanes, with one engine inoperative, it is required that the pilot be able to maintain control down to the so-called minimum control speed, V_{mc}.

The airworthiness requirements (see Appendices A and B in Part VII) stipulate the following relationship between the minimum control speed, V_{mc}, and the stall speed, V_s:

$$V_{mc} < 1.2V_s \qquad (8.16)$$

The stall speed itself is related to the wing loading and the maximum lift coefficient (clean, flaps up and flaps down!):

$$V_s = \{2(W/S)/(\rho C_{L_{max}})\}^{1/2} \qquad (8.17)$$

The minimum control speed (in a takeoff flight condition) for a twin engine airplane can be approximated as follows:

$$V_{mc} = [\{(2T_{TO_e})(y_T)\}/\{(C_{n_{\delta_R}}\delta_{R_{max}})(Sb\rho)\}]^{1/2} \qquad (8.18)$$

where: T_{TO_e} is the take-off thrust of the most critical engine in lbs

y_T is the lateral moment arm of the most critical engine in ft

$C_{n_{\delta_R}}$ is the rudder control power derivative in rad^{-1}

$\delta_{r_{max}}$ is the maximum allowable rudder deflection in rad

S is the wing area in ft^2

b is the wing span in ft

ρ is the atmospheric density in $slugs/ft^3$

Each design parameter in Equation (8.18) itself may depend on a number of other design parameters. For example, the control power derivative, $C_{n_{\delta_R}}$ is a function of:

* vertical tail size
* vertical tail geometry (aspect ratio, sweep angle, taper ratio and airfoil)
* vertical tail location on the configuration
* rudder size, shape and location

The maximum allowable rudder deflection, $\delta_{R_{max}}$, is a function of:

* Vertical tail stall behavior and Reynolds Number
* Rudder geometry
* Vertical tail location on the configuration

When all these design parameters have been decided upon, the V_{mc} constraint defines a minimum allowable vertical tail size. This in turn has an impact on overall airplane drag, weight and center of gravity location. The latter will also impact the stability and control of the airplane!

8.4.1.3 Examples of structural integrity constraints

Examples of structural design constraints are all those factors which impact on the airloads and other loads which act on an airplane as it operates in the air or on the ground. In the preliminary design phase the structural integrity requirements impact the design decision making process through factors such as:

* specified limit load factor

  This has an impact on structural weight and thus on overall weight, center of gravity locations and on cost.

* specified design touchdown velocity

  This has an impact on the design of the landing gear which in turn impacts airplane takeoff weight and the volumetric requirement of the gear upon retraction. The latter can also have an economic impact if the retracted landing gear volume limits payload or fuel carrying capability.

8.4.1.4 Examples of system reliability and redundancy constraints

Whenever the failure of an airplane system is considered to have a major impact on safety, it is called a flight crucial system. The hydraulic system is a flight crucial system if it is the only source of control in flight. That is the case in many transports today.

Whenever a system is defined as flight crucial it must satisfy certain standards of reliability and redundancy. For more information on the subject of safety and redundancy, see Chapters 6 and 13 in Part IV and also Appendix C in Part VII.

Flight crucial systems are 'allowed' to fail in a catastrophic manner at the rate of once per 10^9 flight hours (or sometimes flights). One task of the designer is to estimate the reliability of each such system, in terms of catastrophic failure rate per X flight hours. Once this estimate is available it is easy to compute how many independent systems are required to satisfy the 10^9 requirement. It is not difficult to see that the outcome of these system design considerations can have an impact on airplane weight, airplane layout (signal and power path redundancy!!), airplane complexity, airplane maintenance requirements and airplane cost.

Casting these constraints in mathematical formats suitable for use in optimization is a formidable task.

8.4.2 Design Constraints Related to Operational Safety

Examples of this type of design constraint are:

* Airplane ceiling

  Each airplane has an altitude-weight combination beyond which it is unsafe to operate. The reason for this limit is the proximity of stall speed, buffet limit and maximum allowable Mach number. There are also cases where airplane handling qualities are compromised beyond certain combinations of airplane weight and altitude.

* Airplane operating speed and/or Mach number

  At each altitude, airplanes are limited in minimum and maximum allowable speed for a variety of reasons having to do with engine or airframe limitations or with handling quality limitations.

* Fuel reserves

  When airplanes are flown on a given route, there exist regulations (FAR 135) which specify that flight to alternate airports must be possible in case of weather limitations at the airport of destination. This leads to requirements for fuel reserves. These fuel reserves increase the weight of the airplane. They also require that enough fuel volume is available on board to store it.

* Emergency and evacuation constraints

  These constraints require certain airplanes to be equipped with:

  a) emergency exits of prescribed size and location
  b) evacuation slides
  c) life rafts and floatation equipment
  d) fire safety equipment

  All these requirements translate into weight and volume needs which should be viewed as constraints on the design decision-making process.

These constraints must be translated into functional relations between various airplane design variables. This cannot (in general) be done until detailed data on the aerodynamic and the structural design are available.

8.4.3 Design Constraints Related to Operational Factors Other than Safety

Examples of this type of design constraint are:

* Servicing requirements: commercial

  A jet transport, when arriving at the gate, needs to be serviced by a wide variety of servicing vehicles. Critical here is the so-called 'gate turn-around time' which must be kept to a minimum. Airplanes parked at the gate do not earn money! A discussion of this problem is given in Part III, Chapter 3 (page 82).

  This design constraint implies that there must be sufficient access to various systems of the airplane to make simultaneous servicing possible. This in turn puts constraints on the location of equipment in the airplane and on the size and location of access doors and access covers. These considerations tend to have a significant impact on structural design (strength and fatigue life) and therefore on weight and cost.

* Servicing requirements: military

  A military airplane returning from a combat mission needs te be turned around and readied for its next mission in as short a time as possible.

  This design constraint is similar to that of the jet transport at the gate. However, in the case of military airplanes these functions may have to be carried out by personnel wearing polar and/or chemical warfare clothing. This makes the design of access doors and covers very critical: they tend to be large!

* Landing gear constraints

  The reader is asked to review the landing gear design and disposition criteria of Ch.9 in Part II and in Ch.2 of Part IV. Most of these gear design criteria should be viewed as design constraints.

No attempt has been made here to derive functional relations between these design constraints and the appropriate airplane design variables.

8.4.4 Design Constraints Related to Economics

Examples of design constraints related to economics are:

* Revenue
* Program Cost
* Acquisition cost
* Travel time

A brief discussion of each follows.

* Revenue as a constraint

In certain markets the customer may not be willing to pay more than a certain amount of USD per nm. It is essential that airplane designers understand such limits and incorporate them into their design trade studies to determine the economic viability of a proposed design. Such limits on 'extractable' revenue should be viewed as design constraints.

* Acquisition cost as a constraint

The acquisition cost of an airplane translates into the cost of ownership. There are limitations on the allowable cost of ownership for commercial as well as for military airplanes. Again, such limitations should be viewed as design constraints.

* Program cost as a constraint

In military programs in particular the total allowable program cost should be seen as a design constraint.

These three cost constraints all bear a relation to the 'design-to-cost' problem which is discussed briefly in Section 8.8.

* Travel time as a constraint

Figure 8.8 shows how trip time relates to block-speed for various triplengths (block distances). It is seen that for transpacific distances there is not much payoff, in terms of time gained, for flying beyond Mach 4.0. This design constraint is caused by the 'smallness' of the earth.

8.4.5 Design Constraints Related to the Infra-structure

Examples of design constraints which are related to the infra-structure are:

* Gate space at airports of intended operation

  This may limit the designer to a given wing span. One example of this is the Boeing 747-400. Winglets were added to this airplane largely to avoid greater wing span. In future designs, folding wing tips may have to be considered as well!

* Hangar size

  Airplanes may have to fit into existing hangars. Construction of new hangars is expensive. Some operators may not be willing to spend money to build a new hangar just because the airframer decided to build an airplane which does not 'fit'. This constraint can therefore dictate the outside dimensions of the airplane regardless of any aerodynamic or other considerations!

  Such size constraints are common in the design of carrier based airplanes for the navy: deck spotting, elevator spotting and hangar deck spotting considerations all impose size constraints on the airplane designer.

* Width of existing taxiways

  Construction of airport taxiways involves major expenses. If an airplane manufacturer designs an airplane with a (landing gear) configuration which is not compatible with existing taxiways he may find no buyers for that airplane!

8.4.6 Design Constraints Related to Typical Military Requirements

Examples of such design constraints are:

* Gun integration constraints

  Guns impose a number of serious constraints on the design of an airplane. For a discussion of these, see Chapter 3, Part IV.

* Missile integration constraints

  This may dictate internal and/or external space requirements. It may also dictate a range of design parameters related to aerodynamics, weights and flight controls. For a discussion of weapon integration, see Chapter 3, Part IV.

* Stealth requirements

  Requirements for stealth may dictate the external shape of the airplane and/or the structural arrangement, quite apart from any conventional aerodynamic and/or structural design considerations.

  Prominent examples of such airplanes are the B-2 and the F-117A.

8.4.7 Design Constraints Related to the Environment

Examples of this type of design constraint are:

* Noise

  This design constraint may dictate the size, location and type of powerplant installation. In turn, this has a significant impact on weight, complexity and fuel consumption.

* Fuel dumping

  If a requirement would be formulated against fuel dumping, airplane landing gears would have to be designed for considerably greater weight at touchdown. This would impact airplane weight and landing gear retraction volume requirements.

* Atmospheric pollution

  Requirements to minimize 'visible' atmospheric pollution have already resulted in a redesign of burner systems in jet engines.

  Depending on the outcome of on-going research of potential ozone destruction by high flying supersonic transports, their burner systems may have to be tailored specifically to the prevention of releasing certain combustion products. The effect of these considerations on engine performance is still subject to question.

8.5 RULES FOR EVALUATING COST FUNCTIONS

It will be assumed that at this point the following steps have been taken:

1. One or more cost functions have been selected for optimization. These cost functions must be consistent with the required mission performance of the airplane.

2. The cost functions selected under 1. have been expressed functionally in terms of a range of airplane design variables.

3. Airplane design constraints have been cast in a format which relates them clearly to airplane design variables.

Next, a decision has to be made which of the cost functions are to be maximized or minimized while at the same time observing specific constraints.

NOTE WELL:

A) It is just as important to keep in mind which potential cost functions and which potential design constraints have NOT been considered. Doing this avoids the pitfall of believing that the results of optimization procedures (constrained or not) are somehow absolute.

B) Probably the most difficult aspect of the design optimization process is to decide which cost functions and which constraints should be used.

8.6 METHODS FOR DESIGN OPTIMIZATION

The purpose of this section is to acqaint the reader with reference material in the area of airplane design optimization. A detailed treatment of methods for design optimization is considered beyond the scope of this text.

There exists an impressive amount of literature on the subject of airplane design optimization. References 37 and 38 are recommended as excellent introductions to this subject. Both references contain many citations of other literature pertinent to design optimization.

Optimization methods can be broadly classified as:

1. Optimization within a given discipline

2. Multidisciplinary optimization

In each case, the optimization methods may or may not consider effects of design constraints: constrained or unconstrained optimization.

Most of the early design optimization methods dealt with optimization problems related to a particular design discipline, usually without accounting for constraints.

Examples are methods for optimizing airplane performance and/or aerodynamic parameters. References 39, 40 and 41 fit this category. Typical of the design conclusions drawn from these methods is Kuchemann's conclusion in Ref.39, that the best lift-to-drag ratio which is achievable in supersonic cruise is:

$$(L/D)_{max} = 4(M + 3)/M \qquad (8.19)$$

A problem with optimization within any discipline is that it can result in unrealistic design solutions. For example: induced drag is reduced by allowing the wing aspect ratio to assume very large values. This runs into problems with wing structural weight. Therefore, it is necessary to account for the disciplines of structural and aerodynamic analysis together whenever one impacts the other.

A problem with design optimization without proper regard of constraints is that it also can lead to unrealistic design solutions. Consider the following example:

A supersonic fighter has a requirement for a long subsonic mission leg, followed by a short supersonic dash requirement, followed again by a long subsonic leg to home base.

It is possible, by the application of area-ruling methods, to minimize the supersonic wave drag of this airplane. However, it may be that a result of minimizing wave drag is an increase in wetted area over that needed for efficient subsonic flight. If the fuel used in the short supersonic dash is significantly less than that used during the long subsonic legs, it may be that wave drag minimization makes no sense.

In this case, wave drag should have been minimized subject to the constraint that total mission fuel consumption (subsonic + supersonic) is kept to a minimum.

The subject of airplane design optimization in the presence of design constraints is extensively discussed in References 37 and 38.

Recently, research in airplane design optimization has moved in the direction of multidisciplinary design optimization subject to a multitude of constraints. To become familiar with recent trends in design optimization methods References 42 through 46 should be consulted.

The next section presents some examples of the end product of typical design optimization studies.

8.7 EXAMPLES OF DESIGN OPTIMIZATION STUDIES

The purpose of this section is to present a brief discussion of results of potential design optimization studies. For more details on how to perform such studies references 42 - 46 should be consulted.

Example 1: The economic value of blockspeed

For a given airplane, operated over a given stage length (block distance) and at a given altitude, the question can be asked: at what speed is the operator's profit margin the highest possible?

In this case, a solution can be found by plotting the various operating cost components as a function of blockspeed. Figure 8.9 shows an unscaled example of such a plot. Note the behavior of the fuel cost contribution:

* at low block speeds, fuel usage increases because of the increase in induced drag

* at high block speeds, fuel usage increases because zero-lift drag and compressibility drag increases

Figure 8.9 shows various 'optima' which correspond to different block speeds:

* block speed for minimum fuel cost
* block speed for minimum fuel used per nm per hr
* block speed for minimum DOC
* block speed for maximum profit margin

It is of interest to observe that the block speed for the lowest DOC is lower than the blockspeed for the greatest profit margin!

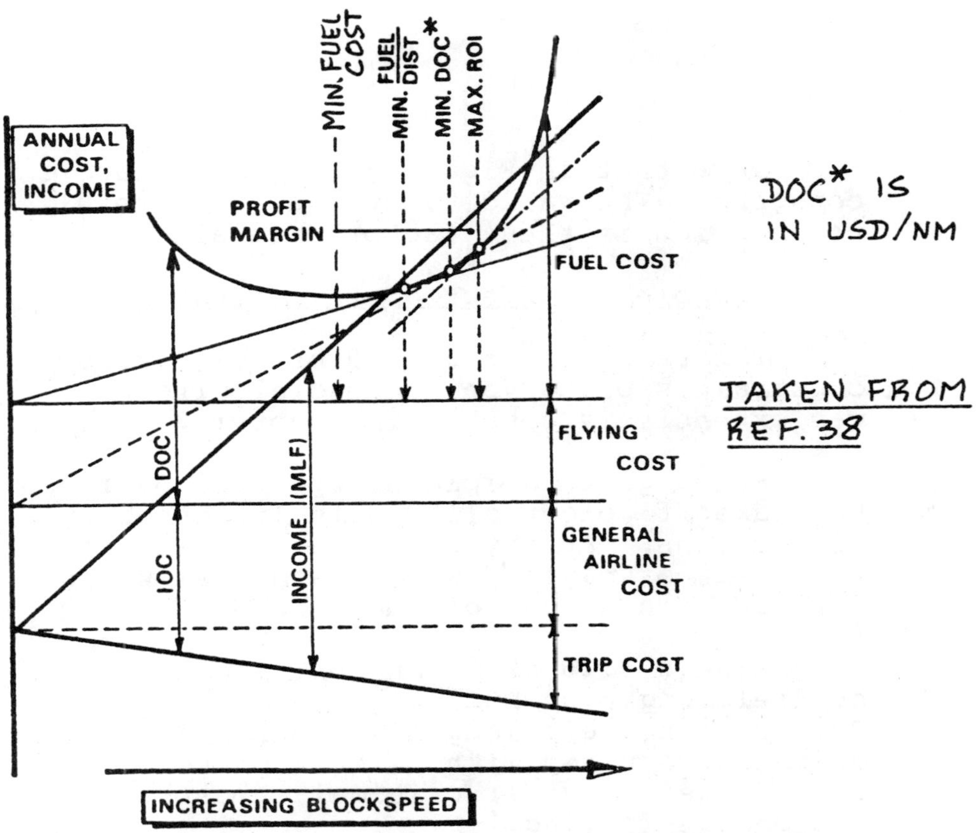

Figure 8.9 The Economic Consequence of Blockspeed

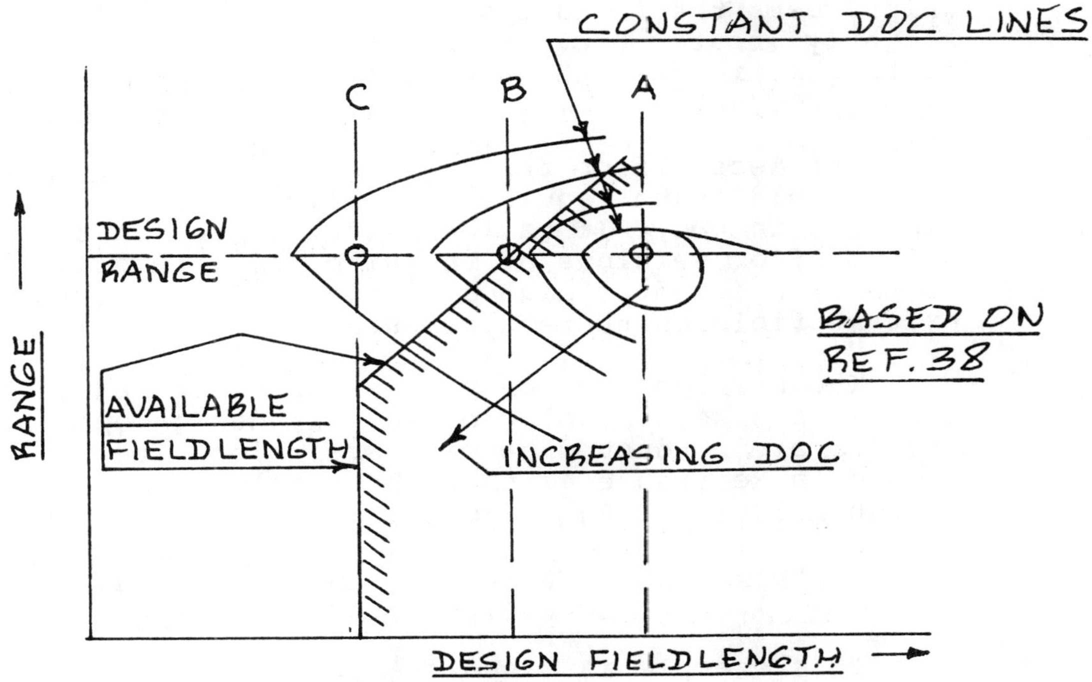

Figure 8.10 Effect of Fieldlength Constraints on Direct Operating Cost (DOC)

Note that in Figure 8.9 various factors were kept constant. The validity of the study results are subject to question if these factors are allowed to vary!

<u>Example 2: Effect of fieldlength constraints on DOC</u>

The reader is reminded of the various fieldlength constraints placed on commercial aircraft. This topic was extensively covered in Chapter 3 of Part I.

For a given airplane, operated over a given range, the takeoff weight allowed is limited by fieldlength considerations. At any given takeoff weight, the useful load (sum of payload weight and fuel weight) is fixed: for long ranges payload must be off-loaded!

Figure 8.10 shows an unscaled example of the effect of fieldlength on DOC, with range as a second parameter. The 'design field length' is defined as that takeoff distance required with maximum passengers and maximum takeoff weight under critical atmospheric conditions and one engine failing.

Point A reflects an airplane which has been 'optimized' for minimum DOC in cruise: it does not meet the fieldlength constraint. Point B represents an airplane which meets the fieldlength constraint exactly, but at slightly greater DOC. Point C reflects an airplane which meets the fieldlength requirement at any range, but at higher DOC yet.

The lesson from this study is that designing for extreme fieldlength requirements may result in serious compromises in DOC. It may be better to design to Point B AND to offer versions of the airplane with slightly different wings and/or high lift devices to operators with extreme fieldlength requirements.

The airplane design process as described in Parts I and II is a sequential process. If design optimization studies are conducted at each step along the way, it is argued in Reference 47 that the result is a sub-optimal design solution. An example of this is given next.

<u>Example 3: Minimizing wing structural weight in a sequential manner</u>

This example is based on Reference 46. Suppose wing structural weight needs to be minimized subject to some performance measure, PM. This performance measure, PM, could be the payload for a given range. Constraints C_1 and C_2 (C_1 could be a fieldlength constraint and C_2 could be an engine-out, rate-of-climb constraint) are also imposed on the problem.

Figure 8.11 shows a possible result where the wing aspect ratio is used as an independent variable. The point P_1 is the constrained optimum for this case study. Suppose that, as a result of this type of study, the wing aspect ratio is frozen at a value A_1. Also suppose that at some later point in the design process a flutter study has been conducted which results in an added constraint, reflected by C_3 in Figure 8.12. As a result, the wing structural weight has to be increased to a value corresponding to point P_2. However, if the flutter study had been conducted earlier, point P_3 might have been selected as the desirable design point!

It is worthwhile to end with an observation first made in Reference 47. The reader is asked to refer to Figure 8.13. Observe that as the time into the design process increases, the knowledge about the design also increases. However, this increasing knowledge about the design implies that more and more design variables are 'frozen'. That in turn limits the potential design freedom. Therefore, the ability to act on the results of design optimization studies gradually disappears as time into the design process increases. Reference 47 refers to this state of affairs as the paradox of sequential design decision making. This 'peril' can be avoided by careful selection of design cost functions and judicious application of design optimization studies.

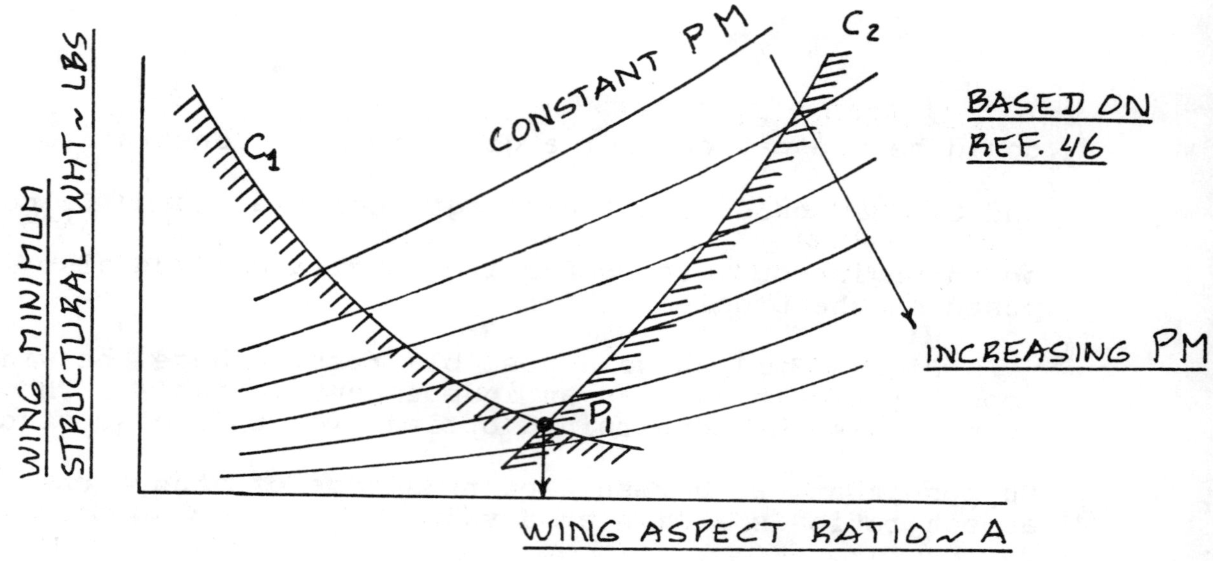

Figure 8.11 Optimization of Wing Weight with Aspect Ratio Subject to Two Constraints

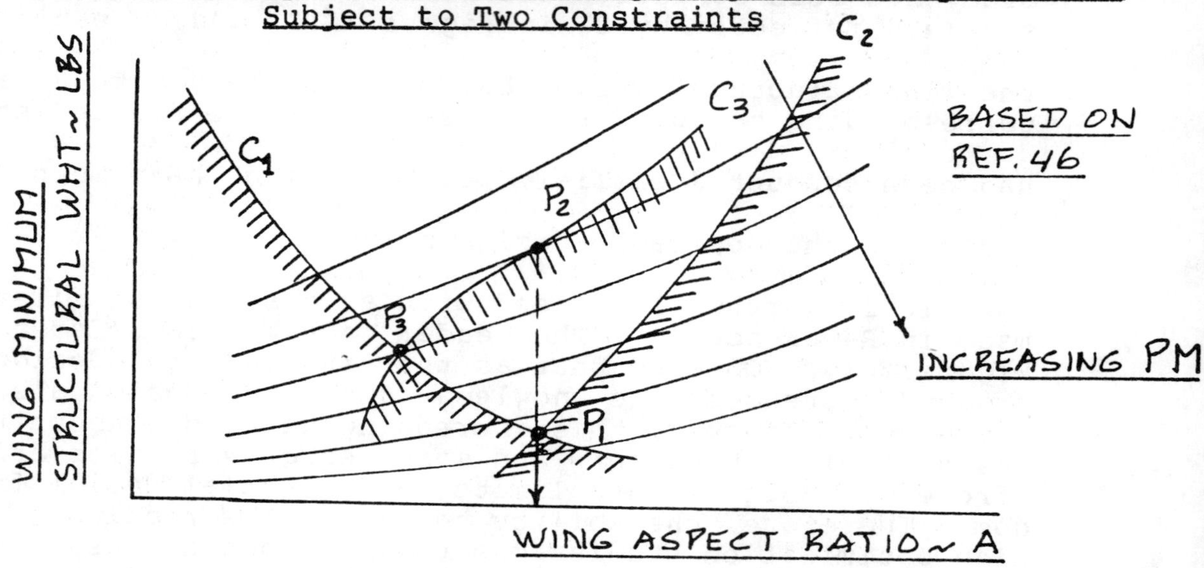

Figure 8.12 Optimization of Wing Weight with Aspect Ratio Subject to an Additional (Later Discovered) Constraint

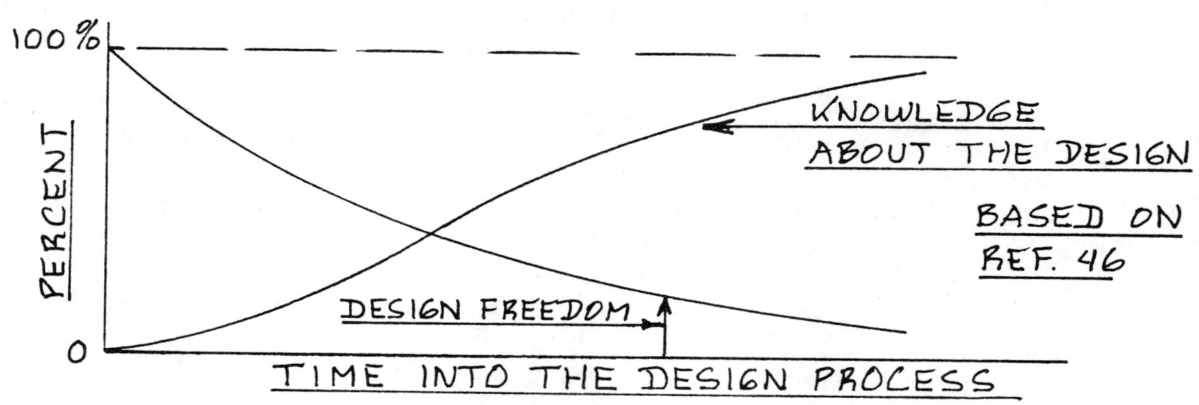

Figure 8.13 Illustration of the Paradox of Sequential Design

8.8 INTRODUCTION TO THE DESIGN-TO-COST PROBLEM

Budgets for airplane programs are always limited. Also, airplane operators cannot afford airplanes if they are too expensive. Interesting questions therefore are the following:

1. Assume that an airplane has been designed to some mission specification. How much is that airplane worth to the operator?

2. Given a certain net worth and given the need for a certain number of airplanes, can a manufacturer design, develop and manufacture such an airplane and make a profit?

The first question is referred to as the so-called 'net worth' problem. Sub-sections 8.8.1 and 8.8.2 give an approach to estimating airplane net worth.

The second question is referred to as the so-called 'design-to-cost' or the 'design-to-price' problem. An approach to this problem is given in sub-section 8.8.3.

8.8.1 The Commercial Airplane Net Worth Problem

Definition: The net worth of an airplane is the price an operator is willing to pay for an airplane because the potential for profit (Return-on-investment, or ROI) is demonstrable.

The net worth of an airplane can be determined via an operational market analysis. Such a market analysis should result in the following data:

* definition of routes, frequencies, passenger demand and/or cargo demand

* definition of sustainable revenue factors such as USD/paxnm and/or USD/tonnm

A preliminary estimate of the sustainable revenues can be obtained from the literature. Figure 8.14 is just one example. From such data it is possible to estimate the annual revenue which can be obtained per airplane. Assuming a desired return-on-investment percentage, the allowable airplane price (net worth) can be recovered from Equation (8.15) by solving it for AEP, the airplane estimated price.

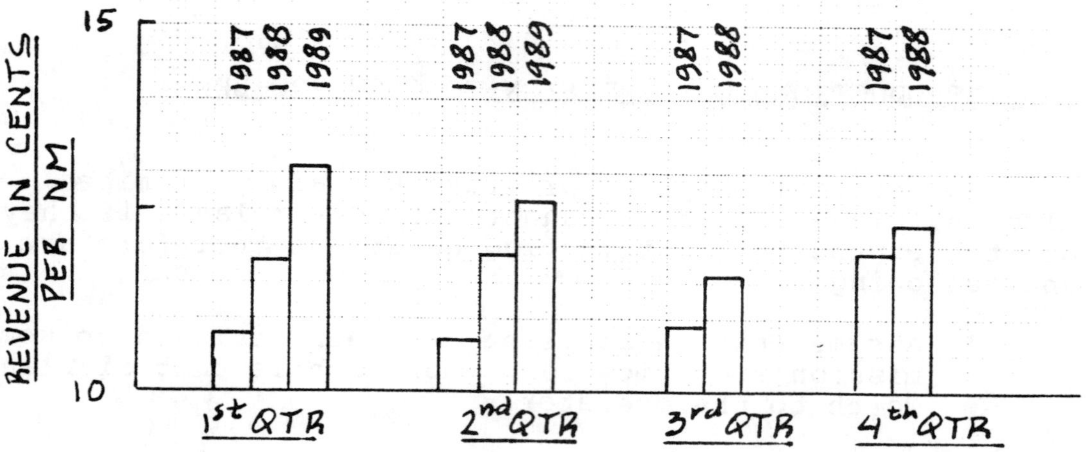

Figure 8.14 Example of Airline Revenues

A problem with this procedure is that the direct operating cost, DOC, itself is a function of AEP, as shown in Chapter 5, Eqns (5.19), (5.40) and (5.41). Solving for AEP is therefore a complicated matter.

The IOC (indirect operating cost in Eqn.(8.15)) can be assumed to be some constant fraction of DOC as suggested in Section 5.3.

8.8.2 The Military Airplane Net Worth Problem

Estimating the net worth of a military airplane is a tricky proposition. For combat airplanes one approach is to estimate the worth of potential targets which have to be destroyed by the airplane in question. At least in theory, it would not make economic sense to develop an airplane which costs considerably more than the targets it is supposed to destroy. This line of thinking thus sets an upper bound on the allowable airplane cost. Reference 48 can be helpful in this regard.

For military training airplanes the problem is a lot simpler. It is not too difficult to determine the cost of training pilots. Knowing the number of pilots which have to be trained an estimate can be made of the allowable cost of the airplane: its net worth. Reference 49 provides more detailed insight into this problem.

8.8.3 The Design-to-Cost Problem

It will be assumed that as a result of the solution to the net worth problem in sub-sections 8.1.1 or 8.8.2, the following information is available:

* Airplane mission specification

* Airplane net worth, ANW (usually assumed to be the same as airplane estimated price, AEP)

* Number of airplanes required in a market, N_{market}

Assuming that the number of airplanes to be built to production standard, N_m, is equal to:

$$N_m = F_m(N_{market}) \qquad (8.20)$$

where: F_m is a factor (> 1.0) which reflects the level of optimism about additional markets for that airplane.

The estimated revenue from this airplane program to the manufacturer is:

$$R_{man} = (N_m)(ANW) \qquad (8.21)$$

If the airplane manufacturer is to make a profit on this airplane program, the amount R_{Man} must equal the sum of airplane acquisition cost, C_{ACQ} (Eqn.(4.2)) and airplane RDTE cost, C_{RDTE}:

$$R_{man} = C_{ACQ} + C_{RDTE} \qquad (8.22)$$

With Eqns (4.2) and (4.19) this yields:

$$R_{man} = C_{MAN}(1 + F_{pro}) + C_{RDTE} \qquad (8.23)$$

Thus, equating (8.21) and (8.23):

$$(N_m)(ANW) = C_{RDTE} + C_{MAN}(1 + F_{pro}) \qquad (8.24)$$

By assuming that C_{RDTE} is a constant fraction of C_{MAN} it is now possible to determine the 'allowable' cost of development and manufacturing. Figure 8.15 shows a typical example of the breakdown of various costs in a commercial jet transport program. Similar data can be extracted from the literature or can be estimated with the methods of Chapters 3 - 7.

Knowing the 'allowable' cost of airplane development and manufacturing, the next problem is to decide if it is feasible to design an airplane which can be profitably built within these cost constraints. This problem can be solved iteratively by using the methods of Chapter 4.

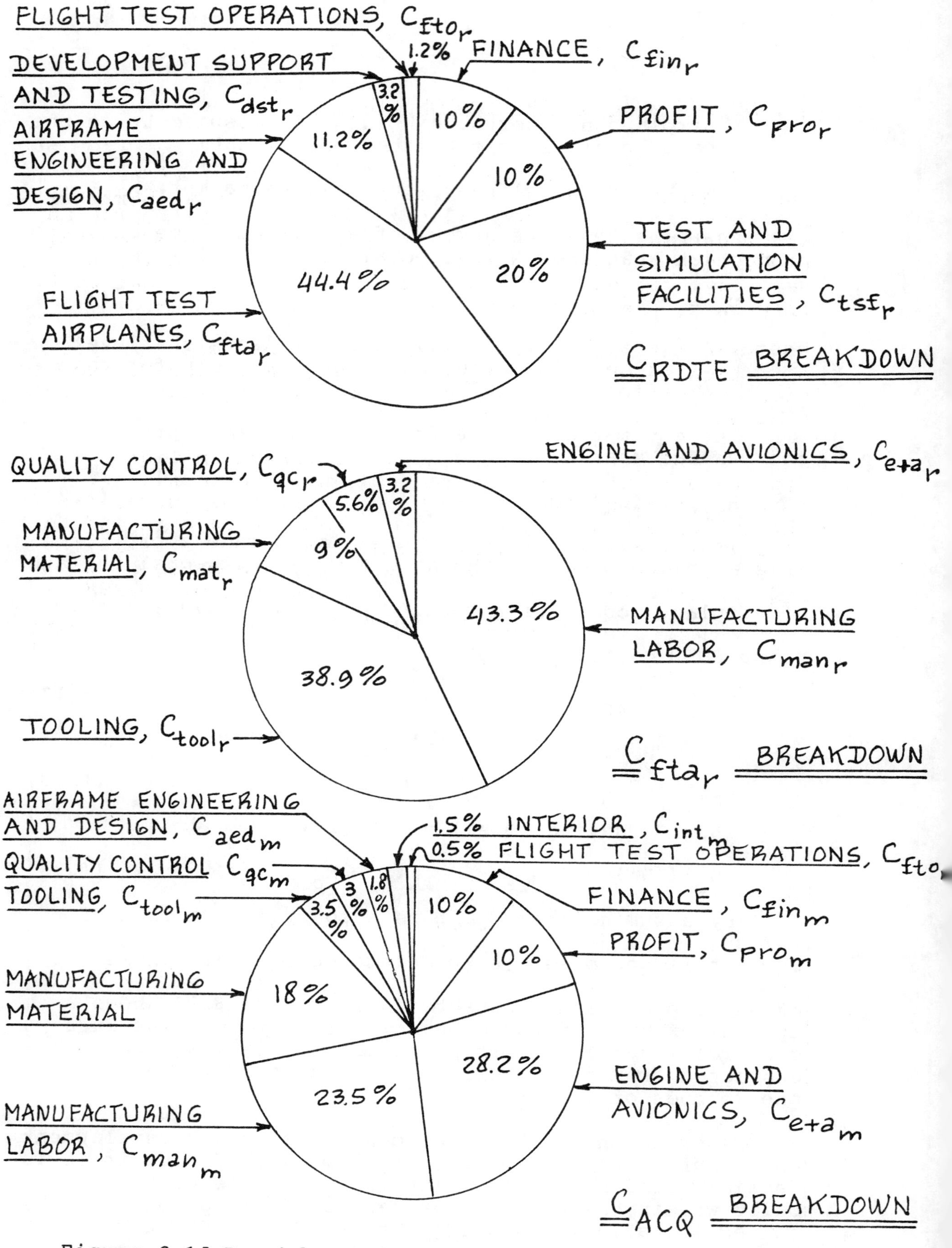

Figure 8.15 Breakdown of Costs in a Jet Transport Program

8.9 DESIGN GUIDELINES FOR LOWERING COST

Even after applying the most thorough of preliminary design methods, including design optimization procedures, many aspects of the detail design of the airplane remain to be decided. Particularly at the system design and at the system design integration level, those detail design decisions can have a significant impact on airplane cost, especially the manufacturing and operating costs.

The purpose of this section is to present a series of design guidelines which, when followed, tend to reduce airplane cost. These guidelines should be carefully considered in any new design.

The 88 design guidelines as presented in Tables 8.4 are the direct result of problems encountered with airplanes designed without the benefit of these guidelines. The author acknowledges the fact that the majority of the 88 guidelines come from research done by Mr. B. Williams, Director of the Center for Reliability, Maintainability and Quality, Air Force Institute of Technology, WPAFB.

SIAI-MARCHETTI S.211 FUEL SYSTEM

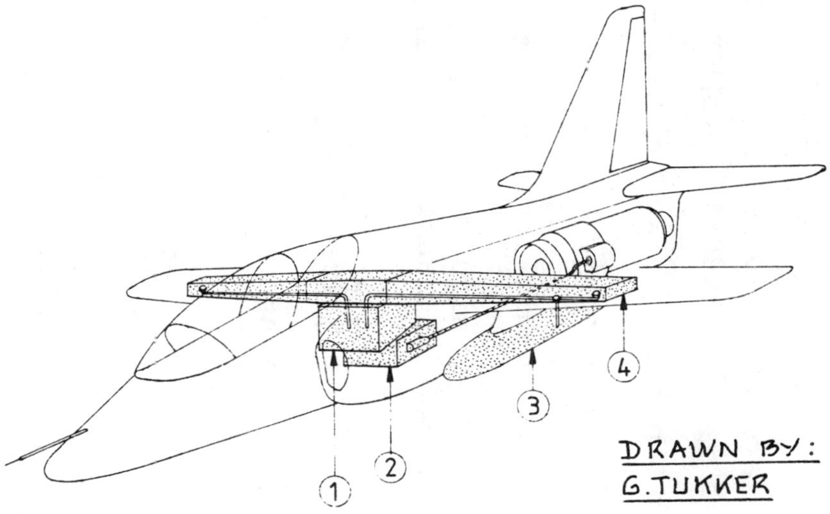

KEY

1. FUSELAGE TANK
2. HOPPER
3. DROP TANK
4. WING TANK

Table 8.4 DESIGN GUIDELINES FOR LOW COST

Guideline	Applicability Commercial	Military
SYSTEMS		
1. Minimize complexity and number of parts associated with all items which require frequent servicing and/or access.	x	x
2. Keep ALL plumbing away from heat sources.	x	x
3. Make most components accessible to people while they stand on the ground. Access step-ups, ladders and platforms are sources of trouble and accidents.	x	x
4. Design so that simultaneous refueling and rearming is possible with the engines running.		x
5. Design so that all required servicing vehicles can access the airplane simultaneously while parked at the gate.	x	
6. Design ALL flight crucial systems with redundant drive and with redundant signal path systems.	x	x
7. Design the installation of identical components for complete interchangeability: no installation should be unique.	x	x
8. Design wire bundles so they will 'fit' only one way, avoiding the possibility of miswiring.	x	x
9. If reversed installation of a component can cause malfunctioning, design the installation for a one way fit.	x	x
10. Design all access panels with CAPTIVE fasteners and with retaining devices.	x	x

Table 8.4 (CONT'D) DESIGN GUIDELINES FOR LOW COST

Guideline	Applicability Commercial	Military
<u>SYSTEMS (Cont'd)</u>		
11. Design access to units which require servicing during turn-arounds so that personnel wearing winter clothing AND gloves can service them.	x	x
12. Design all systems which contain liquids (oil, hydraulics, fuel, glycol and water) so that these liquids can be easily poured in when necessary.	x	x
13. Design oil and hydraulic systems so that they can be completely replenished by ONE person in 5 minutes.	x	x
14. Put servicing instructions on placards in easy to read letters.	x	x
15. Size fluid reservoirs for one full day of operation.	x	x
16. Design avionics systems for fit in integrated, easily removable racks.	x	x
17. Locate avionics racks so that exposure to vibration is minimized.	x	x
18. Locate airplane rate and acceleration sensors so they cannot read false signals due to gun induced vibrations.		x
19. Standardize all micro-integrated circuits.	x	x
20. Design APU systems to provide for adequate cooling on the ground in a hot day environment.	x	x

Table 8.4 (CONT'D) DESIGN GUIDELINES FOR LOW COST

Guideline	Applicability Commercial	Military
SOFTWARE		
21. Establish rigid control over software design AND over software modifications.	x	x
22. Design digital systems so that in-flight computing resources use only HALF of the total system capability (i.e. processing capacity, execution time and memory).	x	x
23. Design software to be compatible with ground systems and with test stations.	x	x
24. Design data management systems to be modular: one update input to update all systems.	x	x
AIRFRAME		
25. Design access doors and panels which require frequent opening with quick opening latches.	x	x
26. Design access doors and panels so that they are easy to close AND so that unlocked doors and panels are easily spotted.	x	x
27. Design door hinges so that if they break, they cannot tear into other primary structure(s).	x	x
28. Design the airframe so that NO structural components have to be removed in order to remove the engine.	x	x
29. Structural areas with a high probability of failure, fatigue or corrosion SHALL be visible and inspectable.	x	x

Table 8.4 (CONT'D) DESIGN GUIDELINES FOR LOW COST

Guideline	Applicability Commercial	Applicability Military
AIRFRAME (Cont'd)		
30. Design airframe jack points so they do not interfere with gear operation, flap operation and other check-out procedures with the airplane on the jacks.	x	x
31. Provide adequate access to integral fuel tanks so that resealing is made possible.	x	x
32. Design for access to all crew station equipment without having to remove a seat.	x	x
33. Ejection seats must be removable without having to disassemble the canopy.		x
34. Install drag chutes in a pre-container to facilitate re-installation.		x
35. All seat components must be inspectable with the seat installed in the airplane.		x
36. Design the nosegear so that the airplane can be towed without having to disconnect steering power and so that the airplane can be towed within the design steering range.	x	x
37. Design landing gear door actuation system as simple as possible.	x	x
38. Wheel and brake assemblies should be separately removable.	x	x
39. Design landing gear so that jacking is NOT needed for scheduled maintenance and inspection events.	x	x

Table 8.4 (CONT'D) DESIGN GUIDELINES FOR LOW COST

Guideline	Applicability Commercial	Military
FLIGHT CONTROLS		
40. Design control surfaces so that they can be replaced without re-rigging the flight control system.	x	x
41. Provide and design for rigging pins with the cockpit controls in the neutral position.	x	x
42. ALL flight control system components must be accessible for all types of maintenance.	x	x
43. Where possible, LIMIT the autopilot authority.	x	x
ENGINES AND INSTALLATION		
44. All engine driven accessories and their electronic controls should be airframe mounted.	x	x
45. Design engine mountings so that each engine can be removed AND replaced within 1.5 hours.	x	x
46. Some engines require access to the top of the engine. Provide easy access in such cases.	x	x
47. Any engine mounted accessories shall be replaceable with the engine installed.	x	x
48. Engine removal shall be possible without jacking or otherwise supporting the airframe.	x	x
49. Design the engine-to-airframe interfaces so they are all located in one area.	x	x

Table 8.4 (CONT'D) DESIGN GUIDELINES FOR LOW COST

Guideline	Applicability Commercial	Applicability Military
<u>ENGINES AND INSTALLATION (Cont'd)</u>		
50. All engine performance sensors must be replaceable within 10 minutes.	x	x
51. Design engine throttle controls so that rigging is not required.	x	x
52. The engine oil system must be designed for operation at prolonged negative 'g's.		x
<u>AUXILIARY POWER UNIT</u>		
53. Ground checkout of the APU must be possible with one access panel.	x	x
54. Design airframe hardpoints to enable quick APU removal and replacement.	x	x
55. It must be possible to operate the APU via an external panel when the airplane is on the ground.	x	x
<u>ELECTRONIC COUNTERMEASURES SYSTEM</u>		
56. Failure of the automatic temperature control system shall NOT preclude manual operation of the system.		x
57. If ECS cooling ducts are located in inaccessible areas DON'T use duct material which deteriorates with age.		x
58. ECS lubrication systems shall be designed so that operation in extreme cold is possible.		x

Table 8.4 (CONT'D) DESIGN GUIDELINES FOR LOW COST

Guideline	Applicability Commercial	Military
ELECTRICAL POWER SYSTEM		
59. Design for over current protection from any single failure.	x	x
60. Design wire bundles for easy removal for modifications.	x	x
61. Design to prevent wire chafing through holes and hinges in the case of long wire bundles.	x	x
62. Keep wire bundles away from heat sources.	x	x
63. Locate the power distribution panel in a non-hazardous area with the engine(s) running on the ground.	x	x
64. All flight station panels must be connected via pendant cables to allow for removal without disturbing ANY adjacent equipment.	x	x
65. Incorporate service loops in the case of sliding chassis or hinged doors.	x	x
66. All grounding receptacles must be in accessible areas AND must be replaceable.	x	x
67. Design the power generation and main battery system so that the main battery is fully charged after one mission.	x	x

Table 8.4 (CONT'D) DESIGN GUIDELINES FOR LOW COST

Guideline	Applicability Commercial	Applicability Military
LIGHTING		
68. Incorporate lights in the wheelwells for maintenance.	x	x
69. Isolate lights from vibration.	x	x
70. All external lights must be rapidly replaceable.	x	x
71. Design strobe light location to avoid disoreintation of flight crew.	x	x
72. Minimize the use of retractable lights and/or use highly reliable retraction/extension mechanisms.	x	x
HYDRAULIC SYSTEMS		
73. Design hydraulic line and hose runs to prevent the inadvertent cross-connection of systems.	x	x
74. Identify the required fluid type CLEARLY at all fluid filler points.	x	x
75. Design all filters, shut-off valves, pressure switches and sub-components for separate removal.	x	x
76. Incorporate isolation valves in each sub-system to allow for maintenance, bleeding, etc.	x	x
77. Incorporate quick disconnect fittings for ground servicing and for connecting to a maintenance cart.	x	x
78. Incorporate a central hydraulic system control panel for maintenance personnel.	x	x

Table 8.4 (CONT'D) DESIGN GUIDELINES FOR LOW COST

Guideline	Applicability Commercial	Military
FUEL SYSTEM		
79. Fuel quantity transmitters should be mounted to accept high 'g' maneuvers.		x
80. Fuel quantity indicators should indicate remaining fuel in a proportional manner.	x	x
81. Design ALL fuel system components so they can be removed WITHOUT draining the fuel tanks.	x	x
82. Any motor operated shut-off valve shall have the motor removable while the valve is in the fuel system.	x	x
83. Design fuel system with common valves and common pumps.	x	x
84. Incorporate a single point refueling and control panel.	x	x
FAULT DETECTION SYSTEM		
85. All individual fault detection systems should report to a master system.	x	x
86. Failure status should be reported to pilot for manual selection of a backup system.	x	x
87. All in-flight failures must be data linked to the ground for maintenance preparations.	x	x
88. The crew station must have a 'mission capability' panel.	x	x

9. FACTORS IN AIRPLANE PROGRAM DECISION MAKING

The purpose of this chapter is to discuss several factors which must be considered before making an airplane 'program-go-ahead' decision. Much of the material included in this chapter is based on ideas which are developed in the series: 'Case Studies in Aircraft Design', published by the American Institute of Aeronautics and Astronautics (AIAA) as References 50 through 53. An excellent overview of the airplane program decision making process is presented by J.E.Steiner in Ref.54. Finally, the reader is urged to familiarize himself with Ref.25 which provides insight into military airplane programs.

The following factors can play a role in a decision to 'launch' or 'not launch' an airplane development and/ or an airplane production program:

9.1 Factors Involving Commercial and/or Financial Feasibility
9.2 Factors Involving Technological Feasibility
9.3 Factors Involving Manufacturing Facilities
9.4 Factors Involving Political and/or Environmental Feasibility
9.5 Lessons Learned from Past Airplane Programs

9.1 FACTORS INVOLVING COMMERCIAL AND/OR FINANCIAL FEASIBILITY

Before embarking on a new airplane program it is necessary to investigate the following commercial feasibility factors:

9.1.1 Market Potential
9.1.2 Development Cost and Time
9.1.3 Availability of a Skilled Work Force
9.1.4 Required Production Investment
9.1.5 Potential for Return on Investment (ROI)

These factors are discussed next.

9.1.1 Market Potential

Market research is an essential component of the decision making process leading to an airplane program go-ahead. The following elements deserve attention:

9.1.1.1 Understand the customer and his organization
9.1.1.2 Identify and compare with the competition
9.1.1.3 Identify voids in the market
9.1.1.4 Forecast the size of the market
9.1.1.5 Forecast the net worth of the airplane
9.1.1.6 Decide on ability to produce at a cost below the airplane net worth
9.1.1.7 Identify and contact customer decision makers

9.1.1.1 Understand the customer and his organization

First, it is necessary to recognize the fundamental differences between commercial and military customers. Figure 9.1 shows the overall relationship between the manufacturer, the customer and the regulator for:

* Commercial Airplane Programs
* Military Airplane Programs

Second, it is essential that those involved in the airplane design decision making process understand the individual customer, his organization and his needs. In the early phases of the design it is of particular importance to know and understand those persons working for potential customers who are in charge of 'technical decision makers'. It should be noted that the technical decision makers are often not the financial decision makers and vice versa. Frequent contact between the customer and the manufacturer is essential, particularly during the early phases of the design.

9.1.1.2 Identify and compare with the competition

Identification of the competition is an essential ingredient in any competitive enterprise. It was also made part of the configuration design development procedures on page 11 of Part II.

The following reference material has been found very useful in identification of competitive designs:

* Jane's All the World Aircraft (British, Annually)
* Business and Commercial Aviation (U.S., Monthly)
* Aviation Week and Space Technology (U.S., Weekly)
* Interavia (Swiss, Monthly)
* Flight International (British, Weekly)
* Sales brochures from competitors

There are many ways to study comparative strengths and weaknesses of competing designs. Table 9.1 and Figures 9.2 and 9.3 show typical presentations from which

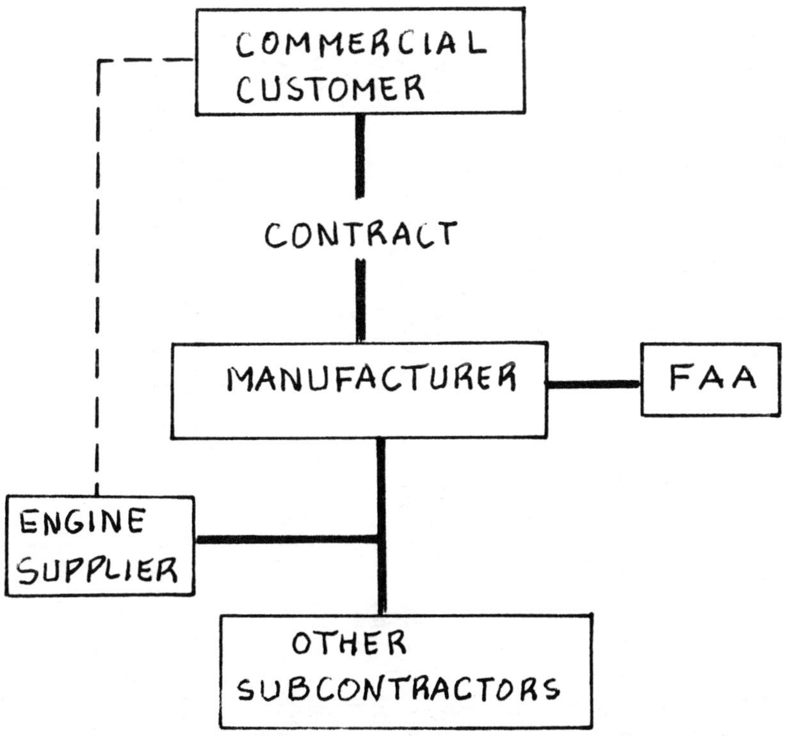

<u>Figure 9.1a Relation Between Manufacturer, Customer and Regulators: Commercial</u>

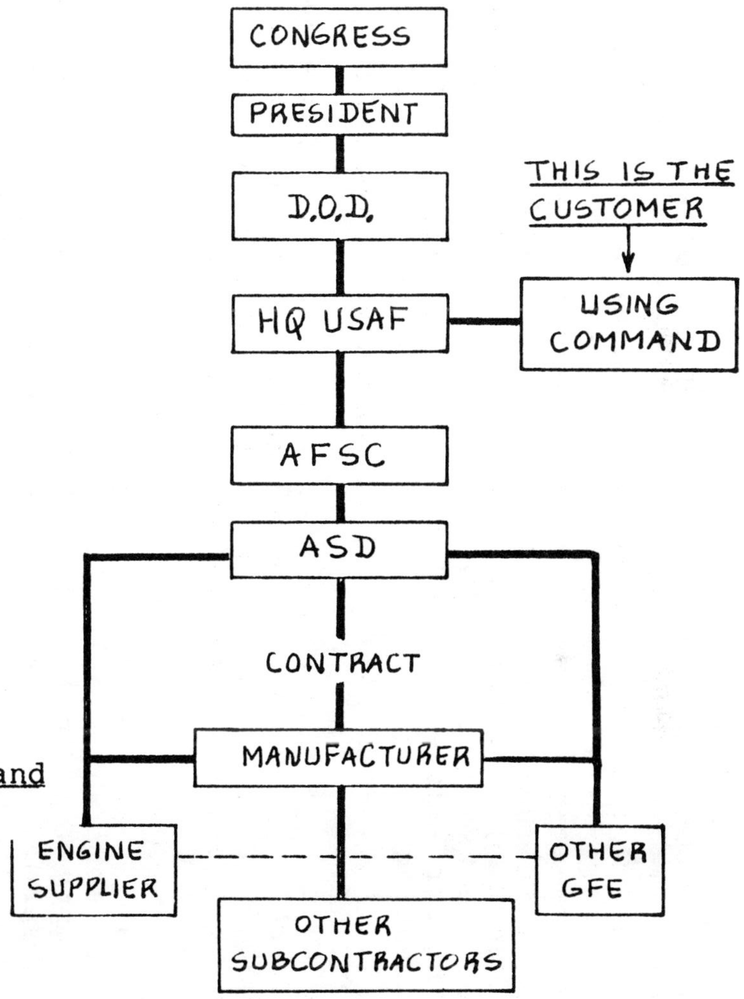

<u>Figure 9.1b Relation Between Manufacturer, Customer and Regulators: Military</u>

TABLE 9.1 COMPARISON OF SIMILAR AIRPLANE DESIGNS

Types:	SAAB 340	Boeing/DH Dash 8	Embraer Brasilia
Specifications			
Max. seating capacity	35	36	30
Power plants (2 each)	CT7-5A2	PW120	PW115
Shp per engine	1,735	1,800	1,600
Max seating capacity	5+35	3+40	3+30
Dimensions			
Length (ft)	64.9	73.0	65.6
Span (ft)	70.3	85.0	64.9
Height (ft)	22.5	24.6	20.8
Wing area (ft^2)	450	585	424.4
Wing aspect ratio	11.0	12.4	9.9
Weights			
Max. takeoff wht (lbs)	27,275	33,000	25,353
Std. empty wht (lbs)	17,415	21,590	15,586
Max. useful load (lbs)	9,860	11,410	9,767
Max. usable fuel (lbs)	5,690	10,160	5,862
(gallons)	843	1,451	875
Max landing wht (lbs)	26,500	32,400	24,802
Max. wing loading (psf)	60.6	56.4	59.8
Performance			
Takeoff fieldlength (ft) (SLS, ISA)	3,960	3,370	4,560
Max. climb rate (ft/min)			
AEO	1,850	1,660	2,250
OEI	500	450	630
Best climb rate speed, AEO (kts)	139	117	NA
Stall speed (kts)	82	72	86
Serv. ceiling AEO (ft)	25,000	25,000	31,000
Serv. ceiling OEI (ft)	16,200	13,400	18,500
Normal cruise speed (kts)	277	265	285
At altitude of (ft)	26,000	15,000	25,000
High speed cruise (kts)	281	265	300
At altitude of (ft)	15,000	15,000	20,000
Fuel flow for:			
Normal cruise (lbs/hr)	948	1,244	832
High sp. cruise (lbs/hr)	1,098	1,244	1,080
Turbulent air penetration speed (kts)	190	180	182
Max IFR range, ISA (nm)*	1,345	1,270	1,240
Range, 4,000 lbs payload	581	1,180	615

*Includes 100 nm diversion plus 45 minute hold

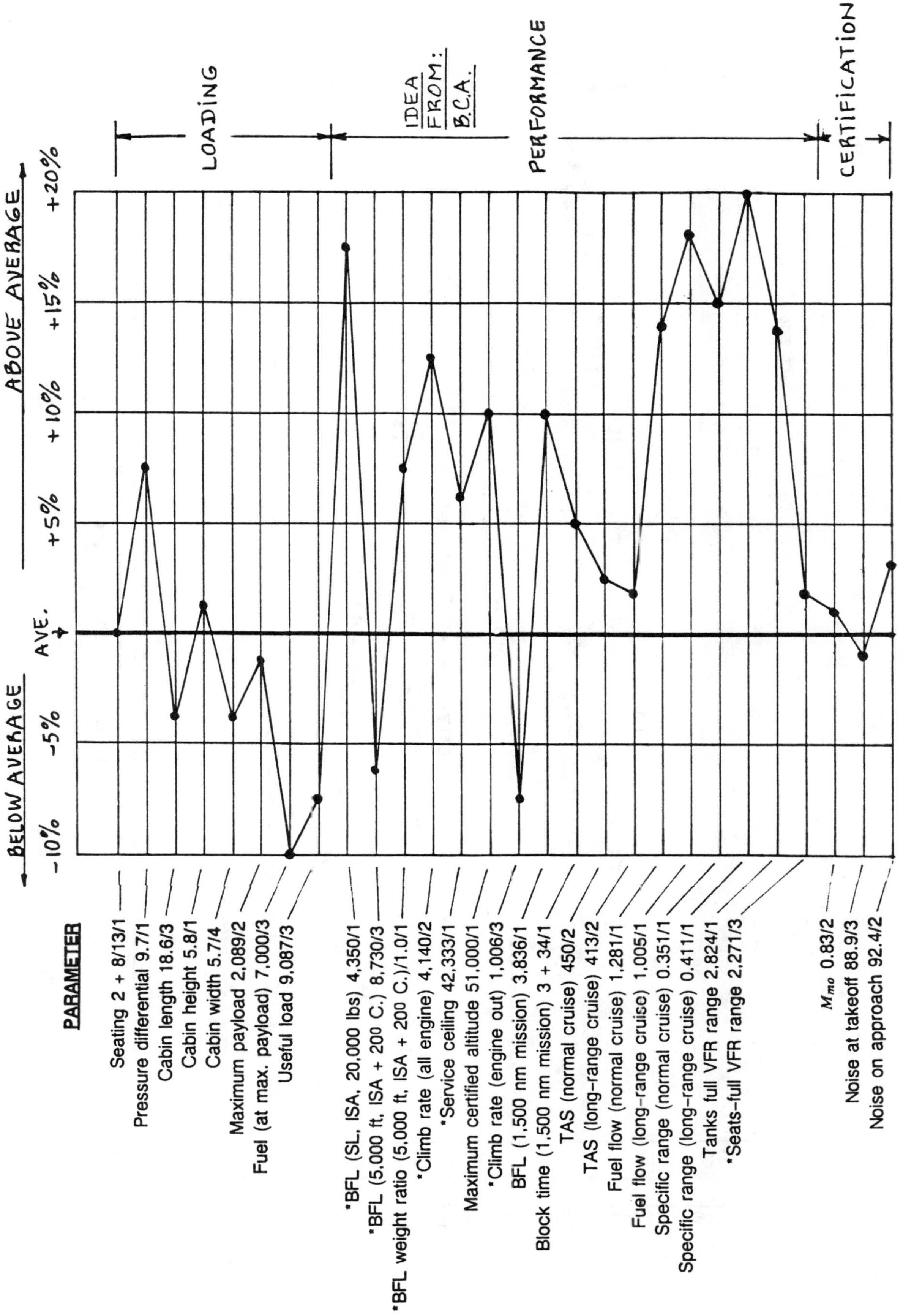

Figure 9.2 Comparison of Similar Airplane Designs

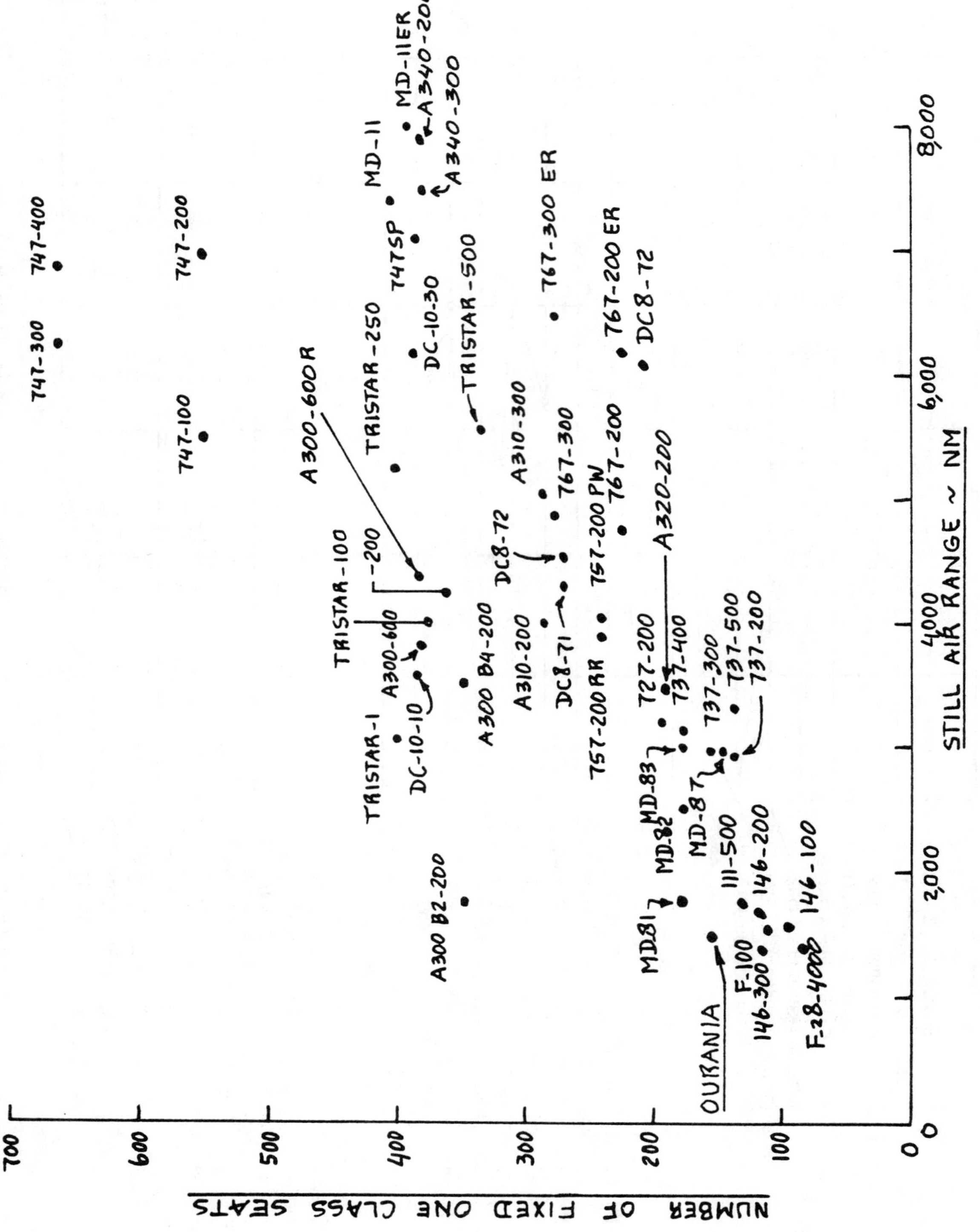

Figure 9.3 Comparison of Similar Airplane Designs

the nature of the competition can be discerned.

When comparing a new design to existing competitive airplanes, ALWAYS assume that the competititon will NOT sit still! Instead, try to imagine various scenarios for the competition to respond. Possible responses from the competition might be:

* Launching of a derivative design

  This could be very cost effective and may be hard to beat with a new design!

* Launching of a competitive new design

  Depending on the cost structure of the competition and on the technology level employed, this may or may not be a threat.

<u>9.1.1.3 Identify voids in the market</u>

Figure 9.3 shows a seat-range diagram for existing jet transports. Figure 9.4 shows that same information but with the airplane identifications left off. It is seen that there are several voids in the diagram. The following questions should be asked:

1.) Are there technical or economic reasons for the voids?

2.) If so, can these reasons be removed by clever and new design approaches?

Take for example the void labeled I in Figure 9.4. Void I would suggest an airplane capable of flying 125-175 passengers over a still air range of 4,000-5,000 nm. The question is: what might be the market for such an airplane? The answer could be: very <u>long</u> but very <u>thin</u> international routes.

* <u>Long</u> means: large distances between city pairs.

* <u>Thin</u> means: passenger demand is relatively small and does not justify the deployment of large wide body airliners.

Table 9.2 shows a list of such city pairs. What must be determined next is how many people might want to travel these routes and what prices might they be willing to pay for their tickets. Market research studies can be conducted to find the answers.

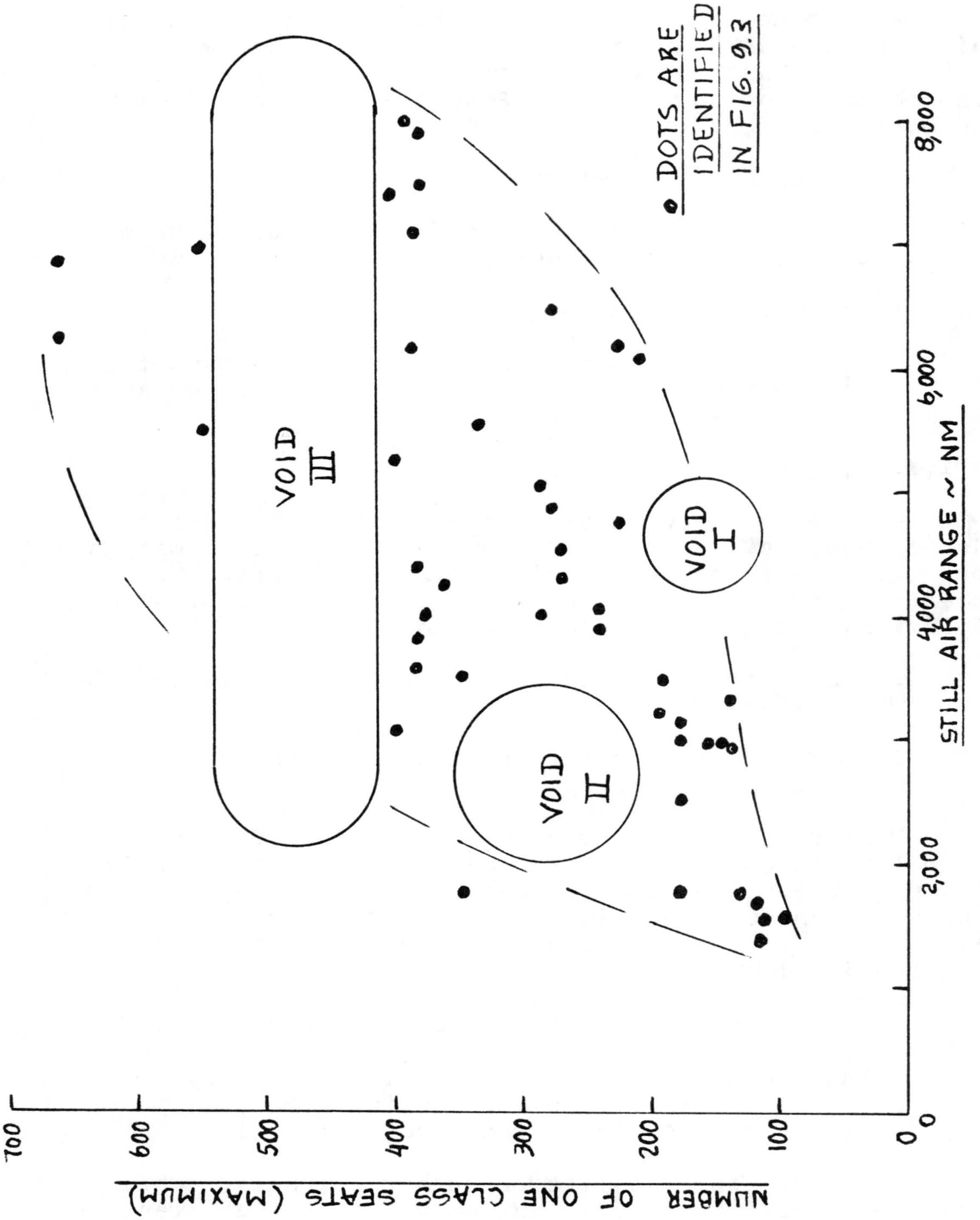

Figure 9.4 Identification of Market Voids

TABLE 9.2 CITY PAIRS WHICH FIT VOID I IN FIGURE 9.4

City Pairs		Great Circle Distance
Kansas City, MO, USA	- Amsterdam, Holland	4,700 nm
Sevastopol, USSR	- Trenton, N.J., USA	5,000 nm
Helena, Montana, USA	- Glasgow, UK	4,200 nm
Baton Rouge, LA, USA	- Rio de Janeiro, Brazil	4,900 nm
Norfolk, VA, USA	- Lima, Peru	3,500 nm
Hamburg, W.Germany	- Dar es Salaam, Tangan.	4,500 nm
Vienna, Austria	- Recife, Brazil	4,900 nm
Osaka, Japan	- Surabaja, Indonesia	3,500 nm
Seattle, WA, USA	- Amsterdam, Holland	4,800 nm
Montreal, Canada	- Abidjan, Ivory Coast	4,800 nm

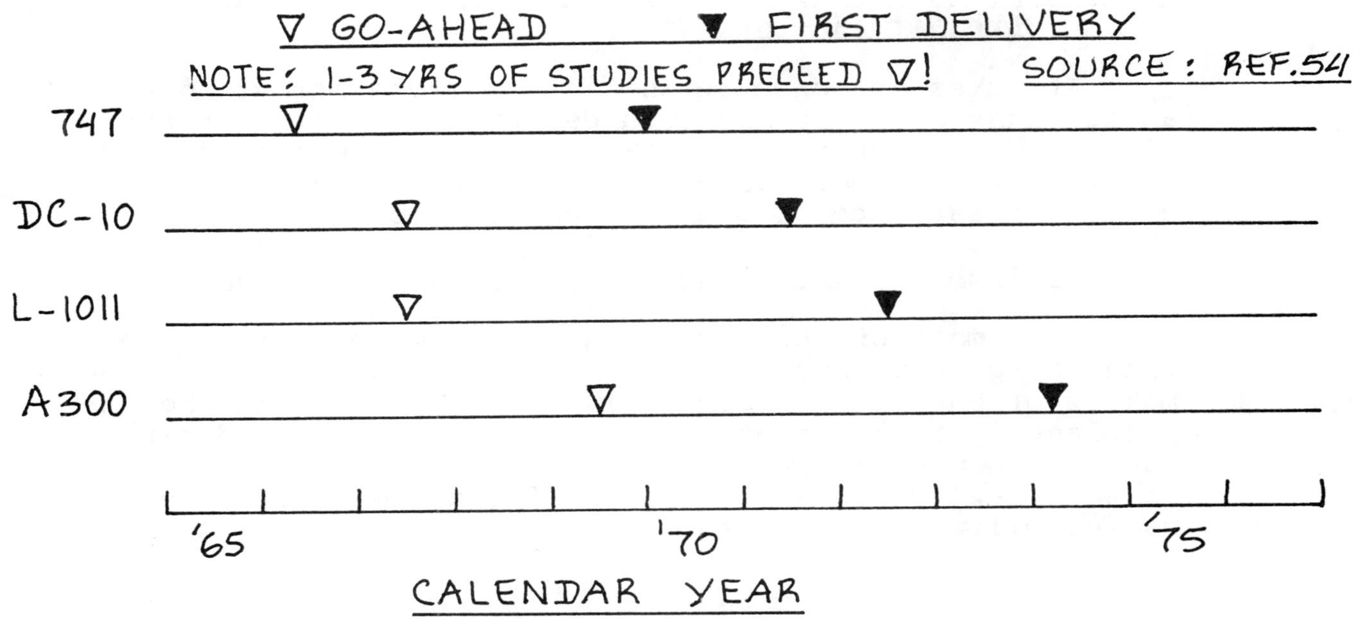

Figure 9.5 Development Times for Four Transport Programs

9.1.1.4 Forecast the size of the market

To make a credible estimate of the size of the market for the new airplane, the information developed in 9.1.1.3 needs to be translated into flight frequencies required to serve example city pairs. Some growth rate is normally accounted for. Typical growth scenarios will postulate that 5 to 10 percent growth occurs annually.

This way the number of airplanes required to serve these new markets can be estimated. With that done, the market share must be estimated. A cautious approach is to assume that a new design might 'catch' 1/3 of the estimated total market.

9.1.1.5 Forecast the net worth of the airplane

The definition of net worth of an airplane is given in Chapter 8 (p.241): it is the price the customer may be willing to pay if there is reasonable assurance of an acceptable return-on-investment (ROI).

A method for determining net worth was also outlined in Chapter 8.

9.1.1.6 Decide on ability to produce at a cost below the airplane net worth

It makes no sense to initiate serious marketing of an airplane until it has been determined that it is possible to manufacture it at a cost well below the airplane net worth. The methods of Chapter 4 should be used to obtain an estimate of the manufacturing cost.

9.1.1.7 Identify and contact customer decision makers

With most of the marketing homework completed, the final step is to try to 'convince' the financial decision makers of one or more customers to commit to ordering the airplane. It should be noted that the financial decision makers are usually not the technical decision makers and vice versa. This tends to be so for commercial as well as for military airplanes.

9.1.2 Development Cost and Time

New airplane development programs can take a very long time. Figure 9.5 illustrates this for several jet transport programs. Also, as shown by the RDTE cost estimation examples of Chapters 3 and 7, the cost of a new airplane development program is extremely high. Management must be prepared to 'sink' a lot of money into the project for many years before any signs of financial returns occur. Figure 9.6 is an illustration of this for the Ourania transport program used in earlier examples.

9.1.3 Availability of a Skilled Work Force

It should not be taken for granted that a skilled work force will always be available to carry out a new airplane development and production program. Historically, aircraft manufacturing employment has gone through many peaks and valleys. Figure 9.7 is an example.

Following major layoffs of production workers, many of these workers will move elsewhere or change job types. Most of these workers may then be 'lost' from that area.

Training of new workers is very expensive but may have to be considered as part of the start-up costs of a new airplane program.

9.1.4 Required Production Investment

Examples of methods for predicting manufacturing costs were given in Chapters 4 and 7. The amount of money involved is usually beyond the capability of an airframer to raise without borrowing money from one or more financial institutions.

The following numbers are recalled:

For the Ourania (150 passenger, medium range jet transport) the estimated program manufacturing cost is:

USD 9,142,000,000 (rounded off from p.61.)

The 'up-front' investment in tooling and engineering cost associated with this manufacturing program is:

USD 552,000,000 (rounded off from pages 60 and 61.)

In addition to these investment needs, there are those for materials, various facilities, engines and avionics. It is true that some of these costs can be

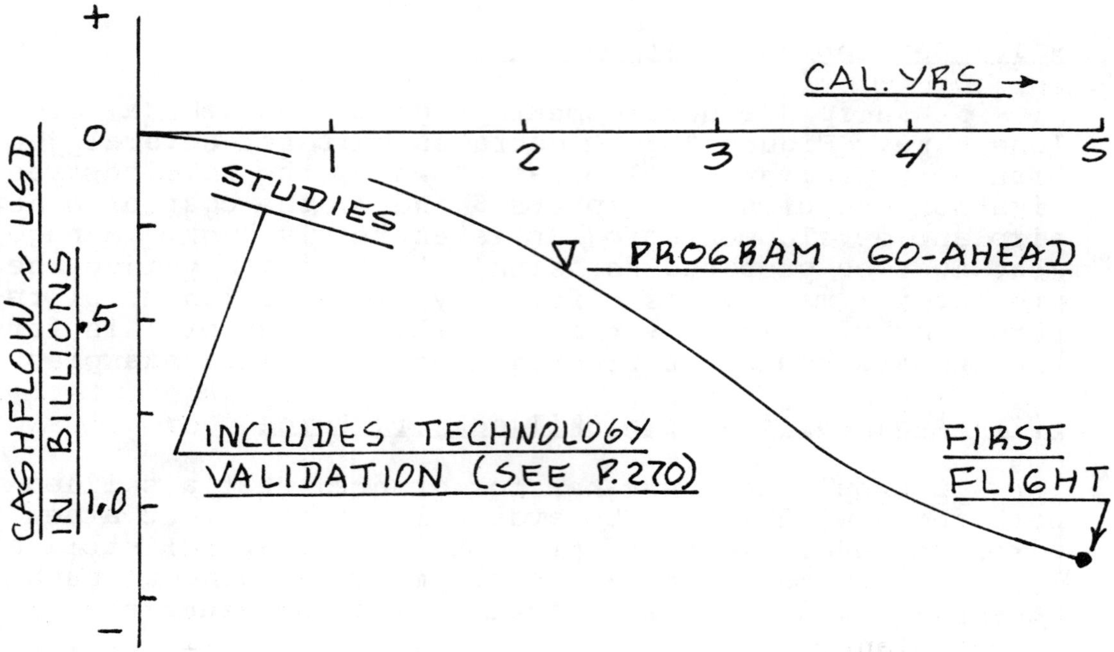

Figure 9.6 Example Cash Sink in a Transport Program

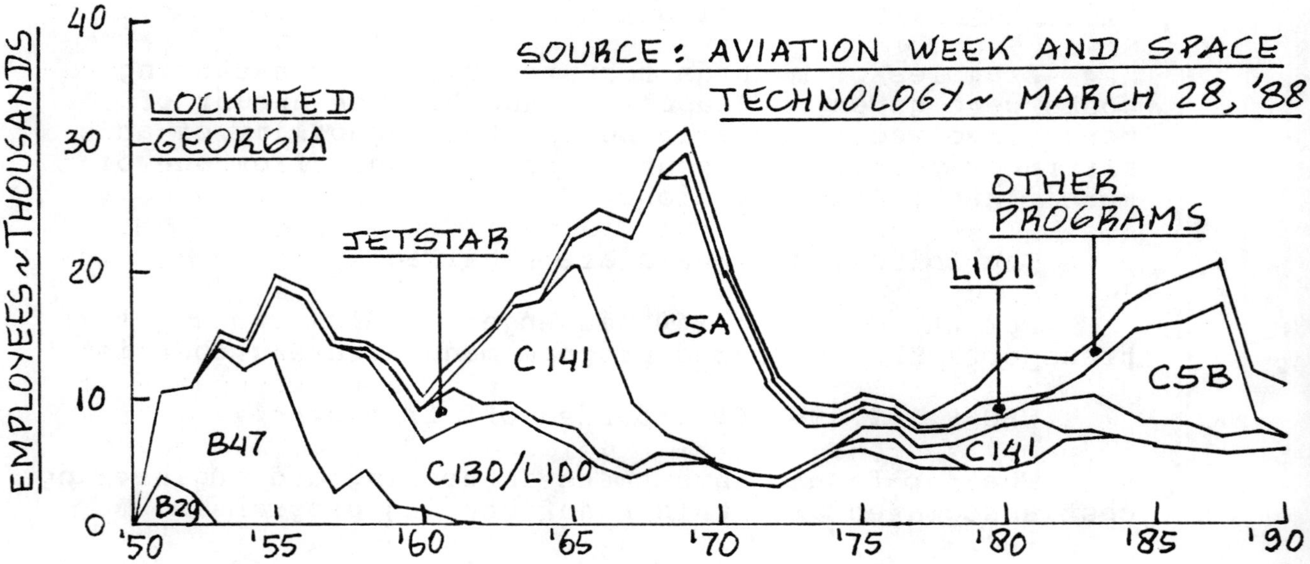

Figure 9.7 Typical Employment Fluctuation in an Airplane Manufacturing Operation

spread out over the duration of the manufacturing program. However, a considerable part has to be invested 'up-front'. The interest over that 'up-front' investment must also be accounted for. If only the tooling and engineering are 'counted' as part of this 'up-front' investment and if the annual interest rate is, say 10 percent, this represents an annual interest payment of USD 55,000,000!!

The need for 'up-front' manufacturing investment money may be so great that risk-sharing arrangements have to be sought with other companies in the USA or abroad.

9.1.5 Potential for Return on Investment (ROI)

At some point well before first deliveries take place, it is customary in the 'airplane business' that the customer begins to make 'progress' payments toward his airplane(s). As a result, some money will start flowing back into the manufacturer's coffers. The difference between program expenditures and program receipts is called the program cash-flow. Figure 9.8 illustrates the cash-flow associated with a typical commercial transport program. Note that the cash-flow does not become positive until 10 years after initiation of the program! However, receipts start coming in after year 3 or 4.

The ultimate profitibility of the program (Return on Investment, or ROI) is determined by the behavior of the cash-flow with time. As it turns out, many airplane programs are viable over the long run ONLY if the manufacturer is willing to adjust his design to new market conditions. This normally results in many 'derivatives' of the original design. However, the design, development, testing and certification of these derivatives is not without cost. Figure 9.9 illustrates the effect of such derivative programs on total program cash-flow.

NOTE TO DESIGNERS: It is important to design the original airplane in such a way that derivatives are in fact possible. For example, if an airplane cannot be grown by lengthening the fuselage because of a landing gear or takeoff rotation limitation it should not have been designed that way in the first place!

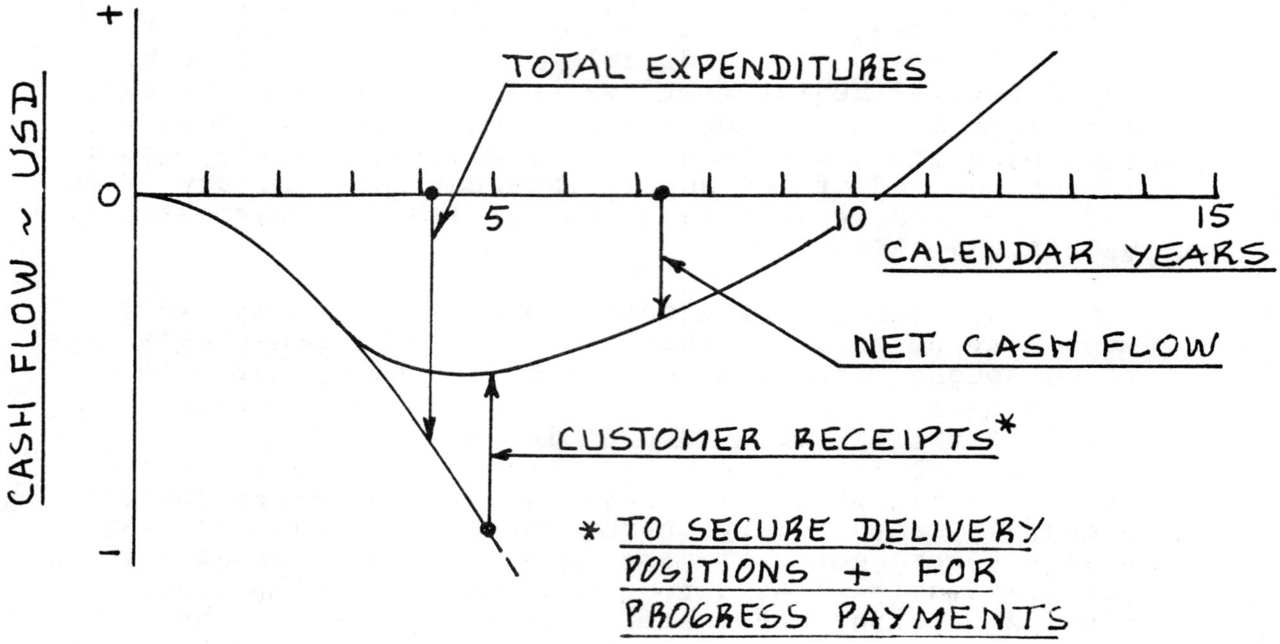

Figure 9.8 Example of Cash Flow in a Commercial Jet Transport Program

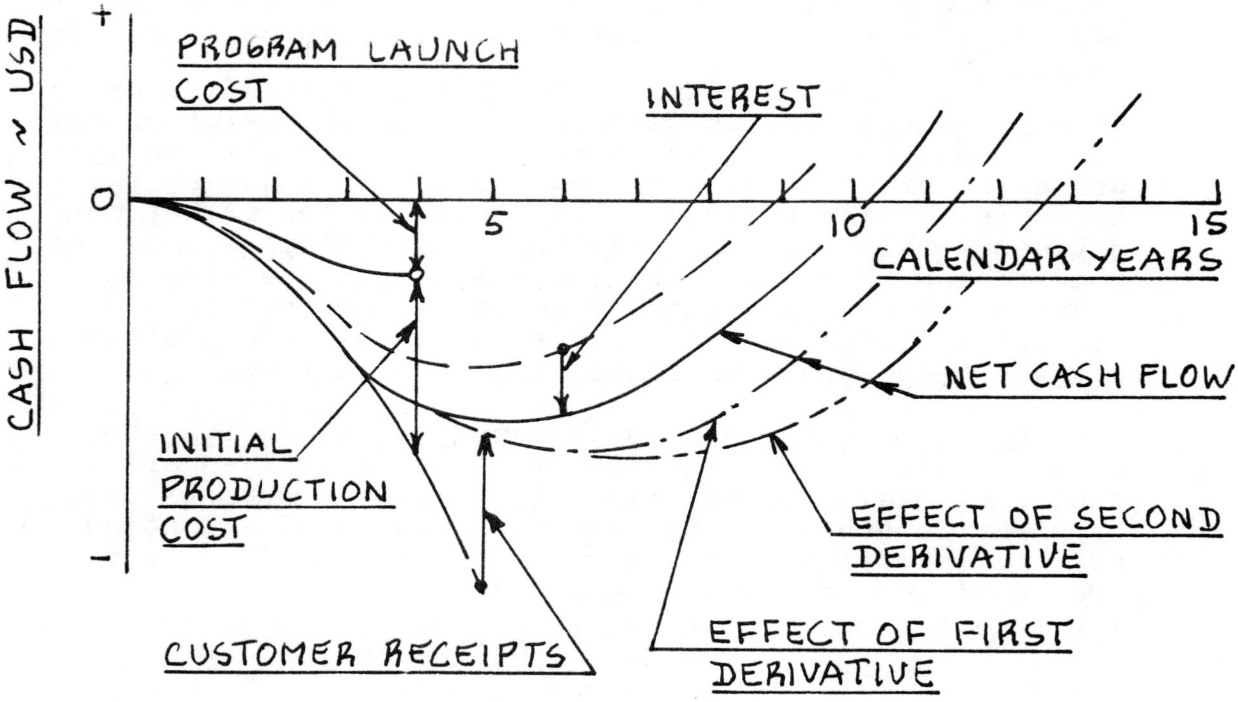

Figure 9.9 Effect of Derivatives on the Cash Flow of a Commercial Jet Transport Program

9.2 FACTORS INVOLVING TECHNOLOGICAL FEASIBILITY

It makes little sense to proceed with new airplane development and/or manufacturing plans without first determining the technological feasibility of the program.

Technological feasibility (or lack thereof) has many facets. To name a few:

* Configuration selection
* Structures
* Flight controls
* Materials
* Aerodynamics
* Propulsion
* Systems
* Manufacturing

The preliminary design methods outlined in Parts I through VII are aimed at establishing the technological feasibility of a new design. Fundamental flaws in the design are supposed to be 'smoked out' as a result of the two preliminary design sequences: P.D. Sequence I and P.D. Sequence II described in Part II.

However, if major advances (innovations) over past airplane designs are to be achieved, it is often required of the preliminary designer to take a risk in one or more of the technological facets mentioned before. Taking a risk implies the extrapolation of currently available and <u>VALIDATED</u> technologies. Examples are:

* Configuration selection: proposing a new configuration which is radically different from existing design practice

* Aerodynamics: assuming that the lift-to-drag ratio in cruise is significantly above that of existing design practice

* Structures: assuming that a new material with much greater strength-to-weight ratio (compared to current design practice) will work out

* Propulsion: assuming that a new engine with much lower s.f.c. (specific fuel consumption) will materialize

* Flight Controls: assuming that a new approach to flight control system design will work and will be certifiable

* Systems: assuming that a new approach to the design of one or more airplane systems will work out

The emphasis in all this is on the word VALIDATED.

NOTE 1: There is nothing wrong with assuming that major advances in technology can be made AND to integrate these advances in a new design during the preliminary design phases.

NOTE 2: There is much wrong with NOT doing the research, development and test work needed to VALIDATE any of these 'extrapolated' technologies.

NOTE 3: It is foolhardy to decide to proceed with the integration of UNVALIDATED technology into the detail design development phase of an airplane.

There are many historical examples of airplane programs which were launched without adequate VALIDATION of one or more aspects of technology used in the design. The consequences of launching an airplane program without technology validation can range from:

* mere embarrassment to a fatal crash, or from:

* minor cost overruns to bankruptcy.

Any unvalidated technology when used in a new airplane design can therefore become a 'profit suppressor' or even a 'show stopper'. It is essential that designers acquaint themselves with all technology factors which may become either a profit suppressor or a show stopper. The purpose of this section therefore is to provide airplane designers with a discussion of a number of technology factors which in past airplane programs have proven to be a profit suppressor or a show stopper. The material is organized as follows:

9.2.1	Drag prediction	9.2.2	Loads prediction
9.2.3	Laminar flow	9.2.4	Range prediction
9.2.5	Maximum speed prediction	9.2.6	VTOL capability
9.2.7	New materials	9.2.8	New engines
9.2.9	Manufacturing processes	9.2.10	De-facto stability
9.2.11	Summary		

This list of possible show stoppers is far from complete. However, it serves to illustrate problems which manufacturers have stumbled into because of excessive optimism or because of insufficient testing.

Examples of each are given without naming the manufacturer or the program involved.

9.2.1 Drag Prediction

Drag prediction is still a 'troublesome' aspect of aeronautical technology, despite all the advances made in computational aerodynamics as well as in wind-tunnel testing techniques. Drag was under-predicted on several recent military airplane programs. The results were:

a) on a military transport program performance was compromised and costly modifications were made

b) on a tactical fighter program performance was compromised to the point where even after costly modifications the effectiveness of the airplane was significantly reduced

c) a military jet trainer program was cancelled well into the flight test program

9.2.2 Loads Prediction

Airframe airload prediction can be a major headache in the transonic speed regime. On a recent military jet transport program the inboard wing airloads were under-predicted. This led to reduced wing fatigue life and costly modifications to the airframe.

9.2.3 Laminar Flow

Laminar flow has the potential to reduce friction drag very considerably. However, the stability of laminar boundary layers is tenuous. Therefore, those factors which are known to affect laminar boundary layer stability must be completely understood. For example, it is well known that high leading edge sweep angles tend to lead to rapid laminar boundary layer transition to turbulent flow which increases friction drag by a considerable amount. Therefore, it might not make sense to use high leading edge sweep if laminar flow is essential to attain the predicted operating speed of a new airplane.

This has resulted in significant downward revision of the cruise speed of a business airplane.

9.2.4 Range Prediction

If an airplane must have very long range capability, the Breguet range equations of page 15, Pt I show that three factors are essential:

1) High average lift-to-drag ratio: $(L/D)_{cr}$

2) Low weight ratio: W_{end}/W_{begin}

3) Low engine fuel specifics: c_p or c_j

The technical program decision makers had better assure themselves that none of these factors are very sensitive to optimism in their prediction. If they are, a directed research program aimed at obtaining these optimistics values may have to be financed.

An early jet transport program ran into this type of a problem resulting in not meeting transatlantic range predictions.

9.2.5 Maximum Speed Prediction

If an airplane must attain very high speed at sea-level, a number of factors bear watching:

1) Drag, including compressibility drag

2) Aeroelastic deformations which affect drag and controllability

3) Flutter, particularly in not yet encountered modal combinations

It takes a major engineering effort up-front to determine whether or not all technical risks in these three areas have been lowered sufficiently to justify launching the program. A number of jet fighter programs were launched without sufficient validation in these areas. The result was either early cancellation or a significant reduction in operational capability.

9.2.6 VTOL Capability

The following factors must be solidly researched and tested before launching a VTOL program:

* Suckdown effects of the configuration

* Adequacy of the reaction controls in hover

* Efficiency of ejectors used

* Reingestion of high temperature gasses

* Foreign object damage

Several VTOL programs of the past failed because one or more of these factors were not adequately researched. In all cases major cost overruns caused the program to be cancelled. In one case a test pilot also lost his life.

9.2.7 New Materials

On several occasions new materials have been introduced without adequate testing. Examples are:

1) Introduction of 7075-ST aluminum in areas of the wing structure where corrosion/crack induced fatigue led to very low structural life.

In a commercial jet transport program this forced the manufacturer into a very costly re-skinning program.

2) Introduction of composites which were found to be subject to (difficult-to-detect) delaminations resulting in compromises to the integrity of the structure.

This caused the manufacturer to use significantly thicker skins than planned. As a result the airplane 'gained' a lot of unplanned weight which reduced the useful load.

3) Introduction of composites without proper manufacturing control over weight. This led to unpredictable weight variability from one production airplane to the next.

This caused the termination of the program and bankruptcy of the manufacturer.

9.2.8 New Engines

History is full of examples of airplanes which did not succeed because the engines around which they were designed never reached either acceptable performance or acceptable reliability.

There are also many programs where engine problems caused major cost overruns, delays in deliveries and corresponding financial headaches for the manufacturer.

9.2.9 Manufacturing Processes

Consider the following case histories:

1) As part of the manufacturing process, a given airplane component may have to be quelched in water. The water used comes from the city water supply. During one season, the city puts some chlorine compounds in their water. The chlorine alters the structural characteristics of that airplane component.

   It took a lot of 'detective' work by engineers to identify this problem.

2) A manufacturer introduces metal-to-metal bonding in small radius skin structures, such as a horizontal tail. Say that in the analysis of the flutter behavior of the tail, all bonding was assumed to be complete. Suppose that as a consequence of local disbonding, the tail develops flutter which causes inflight airframe failure. Now the probability of a minor flaw in manufacturing has major consequences!

9.2.10 De-facto Stability

A new airplane requires a certain inherent static instability to meet certain mission demands. The manufacturer decides to employ a digital flight control system. However, the manufacturer engages a sub-contractor to assist in the development of the flight control laws. Somehow, inadequate provisions are made to simulate the control laws before first flight. The first prototype crashes, causing major program delays and cost overruns.

9.2.11 Summary

There are several lessons in all these examples:

1. PLAY THE 'WHAT-IF' GAME AT ALL LEVELS OF DESIGN DECISION MAKING.

2. LEARN FROM MISTAKES MADE ON PAST PROGRAMS. DO SO BY READING THE TECHNICAL LITERATURE.

3. THERE IS NO SUBSTITUTE FOR THOROUGH ENGINEERING RESEARCH AND TESTING: IN OTHER WORDS, VALIDATE THE TECHNOLOGY

9.3 FACTORS INVOLVING MANUFACTURING FACILITIES

The following examples are offered as illustrations of the importance of considering requirements for manufecturing facilities in early design decision making.

Example 1

Sometimes, to assure technological feasibility, new manufacturing methods and processes must be worked out. There is no guarantee that these will work. Therefore, a certain amount of manufacturing research may have to be performed to ensure that the product will perform as advertised.

Examples of programs where a very intensive program of manufacturing research was carried out are the F-27 and the F-28 transport programs and the SR-71 reconnaissance airplane program.

Example 2

It is desirable from a structural weight viewpoint to have as few joints in any structure as possible. In wing design this may lead to a requirement for very long wing skin panels. If it is not possible to manufacture these panels on existing milling machines new facilities may have to be acquired. The cost of this must be taken into account. An example of this is the 747 program.

Example 3

If laminar flow must be maintained over a large part of the wing chord it is essential that very tight surface tolerances be maintained. This has been found relatively easy to accomplish when using composite skins because of their inherent high degree of stiffness.

In metallic structures, maintaining large laminar flow runs is critically dependent on the number of joints and fasteners used. Two methods are used to minimize the number of joints and fasteners:

A) Manufacturing the wing torque box out of two sections which are joined on the neutral axis. Figure 9.10 shows an example of that method. This method requires the availability of sophisticated machining equipment.

B) Manufacturing external surfaces with 'outside-in' instead of the more conventional 'inside-out' method.

Figure 9.10 Manufacturing a Wing Torquebox in Two Pieces

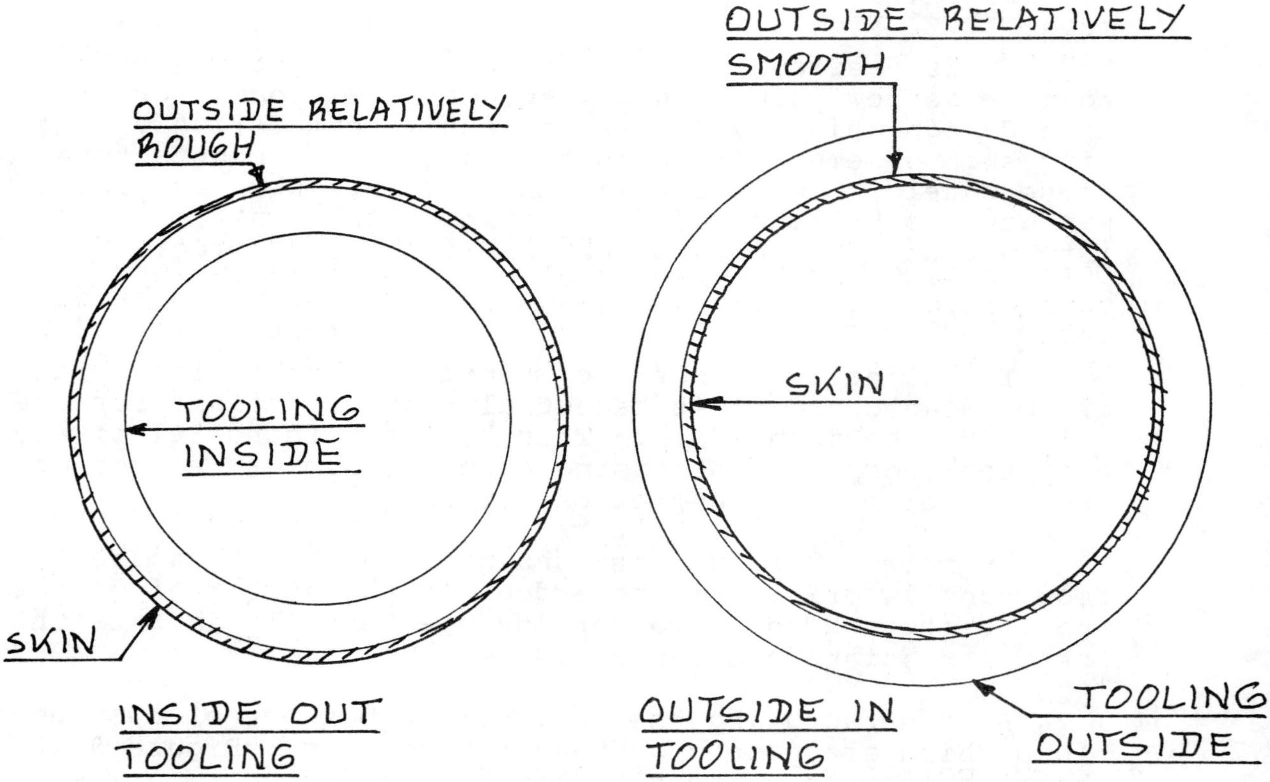

Figure 9.11 Inside-out (Old Style) and Outside-in Method (New Style) of Airframe Manufacturing

Figure 9.11 shows an example of this method. It is also possible to manufacture wings in that manner. This method requires a new approach to tooling design.

All this emphasizes the need for preliminary design organizations to discuss their designs with manufacturing experts at the earliest possible time. In this manner it is possible to prevent poor design decisions to go unchallenged before it is too late. In addition, this can help to identify the need for new manufacturing equipment and new manufacturing facilities early in the design program. This way, the cost consequences of design deciions which affect manufacturing receive early scrutiny.

9.4 FACTORS INVOLVING POLITICAL AND/OR ENVIRONMENTAL FEASIBILITY

There are situations where political and/or environmental pressures cause airplane programs to be terminated or altered. The purpose of this section is to acquaint the reader with some of these pressures. Three examples will be discussed.

Example 1: Nuclear Power

In the fifties, major studies were conducted to determine the technological and financial feasibility of using nuclear power for the propulsion of large intercontinental bombers and transports. Flight tests of at least one small nuclear reactor were in fact carried out in a B-36 airframe. This program was eventually terminated for a number of reasons. Questions were raised relative to the technological and financial feasibility (i.e. the projected high cost).

But there were other questions such as: what happens in the case of a crash with perhaps an explosion and/or a fire? How much nuclear contamination of the environment will occur? How many people might get killed due to the radiation effects? What is the probability of such a crash occurring?

These questions all beg for answers because they affect political and environmental issues.

It is the lack of acceptable answers to these questions which, probably more than any technical or financial issue, led to the termination of this and other nuclear aircraft propulsion projects.

But there is another issue here: ETHICS.

Should engineers push for starting a flight hardware project if there are serious questions about safety and about the environment?

The author believes that engineers SHOULD NOT DO SO! Engineers have an ethical commitment to society not to engage in activities which lead to major safety problems and/or problems with the environment.

Example 2: Supersonic Transports

Consider the following question: Do large fleets of supersonic transports have the potential of causing significant damage to the earth's ozone layer?

Major study programs of the chemical interactions between engine exhausts and the upper atmosphere are now being conducted by NASA to determine the answer to this question.

Until the answer to this question is available and widely accepted by the scientific and political community there SHOULD NOT be a commercial supersonic transport program.

The reader will remember the cancellation of the US-SST program in the late sixties. That program was cancelled mostly because of environmental pressures. These pressures prevailed because the aeronautical community failed to come up with scientifically sound arguments to defuse them.

The reader may ask: what about military supersonic flights? The answer has to do with relative levels of pollution. Consider the following argument.

Assume that there are, worlwide, 5,000 military airplanes able to sustain supersonic cruising flight. These airplanes fly on the average 350 hours per year. However, of those 350 hours, they probably spend less than 10 percent of their flying time in supersonic, high altitude flight conditions. This represents 175,000 hours of flight in the ozone layer.

Now consider a worldwide fleet of 1,000 commercial supersonic transports. This fleet could consist of a mixture of business jets and airliners. Say that these airplanes fly an average of 3,000 hours per year. About 70 percent of this time will be spent in a supersonic flight condition. The average takeoff weight of this fleet of airplanes will probably be about 500,000 lbs, compared to at most 50,000 lbs for their military counterparts. The reason why takeoff weight is important has to do with the fact that the cruise thrust (and thus the engine massflow) is proportional to takeoff weight. This represents a factor 10 difference.

This commercial fleet of SST's will therefore produce the military equivalent of (1,000)(0.7)(3,000)(10) = 21,000,000 hours of flight in the ozone layer. This represents a factor of 120 difference. That is more than TWO ORDERS OF MAGNITUDE!! That is why commercial supersonic flying might be a problem whereas military supersonic flying is thought to be insignificant in terms of ozone destruction.

Example 3: Noise

During the late sixties and early seventies serious questions were raised about airplane noise, particularly the following types of noise:

1) Airport and community noise caused by takeoff operations

2) Airport and community noise caused by approach and landing operations

3) Community noise and damage caused by sonic booms from supersonic overflights

The noise levels in many instances had become intolerable to many people. Political action followed. This in turn resulted eventually in legislation under which:

1) Takeoff noise is regulated in FAR 36

2) Approach and landing noise is regulated in FAR 36

3) Supersonic overflight (other than in certain restricted, military areas) is prohibited by law

As a result, airplane propulsion systems must all be designed so that they meet the minimum noise standards of FAR 36. Many older transport jets were (and are still) being phased out because their modification to the latest FAR 36 standards is not economical.

9.5 LESSONS LEARNED FROM PAST AIRPLANE PROGRAMS

The purpose of this section is to present a listing of lessons learned (the hard way) on past airplane programs. The 19 lessons were taken from Reference 54 with minor revisions.

Lesson 1: Engines

Engines usually take at least a year longer for development than airplanes. In planning an airplane program this must be accounted for.

It is advisable to design the airframe around at least two engines from different engine manufacturers. If not, one knowingly takes a large degree of program risk. Figure 9.12 shows the engine options available on the 747 in 1982.

Lesson 2: Being there First

Being the first manufacturer to go ahead with a new, advanced airplane does not assure success. History shows that the eventual prize may go to the manufacturer who is second, provided that one is fast on his feet and is technically and financially sound.

Lesson 3: Losing a Competition

A competitive loss can motivate the loser more than the winner. This possibility should be carefully considered by management!

Lesson 4: Validated Technology

Generally speaking, the highest VALIDATED state of technology has produced the longest term, and thus, the most financially rewarding production program. Note the emphasis on the word VALIDATED.

Management should always weigh the risks and unanticipated costs associated with advanced technology. It should be ready to take program actions to minimize these.

Lesson 5: Hidden Conservatisms

Engineering should not be allowed to make conservative design assumptions AND keep these hidden from program management.

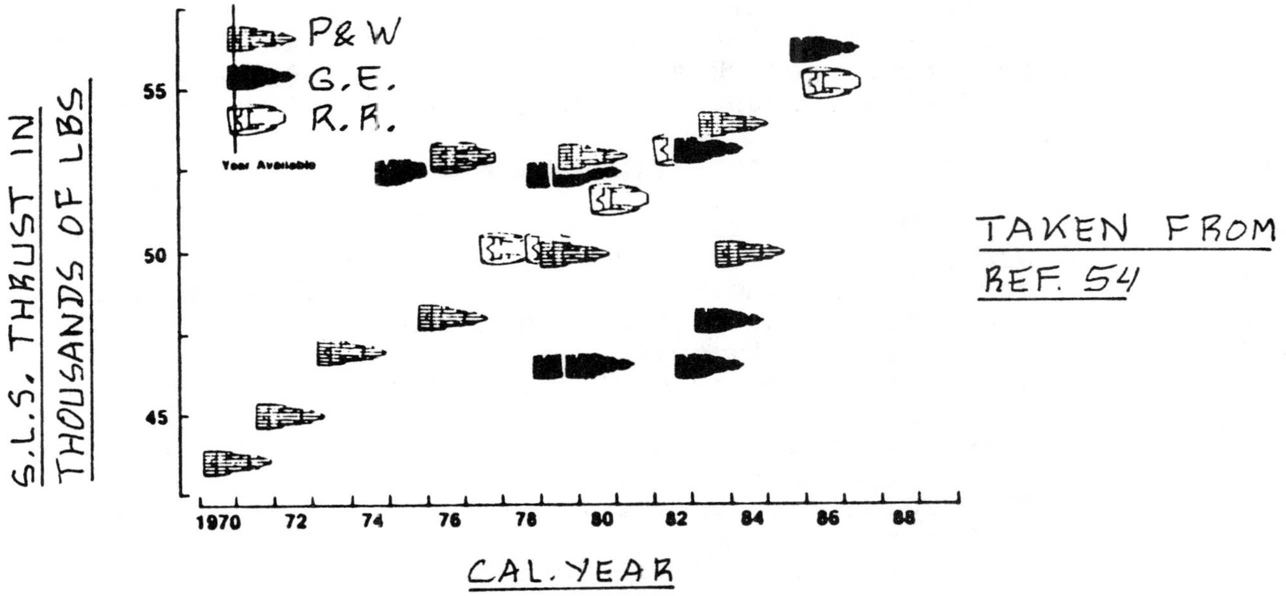

Figure 9.12 Engine Options Available in the 747 Program

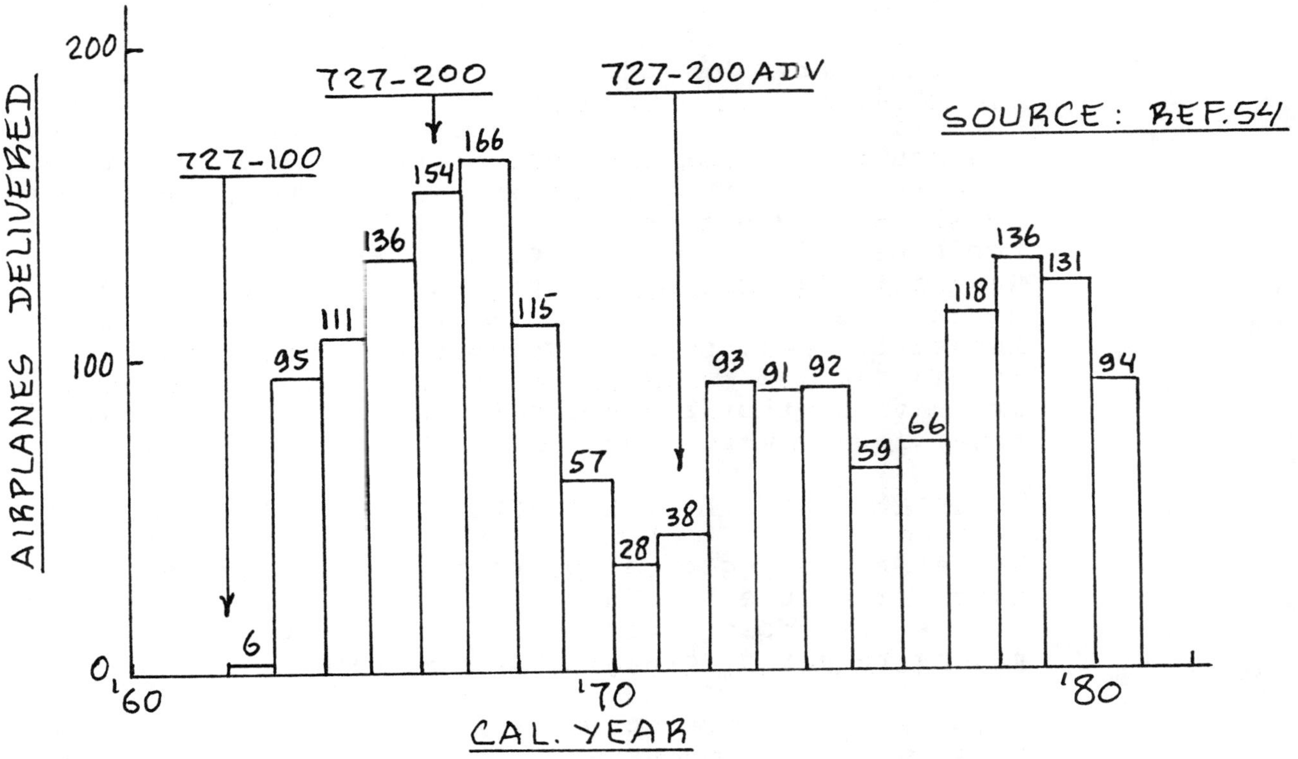

Figure 9.13 Early Delivery History of the 727 Program

A consequence of not heeding this lesson may be that the design is invalidated and/or that growth is not felt to be possible. This could stop an actually promising design from being launched!

Lesson 6: Non-recurring Investment

The RDTE cost associated with airplane development programs is also referred to as a 'non-recurring' cost. If a company wants to be in the commercial airplane market it should be prepared for investments of roughly twice the original RDTE investment cost. These costs may cover a period of 3-10 years.

Lesson 7: Staying power

Commercial airplane programs require that the manufacturers have 'staying power' to carry them over temporary market depressions or unexpected actions taken by the government.

Figure 9.13 shows the early delivery history of the Boeing 727, clearly indicating the need for staying power.

Lesson 8: Red-White-and-Blue Team Approach

To arrive at the 'best possible' design solution for a given mission it has been found to be cost effective to deploy two or three design teams. Each team should be given access to the same budget and talent pools as well as other required resources. This way, two to three different design approaches to a given mission can be compared and the 'best' one selected for full scale development.

It is vital that all three teams know and understand the rules by which what is 'best' will be decided.

Lesson 9: Up-Front Investment

To get a truly good airplane requires a large 'up-front' investment. This investment is to cover cost for new technology development, new machinery which may be required, training and equipment.

In commercial developments, management is free to decide on such investments AND to implement such decisions.

In U.S. government acquisition procedures there is no provision to adequately accommodate this.

Lesson 10: Producing Paper Airplanes

Government paper competitions which start airplane production programs without prototypes MAY result in selection of the bravest and least informed winner, with uncorrectable consequences.

Lesson 11: Readying Technology

Readying new technology developments for application to a production program can be a long and very expensive process. This process will frequently warrant funding stimulation. The payoff, when the need and technology readiness match, can be enormous.

Lesson 12: Sizing a New Airplane

Past decisions that have resulted in airplanes which were improperly sized indicate that designers should moderate their thinking to find a middle ground between 'highs' and 'lows' in projected market growth, competitor proliferation and airport constraints.

Lesson 13: Need for Maturation

A good airplane program requires at least two years of concentrated 'pre go-ahead' study and planning.

The penalty for omitting this phase is greater exposure to large changes late in the development program.

Lesson 14: Understanding Real Customer Needs

There is no substitute for understanding the real needs of a customer. This applies to commercial as well as to military programs. The job of the airplane designer is to design the airplane which the customer wants WHILE still meeting or providing for product capabilities the customer will want five years after all the basic design decisions have been made.

Lesson 15: Definition of Program Success

Management should clearly have in mind what is meant by ultimate program success. Designing two (or even more) airplanes 'head on' for the same market may result in financial failure of one manufacturer unless the market is very large or segmented.

Lesson 16: Catering a Design to One Customer

Most good airplanes are designed to meet a broad spectrum of market requirements. Therefore, design compromises are ESSENTIAL. In the commercial market case, designing to the detailed requirements of one customer can result in an unsatisfactory program. Starting a program with one customer, while designing to a broader market need, can be satisfactory.

Lesson 17: Product Support

Product support after delivery is just as important as is successful design. Product support represents a growing financial burden with time in a successful airplane program. Without it, a manufacturer will soon find himself losing customers.

Lesson 18: World Competition

The airplane business is an international business.

In the race to acquire the most rewarding aeronautical technology, the prize will go to the nation and/or to the manufacturer who runs the fastest. The nation and/or the manufacturer who stops running will lose, regardless of lead or government protection.

Lesson 19: Design Emphasis

In future designs, the most constructive emphasis is to build a superior product at the lowest manufacturing cost.

10. REFERENCES

1. Roskam, J., Airplane Design: Part I, Preliminary Sizing of Airplanes.

2. Roskam, J., Airplane Design: Part II, Preliminary Configuration Design and Integration of the Propulsion System.

3. Roskam, J., Airplane Design: Part III, Layout Design of Cockpit, Fuselage, Wing and Empennage: Cutaways and Inboard Profiles.

4. Roskam, J., Airplane Design: Part IV, Layout Design of Landing Gear and Systems.

5. Roskam, J., Airplane Design: Part V, Component Weight Estimation.

6. Roskam, J., Airplane Design: Part VI, Preliminary Calculation of Aerodynamic, Thrust and Power Characteristics.

7. Roskam, J., Airplane Design: Part VII, Determination of Stability, Control and Performance Characteristics: FAR and Military Requirements.

Note: These books are all published by: Roskam Aviation and Engineering Corporation, Rt4, Box 274, Ottawa, Kansas, 66067, Tel. 913-2421624.

8. Blanchard, B.S., Design and Manage to Life Cycle Cost, M/A Press, Portland, Oregon, 1978.

9. Anon., World Aviation Directory, Published annually by Ziff-Davis Publishing Company, Washington, D.C.

10. Earles, M.E., Factors, Formulas and Structures for Life Cycle Costing, Eddins-Earles, 89 Lee Drive, Concord, Mass, 01742.

11. Levenson, G.S., Boren, H.E., Tihansky, D.P. and Timson, F., Cost Estimating Relationships for Aircraft Airframes, Rand Report R-761-PR, February 1972, The Rand Corporation, Santa Monica, California.

12. Reel, R.E., Totey, C.E. and Johnson, W.L., Weapon System Support Cost Reduction Study (U), ASD/XR Report 72-49, ASD/XRG, WPAFB, Ohio, June 1972.

13. Johnson, W.L. and Reel, R.E., Maintainability/Reliability Impact on System Support Costs, AFFDL-TR-73-152, AFFDL/PTC, WPAFB, Ohio, December 1973.

14. Nicolai, L.M., Fundamentals of Aircraft Design, METS Inc., 6520 Kingsland Court, California, 95120.

15. Anon., Standard Method of Estimating Comparative Direct Operating Costs of Transport Airplanes, Air Transport Association of America report, June 1960.

16. Anon., Standard Method of Estimating Comparative Direct Operating Cost of Turbine Powered Transports, Air Transport Association of America report, December 1967.

17. Anon., Standard Method for the Estimation of Direct Operating Costs of Aircraft, Society of British Aircraft Constructors report, November 1959.

18. Maddalon, D.V., Estimating Airline Operating Costs, p.849-870, CTOL Transport Technology, NASA N78-29046, 1978.

19. Anderson, J.L. and Andrastek, D.A., Operating Cost Model for Local Service Airlines, SAWE Paper 1098, Society of Allied Weight Engineers, 1976.

20. Coykendall, R.E. et al, Study of Cost/Benefit Tradeoffs for Reducing the Energy Consumption of the Commercial Air Transportation System, NASA CR-137891, 1976.

21. Vos, R. et al, United States Air Transportation 1980, NASA/WVU Summer Predoctoral Program in Engineering Systems Design, West Virginia University, Morgantown, West Virginia, 1980.

22. Ball, R.E., The Fundamentals of Aircraft Combat Survivability and Design, AIAA Education Series, 1985, American Institute of Aeronautics and Astronautics, Washington, D.C.

23. Large, J.P., Campbell, H.G. and Cates, D., Parametric Equations for Estimating Airframe Costs, Rand Report R-1693-1-PA and E, February 1976, The Rand Corp., Santa Monica, California.

24. Raymer, D.P., Aircraft Design: A Conceptual Approach, AIAA Education Series, 1989, American Institute of

Aeronautics and Astronautics, Washington, D.C.

25. Augustine, N., Augustine's Laws, American Institute of Aeronautics and Astronautics, Wash., D.C., 1983.

26. Gebman, J., Avionics Maintainability: More Important Than Reliability, AIAA Paper 89-2096, 1989.

27. Kemp, A.H., Effective Integration of Supportability Design Criteria into Computer Aided Design for the Conceptual Design Phase, AIAA Paper 88-4473, 1988.

28. Lerner, E.J., Avionics Unreliability Turns Fighters into Shop Queens, Aerospace America, August 1985.

29. Weinstock, R., Calculus of Variations, McGraw Hill, N.Y., 1952.

30. Leitmann, G., Optimization Techniques with Applications to Aerospace Systems, Academic Press, N.Y., 1962.

31. Roskam, J., What Drives Unique Configurations?, SAE Paper 881353, Aerospace Technology Conference and Exposition, Anaheim, CA, Oct.3-6, 1988.

32. Roskam, J. and Malaek, S., Automated Aircraft Configuration Design and Analysis, SAE Paper 891072, General Aviation Aircraft Meeting and Exposition, Wichita, KS, April 11-13, 1989.

33. Bisplinghof, R.L., Ashley, H. and Halfman, R., Aeroelasticity, Addison Wesley, 1955.

34. Kenyon, R.E. and Mueller, R.N., Cost-Weight Interface in Aircraft Design, Paper No.969, 32nd Annual Conference of the Society of Allied Weight Engineers, London, England, June, 1973.

35. Roskam, J., Airplane Flight Dynamics and Automatic Flight Controls, Pt.I, Roskam Aviation and Engineering Corp., Rt.4, Box 274, Ottawa, KS, 66067, 1982.

36. Roskam, J., Airplane Flight Dynamics and Automatic Flight Controls, Pt.II, Roskam Aviation and Engineering Corp., Rt.4, Box 274, Ottawa, KS, 66067, 1982.

37. Sliwa, S.M., Use of Constrained Optimization in the Conceptual Design of a Medium-Range Subsonic Transport, NASA Technical Paper 1762, December 1980.

38. Torenbeek, E., Fundamentals of Conceptual Design Optimization of Subsonic Transport Aircraft, Report LR-292, Delft University of Technology, The Netherlands, August, 1980.

39. Küchemann, D., The Aerodynamic Design of Aircraft, Pergamon Press, Oxford, England, 1978.

40. Sears, W.R., On Projectiles of Minimum Wave Drag, Quarterly of Applied Mathematics, Vol.4, No.4, 1947.

41. Jones, R.T., The Minimum Drag of Wings in Frictionless Flow, Jornal of the Aeronautical Sciences, No.18, pp 75-81, 1951.

42. Sliwa, S.M. and Arbuckle, D., OPDOT: A Computer Program for the Optimum Preliminary Design of a Transport Airplane, NASA TM 81857, September, 1981.

43. Sobieszczanski-Sobieski, J., Berthelemy, J.F. and Riley, K.M., Sensitivity of Optimum Solutions to Problem Parameters, AIAA Journal, Volume 20, No.9, p.1291, September 1982.

44. Wrenn, G.A. and Dovi, A.R., Multilevel Decomposition Approach to the Preliminary Sizing of a Transport Aircraft Wing, Journal of Aircraft, Volume 25, No.7, pp 632-638, July, 1988.

45. Sobieszczanski-Sobieski, J., On the Sensitivity of Complex, Internally Coupled Systems, AIAA Paper 88-2378, 29th Structures, Structural Dynamics and Materials Meeting, April 18-20, Williamsburg, VA, 1988.

46. Sobieszczanski-Sobieski, J., Multidisciplinary Optimization for Engineering Systems: Achievements and Potential, DFVLR Symposium on Optimization, Bonn, West Germany, June 1989.

47. Sobieszczanski-Sobieski, J., Barthelemy, J.F. and Giles, G.L., Aerospace Engineering Design by Systematic Decomposition and Multilevel Optimization, Int'l Council of Aeronautical Sciences, 14th Congress, Toulouse, France, September 1984, ICAS-84-4.7.3.

48. Bennett, J.H., Kills, Sorties and Dollars, AIAA Paper 89-2073, AIAA/AHS/ASEE Aircraft Design, Systems and Operations Conf., Seattle, WA, July 31 - Aug.2, 1989.

49. Bach, D., United States Air Force Trainer Masterplan, SAE Paper 881466, Aerospace Technology Conference and

Exposition, Anaheim, CA, October 3-6, 1988.

50. Steiner, J.E. et al, Case Study in Aircraft Design: The Boeing 727, AIAA Professional Study Series, September, 1978.

51. Fozard, J.W., Case Study in Aircraft Design: The British Aerospace Harrier, AIAA Professional Study Series, July, 1978.

52. Stuart, W.G., Case Study in Aircraft Design, The Northrop F-5, AIAA Professional Study Series, September, 1978.

53. Garrard, W.C., Case Study in Aircraft Design: The Lockheed C-5, AIAA Professional Study Series, not dated.

54. Steiner, J.E., How Decisions Are Made: Major Considerations for Aircraft Programs, AIAA 1982 Wright Brothers Lectureship in Aeronautics, Seattle, WA, 1982.

55. Anderson, J.L., A Parametric Determination of Transport Aircraft Price, SAWE Paper 1071, 34th Annual Conference of the Society of Allied Weight Engineers, Seattle, Washington, May 5-7, 1975.

56. Muehlbauer, J.C. et al, Turboprop Cargo Aircraft Systems Study, Phase I, NASA CR 159355, Lockheed Georgia Company, November 1980.

57. Lays, E.J. and Murray, D.L., General Aviation Turbine Engine (GATE) Study, Final Report, NASA CR 159603, Williams Research Corporation, June, 1979.

58. Renze, P.P. and Terry, J.E., Application of Advanced Technologies to Derivatives of Current Small Transport Aircraft, Final Report, NASA CR 166197, Beech Aircraft Corporation, July, 1981.

59. Keiter, I.D., Advanced Propeller Technology Examined, Automotive Engineering, Volume 89, Number 7, 1981.

60. Anon., Cost Trends for Fighter Avionics, Advertisement in: Aviation Week and Space Technology, p.132, December 2, 1985.

61. Spearman, L., Some Fighter Aircraft Trends, AIAA Paper 84-2503, AIAA Aircraft Design, Systems and Operations Meeting, Oct./Nov. 1988, San Diego, CA.

Airplane Servicing Arrangement 757

Flight Turnaround

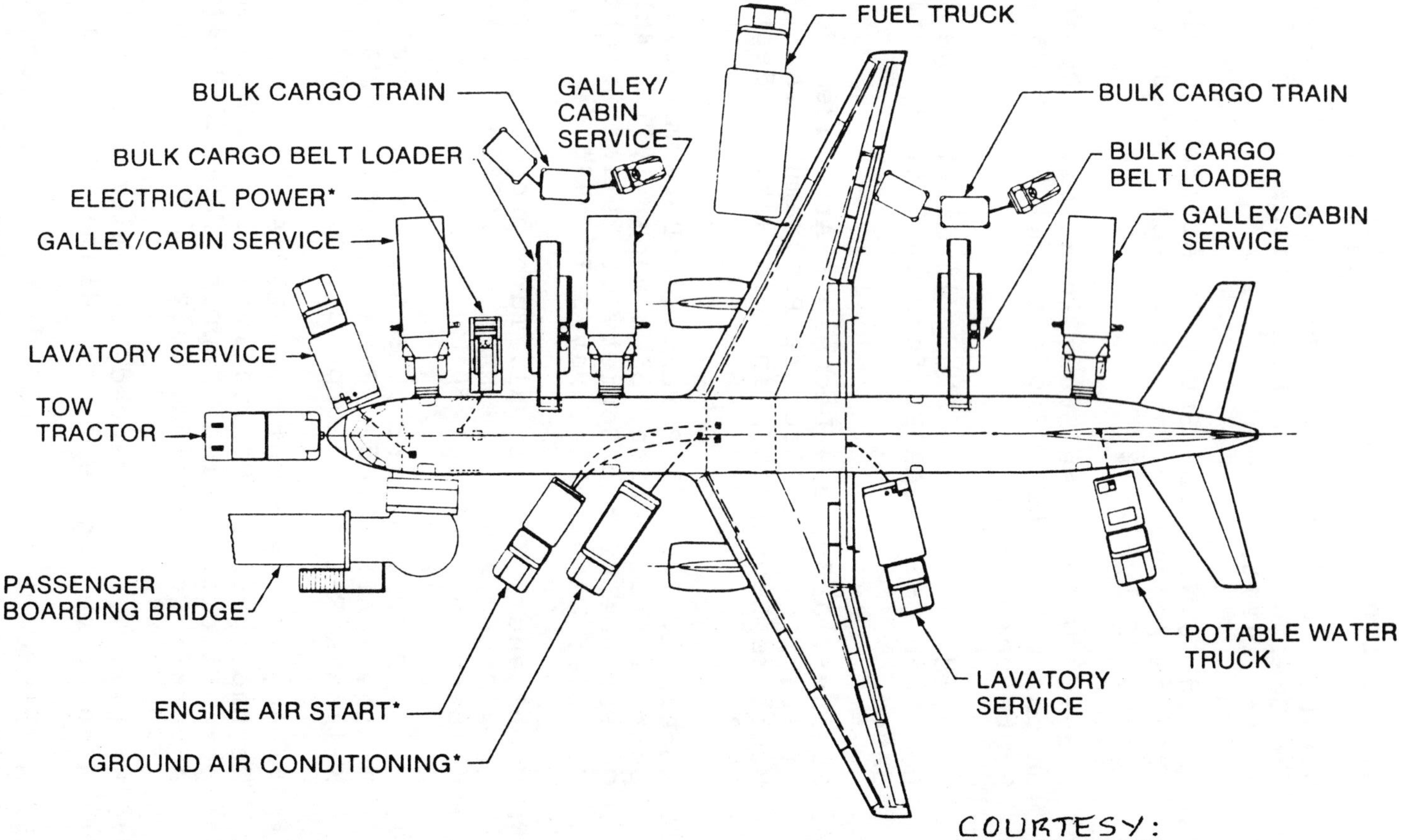

*ONLY REQUIRED WHEN APU IS NOT BEING USED

COURTESY:
BOEING

11. INDEX

Nacelle Installation 757

CF6-32C

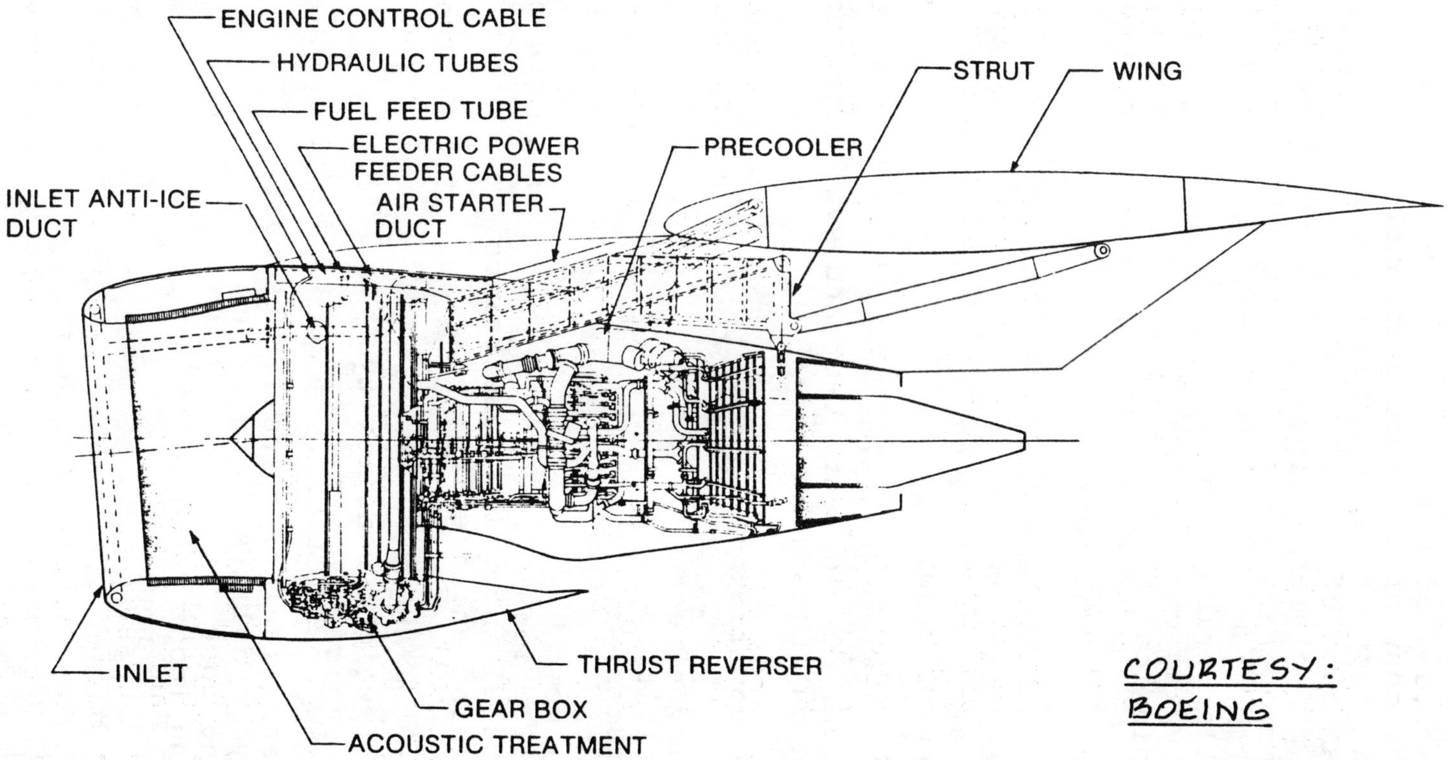

COURTESY: BOEING

ACOUSTIC TREATMENT

APPENDIX A: AIRPLANE PRICE DATA

The purpose of this appendix is to provide airplane price data as well as a rapid method for 'ball-parking' airplane prices for future designs. Most of the information presented was obtained from the following sources:

1. Business and Commercial Aviation: this monthly magazine publishes an annual Planning and Purchasing Handbook, usually as part of the April or May issue.

2. Aviation Week and Space Technology: this weekly magazine publishes an annual Aerospace Forecast and Inventory issue, usually as part of the March issue.

3. Flight International: this British weekly magazine publishes regular overviews of airplane price data.

4. AOPA Pilot: this monthly magazine publishes a general aviation aircraft directory usually as part of its March issue.

5. Professional Pilot Magazine: this monthly magazine publishes regular listings of airplane prices.

6. Interavia: this Swiss monthly magazine contains regular overviews of airplane price data.

Definition: The amount of dollars charged to an airplane operator to acquire that airplane is called the airplane market price: AMP.

Airplane prices vary with the following factors:

* Takeoff weight
* Performance
* Technology features
* Manufacturing cost
* Calendar time (if only because of cost escalation)
* Market conditions (demand and competition)
* Manufacturer sales strategies
* Government rules and regulations (certification standards, politics)
* Legal issues (product liability)

Except for the effect of cost escalation, no easy

rules for predicting the effect of these factors on airplane price can be given. Since all airplane price data are 'dated' by calendar year, the following equation is suggested to allow for cost escalation:

$$AMP_{19XX} = AMP_{then\ year}(CEF_{19XX}/CEF_{then\ year}) \qquad (A1)$$

where: AMP_{19XX} is the desired airplane price in 19XX

$AMP_{then\ year}$ is the given airplane price in 'then year' dollars

CEF is the cost escalation factor obtained from Figure 2.7. The ratio $(CEF_{19XX}/CEF_{then\ year})$ is called the CEF-ratio.

Sources 1-6 provide price data for many different calendar years. Therefore, Eqn.(A1) was used to 'correct' all data to 1989 dollars.

Airplane prices can be correlated with many factors. The most common factors used are the takeoff weight and the number of seats. In this appendix airplane prices are correlated with takeoff weight only. A brief discussion of other price parameters is given in Section A11. The airplane price data are organized as follows:

A1 SAILPLANE PRICE DATA

A2 ULTRALIGHT AIRPLANE PRICE DATA

A3 AGRICULTURAL AIRPLANE PRICE DATA

A4 SINGLE ENGINE PISTON AIRPLANE PRICE DATA

A5 MULTIENGINE PISTON AIRPLANE PRICE DATA

A6 MULTIENGINE TURBOPROP AIRPLANE PRICE DATA

A7 BUSINESS JET AIRPLANE PRICE DATA

A8 TURBOPROP COMMUTER AIRPLANE PRICE DATA

A9 COMMERCIAL TRANSPORT PRICE DATA

A10 MILITARY AIRPLANE PRICE DATA

A11 A GENERAL LOOK AT AIRPLANE PRICE DATA

A1 SAILPLANE PRICE DATA

Sailplane prices are plotted as a function of their takeoff weight in Figure A1. These data are based on information from the March, 1985 issue of AOPA Pilot Magazine. The data were adjusted to 1989 dollars with Equation (A1) and a CEF ratio of 1.082 obtained from Fig.2.7.

It is seen that there is a large spread in the data. The trend line is seen to 'fit' the data (+/- 25 percent) for unpowered as well as powered sailplanes. Based on this trend line the following equation is suggested for 'ballparking' sailplane prices:

$$AMP_{1989} = \text{invlog}\{1.1592 + 1.0705(\log W_{TO})\} \quad \text{(A2)}$$

where: AMP_{1989} is the predicted market price for sailplanes in 1989 dollars.

This equation applies to the following range of takeoff weights: 600 lbs $< W_{TO} <$ 2,000 lbs and should NOT be used outside that range.

A2 ULTRALIGHT AIRPLANE PRICE DATA

The March, 1985 issue of AOPA Pilot Magazine contains a lengthy listing of ultralight prices. Because these aircraft are usually delivered in the form of a kit (raw materials or pre-cut) most prices apply to the unassembled vehicle. There did not seem to be a reasonable correlation between prices and takeoff weights. Based on the aforementioned data, the following ranges are suggested for 'ballparking' ultralight airplane prices in 1989 dollars:

Pre-cut kits: $AMP_{1989} = 5(W_{TO})$ to $10(W_{TO})$ (A3)

Raw material kits: $AMP_{1989} = 3(W_{TO})$ to $10(W_{TO})$ (A4)

According to the data, ultralight takeoff weights range from 400 to 800 lbs. Note: ultralights are not certified to FAR standards.

A3. AGRICULTURAL AIRPLANE PRICE DATA

Prices for agricultural airplanes are plotted as a function of takeoff weight in Figure A2. These data are based on information from the March, 1984 issue of AOPA Pilot Magazine and from the October, 1984 issue of Agricultural Aviation Magazine. The data were adjusted to 1989 dollars using Equation (A1). The CEF ratio of 1.119 was obtained from Figure 2.1.

Some of the price data were judged to be unrealistically low by the author. These 'suspect' data are identified in Figure A2. After disregarding these data, a regression analysis was performed. This resulted in the following formula for 'ballparking' prices of agricultural airplanes:

$$AMP_{1989} = \text{invlog}\{-0.6681 + 1.5799\log(W_{TO})\} \qquad \text{(A5)}$$

where: AMP_{1989} is the predicted market price for agricultural airplanes in 1989 dollars.

This equation applies to the following range of takeoff weights: 3,000 lbs $< W_{TO} <$ 15,000 lbs and should NOT be used outside that range.

It is interesting to observe that engine type does not seem to have a significant effect on agricultural airplane price trends.

A4. SINGLE ENGINE PISTON AIRPLANE PRICE DATA

The price of single engine piston airplanes depends on the following factors:

* engine aspiration: normal or turbocharged
* cabin pressurization (most are not!)
* takeoff weight and performance
* product liability insurance

Figure A3 shows the calendar time history of a single engine piston training airplane compared to the CEF. The price of this airplane compares reasonably with CEF trends until about 1982 when product liability became a major factor in airplane pricing. In 1985 Cessna decided to discontinue the production of this type: the price had become unreasonably high!

Price data for single engine piston airplanes were taken from the March, 1989 Planning and Purchasing Handbook issue of Business and Commercial Aviation Magazine. The data are plotted in Figure A4. Two trend lines suggest themselves: one for airplanes with normally aspirated engines and the other for airplanes with turbocharged engines. Note, that the pressurized Piper Malibu follows the latter trend line. Also note, that the only (in 1989) certified single engine turboprop airplane (the Cessna Caravan I) follows the normally aspirated trend line.

<u>Note:</u> the Lake-4 amphibian and the Beech BE-A36 were excluded from the data base because of the 'specialty nature of these airplanes. The Piper Super Cub was added to the data as follows: W_{TO} = 1,750 lbs at USD 45,000.

A regression analysis was performed to determine the trend line equations which may be used for 'ballparking' single engine piston and single engine turboprop prices.

For single engine airplanes with <u>normally aspirated piston engine or with a turboprop engine</u>:

$$AMP_{1989} = \text{invlog}\{-1.2435 + 1.8459(\log W_{TO})\} \qquad (A6)$$

For single engine airplanes with <u>turbocharged piston engines</u> (with and without cabin pressurization):

$$AMP_{1989} = \text{invlog}\{-1.1174 + 1.8459(\log W_{TO})\} \qquad (A7)$$

where: AMP_{1989} is the predicted airplane market price for single engine airplanes in 1989 dollars.

These equations apply inside the following range of takeoff weights: $1,500 < W_{TO} < 10,000$ and should NOT be used outside that range.

A5. MULTIENGINE PISTON AIRPLANE PRICE DATA

Airplane price data for 1989 and 1980 were obtained from the Planning and Purchasing Handbook (March issue of Business and Commercial Aviation Magazine). The data for 1980 were adjusted to 1989 with a CEF-ratio of 2.045 obtained from Figure 2.7. The resulting price data are plotted in Figure A5. Observe that the effect on price of turbocharging (over normally aspirated engines) and of cabin pressurization (over unpressurized) is negligible.

The author decided to eliminate the relatively very

highly priced Partenavia airplanes from the data before determining the trend line (See Figure A5) with a regression analysis. As a result, the following equation can be used for 'ballparking' prices for multiengine piston airplanes:

$$AMP_{1989} = \text{invlog}\{-0.8526 + 1.7413(\log W_{TO})\} \qquad \text{(A8)}$$

where: AMP_{1989} is the predicted market price for multiengine piston airplanes in 1989 dollars.

This equation applies to the following range of takeoff weights: 3,000 lbs < W_{TO} < 8,000 lbs and should NOT be used outside that range.

A6. MULTIENGINE TURBOPROP BUSINESS AIRPLANE PRICE DATA

This airplane category consists of airplanes which are configured for business and NOT for airline use. The 1989 Planning and Purchasing Handbook of Business and Commercial Aviation Magazine (March issue) was used to plot the price data of Figure A6. A regression analysis was performed on these data. From this, the following equation was derived, which may be used for 'ballparking' airplane prices:

$$AMP_{1989} = \text{invlog}\{1.9153 + 1.1115(\log W_{TO})\} \qquad \text{(A9)}$$

where: AMP_{1989} is the predicted airplane market price for multiengine turboprop business airplanes in 1989 dollars.

This equation applies to the following range of takeoff weights: 8,000 lbs < W_{TO} < 50,000 lbs and should NOT be used outside this range.

A7 BUSINESS JET AIRPLANE PRICE DATA

This airplane category consists of business airplanes, NOT airline configured airplanes. Price data were taken from the March 1989 issue of the Planning and Purchasing Handbook of Business and Commercial Aviation Magazine. Figure A7 shows a plot of these data. As seen, prices of the Gulfstream GIV and several Boeing airplane adaptations for business usage do not fit a linear trend. After eliminating these airplanes from the data a regression analysis was performed. The result is the following equation which may be used to 'ballpark' airplane prices for this category:

$$AMP_{1989} = \text{invlog}\{0.6570 + 1.4133\log(W_{TO})\} \qquad \text{(A10)}$$

where: AMP_{1989} is the predicted airplane market price for business jets in 1989 dollars.

This equation applies to the following range of takeoff weights: 10,000 lbs $< W_{TO} <$ 60,000 lbs and should NOT be used outside this range.

A8 TURBOPROP COMMUTER AIRPLANE PRICE DATA

Price data for turboprop commuter airplanes were obtained from the March 1989 issue of the Planning and Purchasing Handbook of Business and Commercial Aviation Magazine. The data are plotted in Figure A8. A regression analysis was performed on the data. This resulted in the following equation, which can be used to 'ballpark' airplane prices for this category:

$$AMP_{1989} = \text{invlog}\{1.1846 + 1.2625(\log W_{TO})\} \qquad \text{(A11)}$$

where: AMP_{1989} is the predicted airplane market price for turboprop commuters in 1989 dollars.

These equations apply to the following range of takeoff weights: 6,000 lbs $< W_{TO}$ 50,000 lbs and should NOT be used outside these ranges.

A9 COMMERCIAL JET TRANSPORT PRICE DATA

Commercial jet transport price data were taken from the October 10, 1987 issue of Flight International (British weekly aviation magazine). A CEF ratio of 1.0237 was applied to adjust the data to 1989 dollars. The adjusted data are plotted in Figure A9. The trend line was determined with a regression analysis which yielded the following equation:

$$AMP_{1989} = \text{invlog}\{3.3191 + 0.8043(\log W_{TO})\} \qquad \text{(A12)}$$

where: AMP_{1989} is the predicted airplane market price for jet transports in 1989 dollars.

This equation can be used to obtain a 'ballpark' estimate of jet transport prices. The equation applies to the following range of takeoff weights:

$60,000 \text{ lbs} < W_{TO} < 1,000,000 \text{ lbs}$ and should NOT be used outside this range.

A10 MILITARY AIRPLANE PRICE DATA

Prices for military airplanes vary with the following factors:

* takeoff weight and performance
* types of technology included
* mission orientation (weapons capability)
* procurement politics

Price data for military airplanes are published on occasion in sources such as magazines number 2, 3 and 6 mentioned on page 297. The author has collected price data (in the form of unit fly-away prices without spares) from these sources and applied the appropriate CEF-ratio to adjust these data to 1989 dollars. Figure A10 represents a plot of these data. It is noted that military airplane prices vary 'all over the lot' when plotted in this manner. After eliminating the price data for what the author considers 'old technology' fighters, a regression analysis was performed and a trend line established. Except for the T-45 (which appears to be extremely expensive) the trend line follows the data rather well. Thus, the following equation is suggested for 'ballparking' prices for military airplanes:

$$AMP_{1989} = \text{invlog}\{2.3341 + 1.0586(\log W_{TO})\} \qquad \text{(A13)}$$

where: AMP_{1989} is the predicted airplane market price for military airplanes in 1989 dollars.

This equation applies within the following range of takeoff weights: $2,500 \text{ lbs} < W_{TO} < 1,000,000 \text{ lbs}$ and should NOT be used outside that range.

A11 A GENERAL LOOK AT AIRPLANE PRICE DATA

Figure A11 presents a plot of all price-versus-takeoff-weight equations discussed in Sections A1 - A10. The suggested ranges of takeoff weight values are indicated with large dots. Note the two <u>overall trend</u> lines drawn in Figure A11. As seen, business jets and military airplanes differ from these trend lines significantly. It seems that, roughly speaking, the business jets and the military airplanes are 2 to 3.5 times as expensive as all other types of airplane.

An interesting question is: How have airplane prices varied with calendar years in 'then yaer' dollars? The answer may be found in Figures 9, 10 and 15 of Ref.25. The author has taken the trend lines from Figures 9, 10 and 15 of Ref.25 and derived the following equations:

<u>For tactical airplanes (Fig.9, Ref.25):</u>

$$(AMP_{then\ year}) = \quad (A14)$$

$$= invlog\{-104.8212 + 0.056604(19XX)\}$$

<u>For bombers (Fig.15, Ref.25):</u>

$$(AMP_{then\ year} = \quad (A15)$$

$$= invlog\{-138.4443 + 0.074074(19XX)\}$$

<u>For commercial transports (Fig.10, Ref.25):</u>

$$(AMP_{then\ year}) = \quad (A16)$$

$$= invlog\{-99.2972 + 0.054054(19XX)\}$$

where: $(AMP_{then\ year})$ is the airplane market price in 'then year' dollars.

19XX is a calendar year anywhere between 1910 and 1990.

Figure A12 presents plots for each equation. Plots of the U.S. annual defense budget and the gross national product are superimposed. Note the following:

I. All airplane prices tend to increase with calendar time. The logarithmic slope of the price trends is larger than a similar slope of economic trends. If that trend is not reversed, airplanes will become unaffordable!

II. The logarithmic slope of bomber price trends is 50 percent greater than that of commercial transports!

In comparing airplane prices with one another, the following parameters are often used:

* Dollars price per pound of takeoff weight
* Dollar price per seat (civilian airplanes only)

Table A1 tabulates the prices per lbs of takeoff weight for all types of airplanes considered in this appendix. Note that, except for the commercial jet transports, the price per lbs increases for all types of airplanes with increasing takeoff weight. The slope constant in Eqns (A2) through (A11) and Eqn.(A13) are all larger than 1.0, but the slope constant for commercial transports (see Eqn.(A12)) is less than 1.0! An explanation for this may be that commercial transports represent a relatively 'mature' technology.

Note again that business jets and military aircraft at the high end of their weight ranges cost about the same in terms of dollars per lbs of takeoff weight!

Table A2 shows the cost per seat for civilian airplanes. It is of interest to observe that the price per seat for transport type airplanes falls in a relatively narrow range: 160,000 to 360,000 dollars/seat. Also note the very high price per seat for the upper end of the business jets!

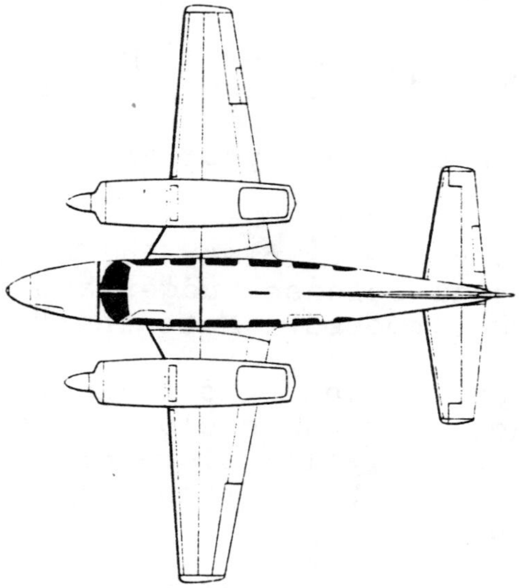

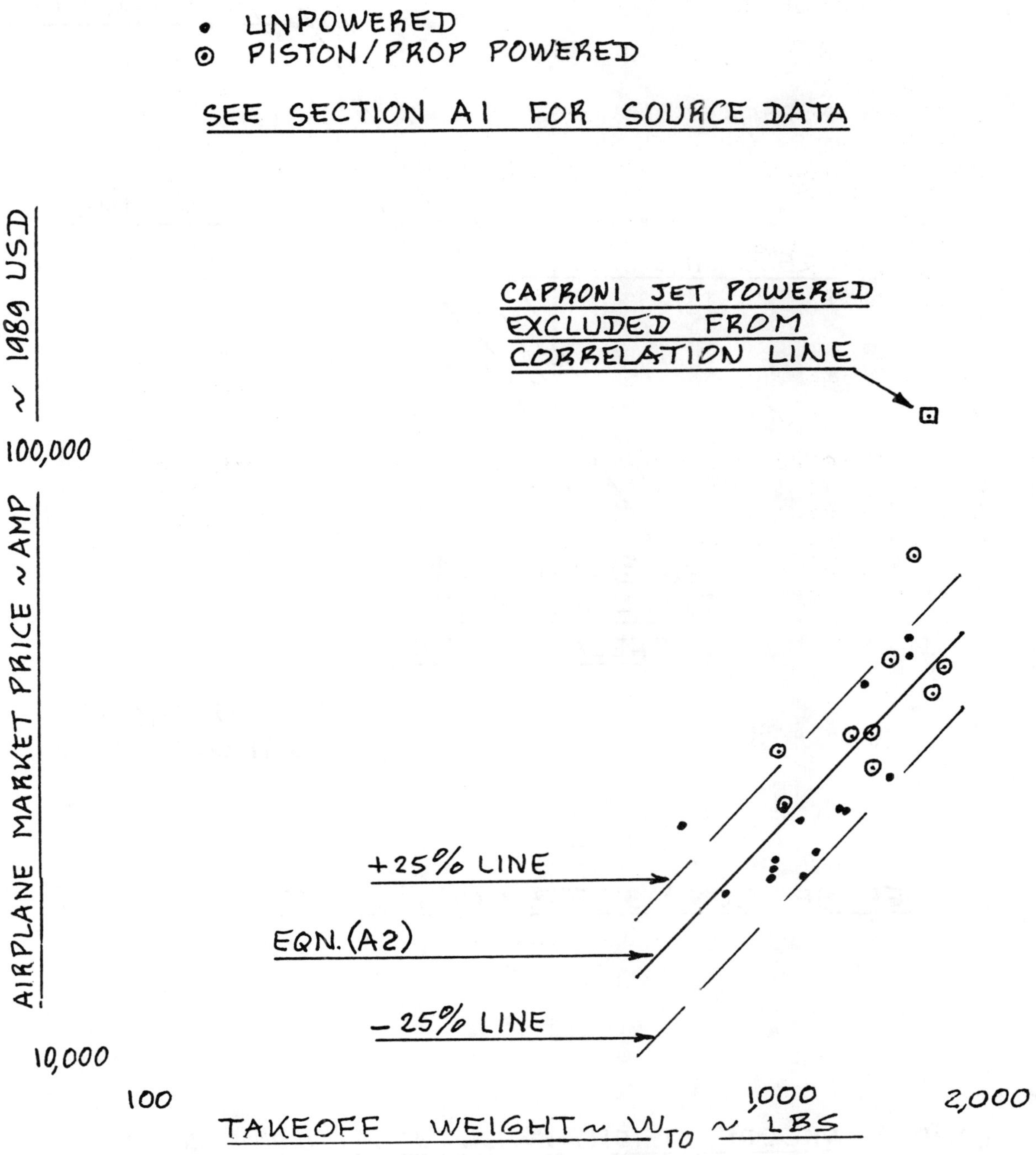

Figure A1 Effect of Takeoff Weight on the Price of Sailplanes

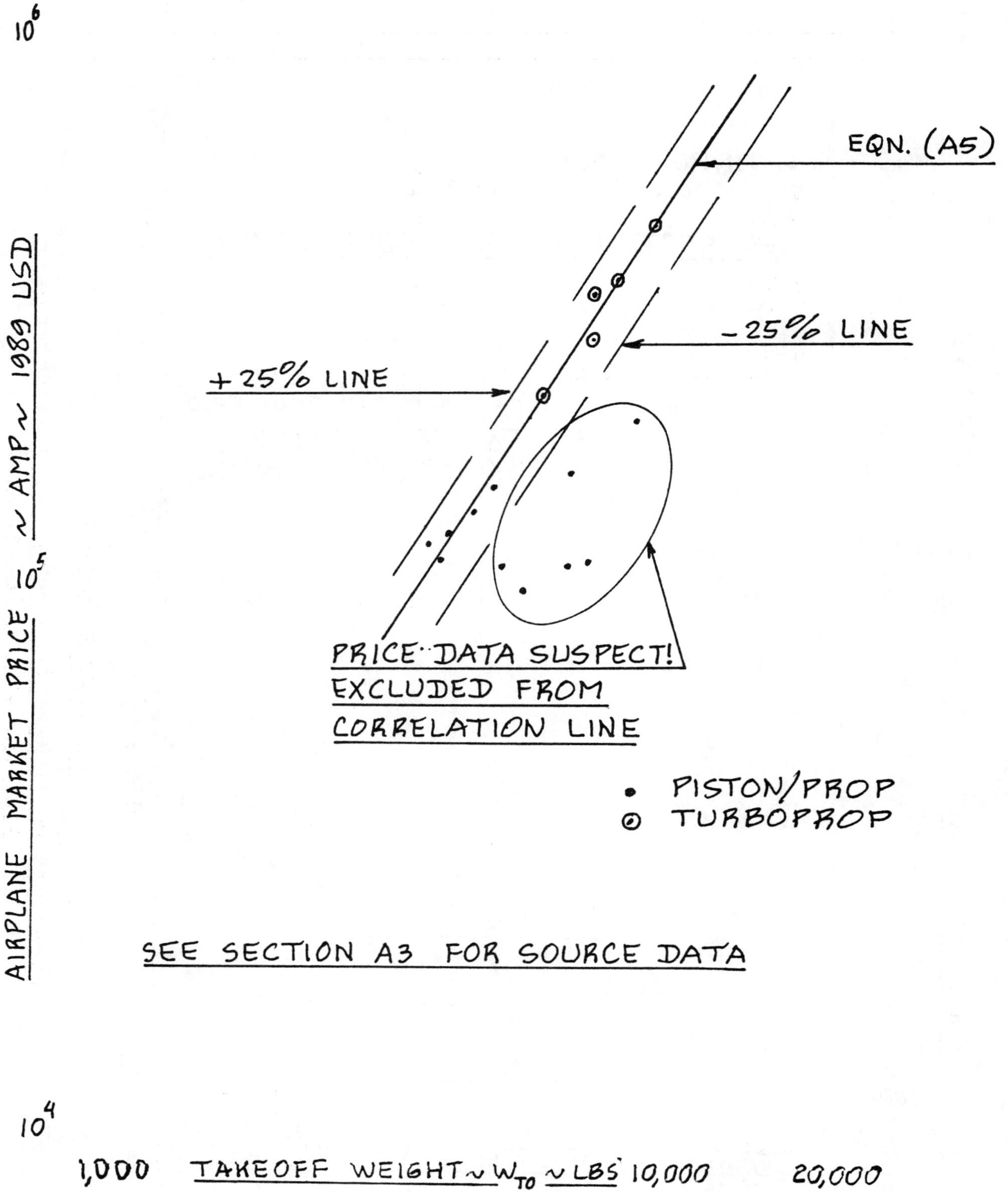

Figure A2 Effect of Takeoff Weight on the Price of Agricultural Airplanes

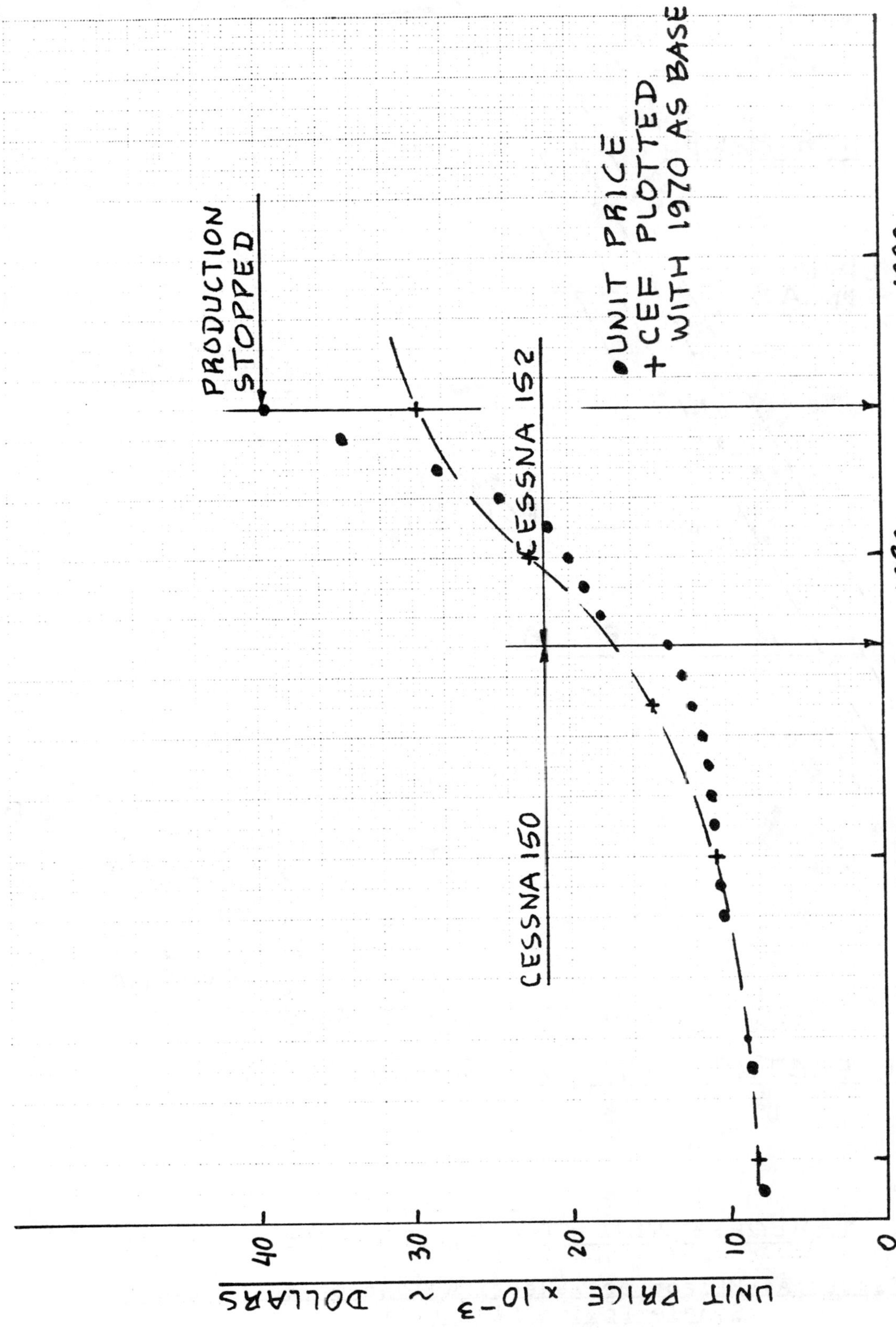

Figure A3 Variation of the Price of a Light Trainer Airplane with Calendar Year

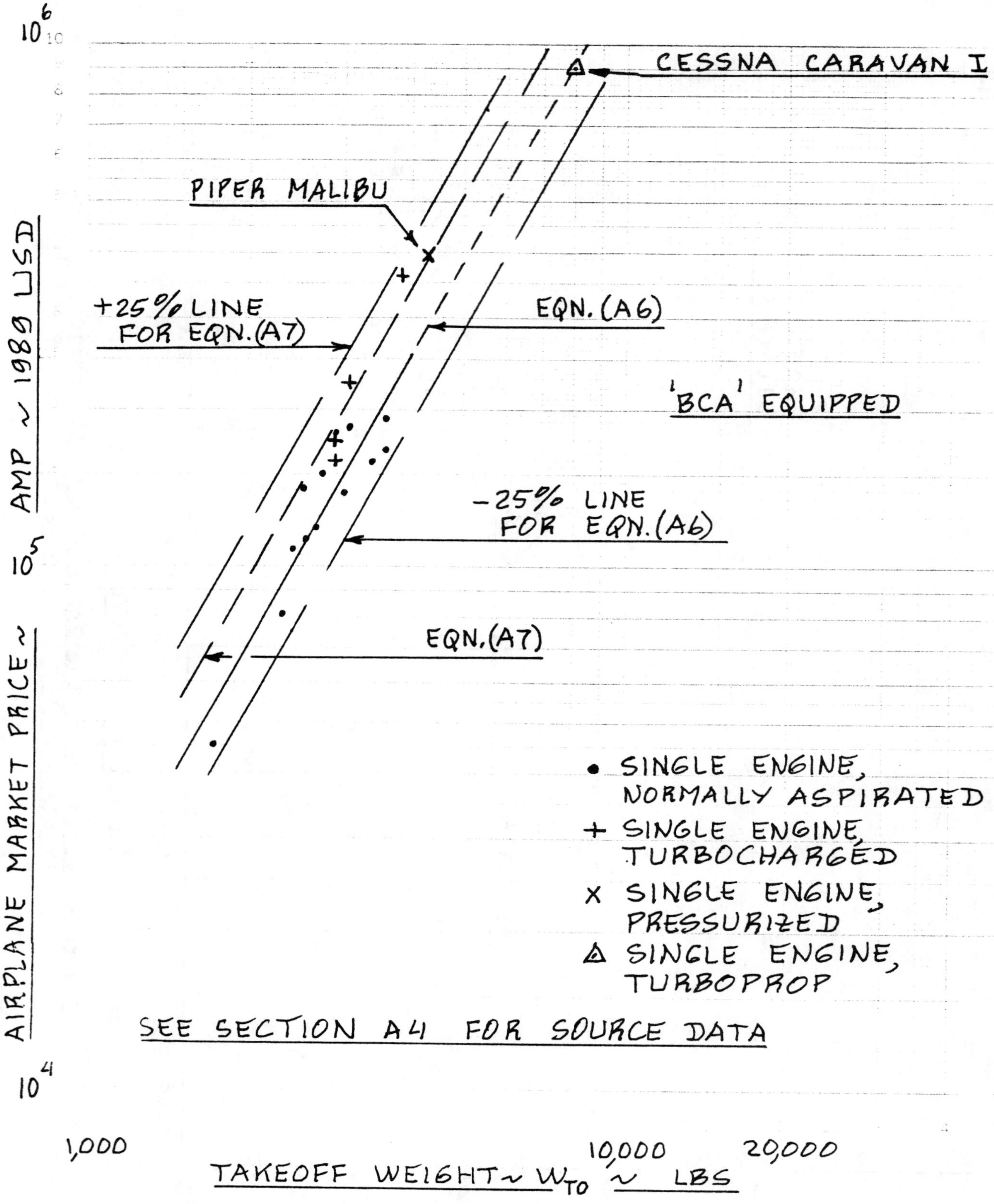

Figure A4 Effect of Takeoff Weight on the Price of Single Engine Piston Airplanes

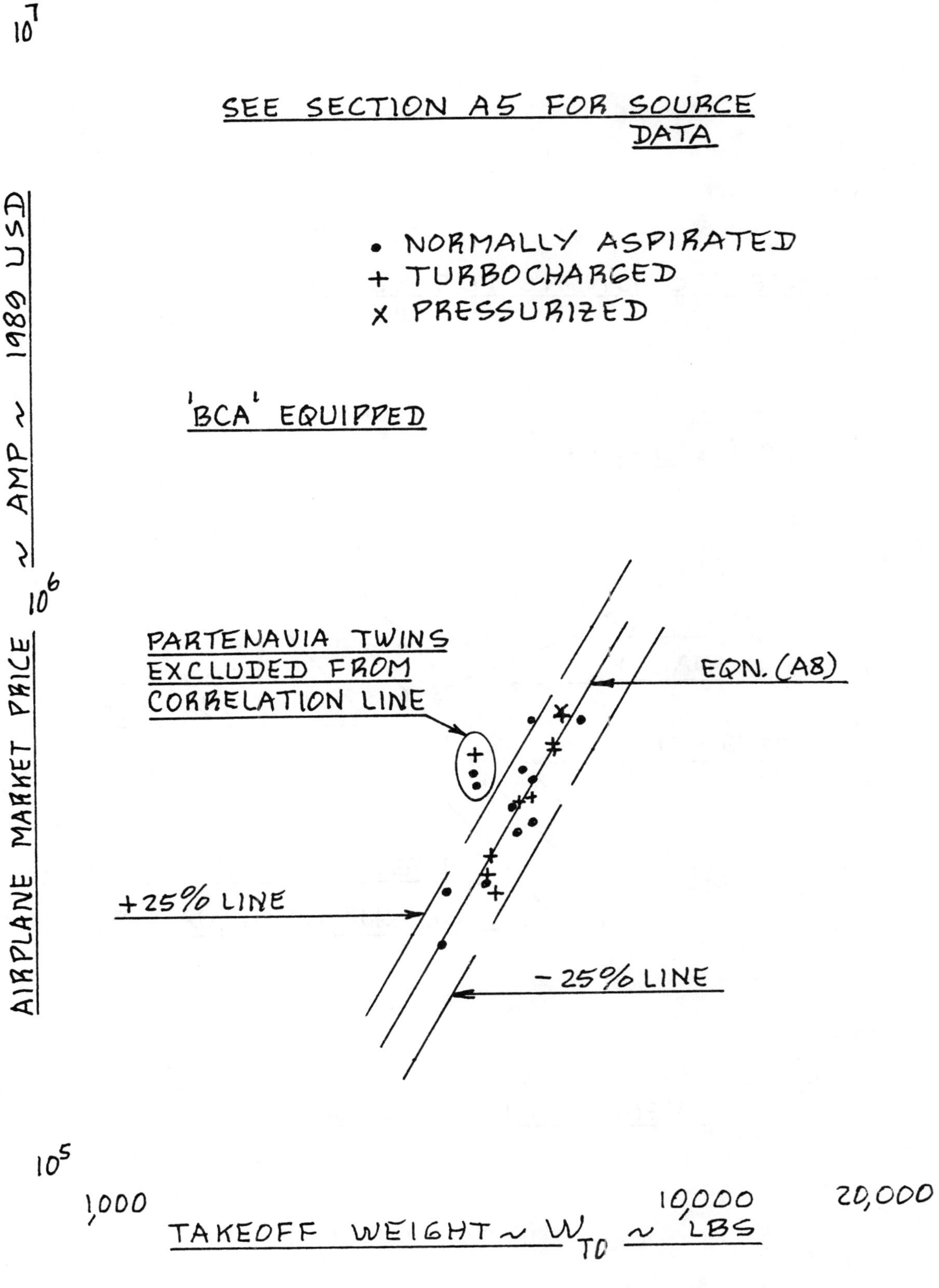

Figure A5 Effect of Takeoff Weight on the Price of Multiengine Piston Airplanes

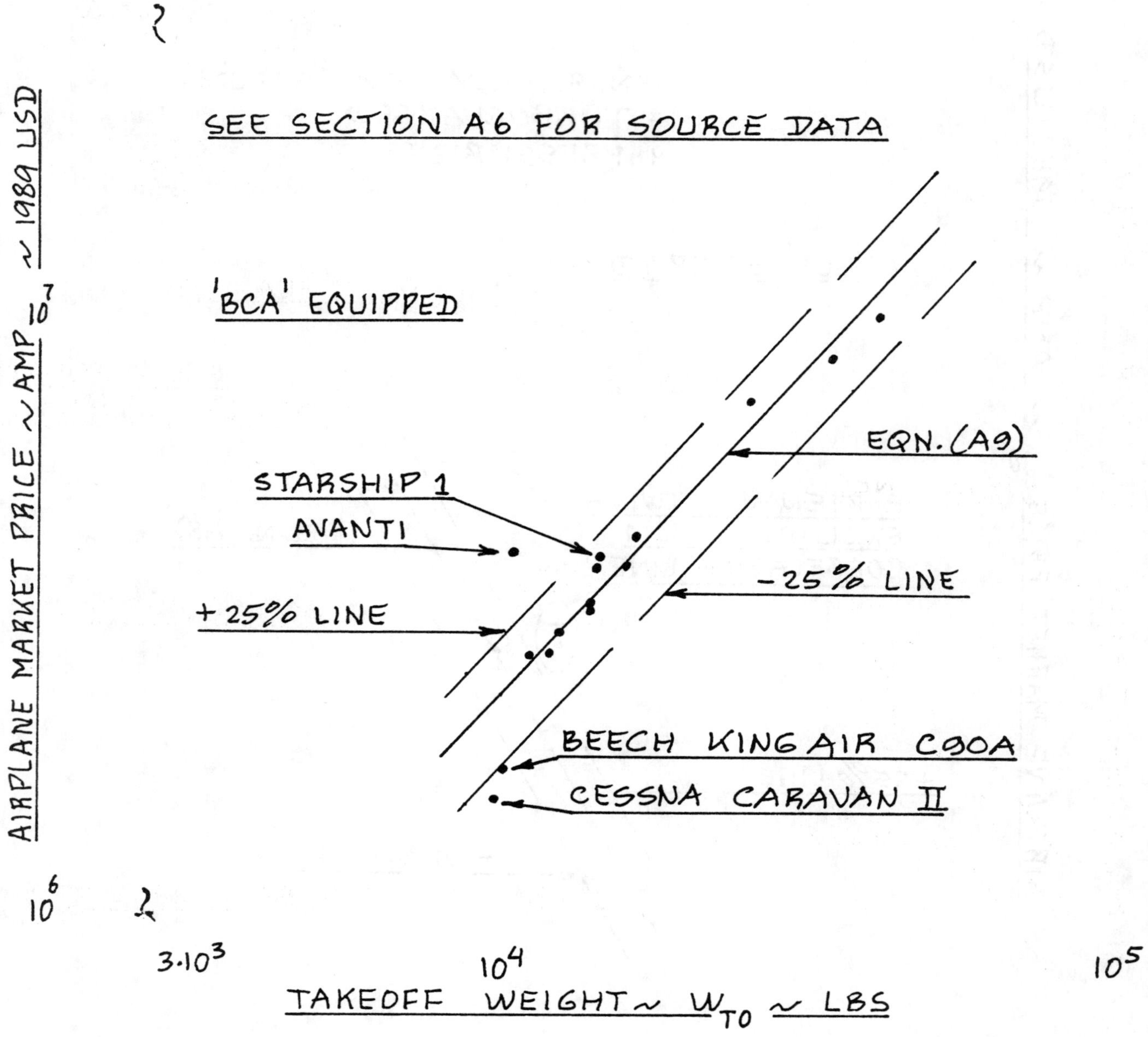

Figure A6 Effect of Takeoff Weight on the Price of Multiengine Turboprop Business Airplanes

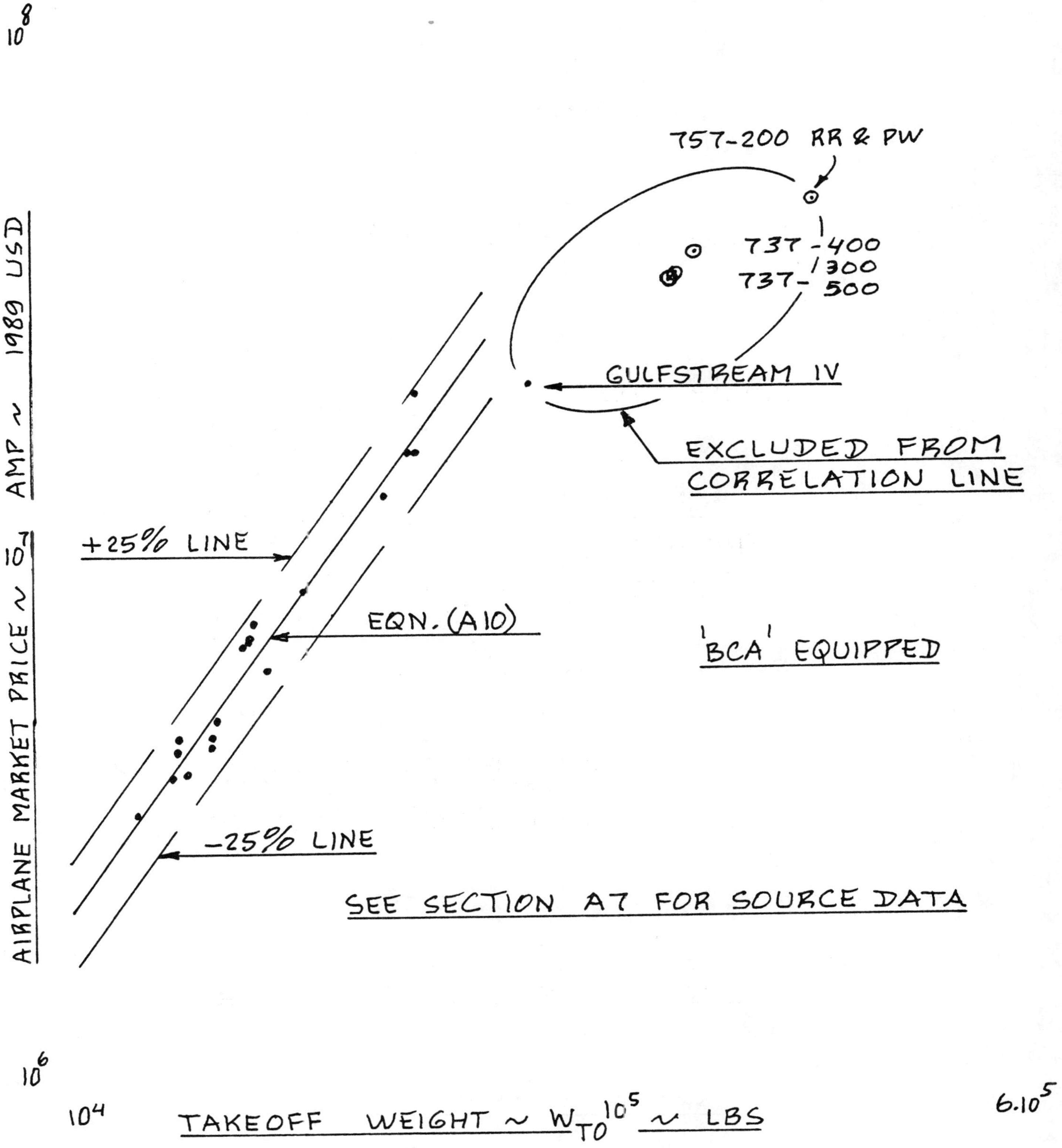

Figure A7 Effect of Takeoff Weight on the Price of Business Jets

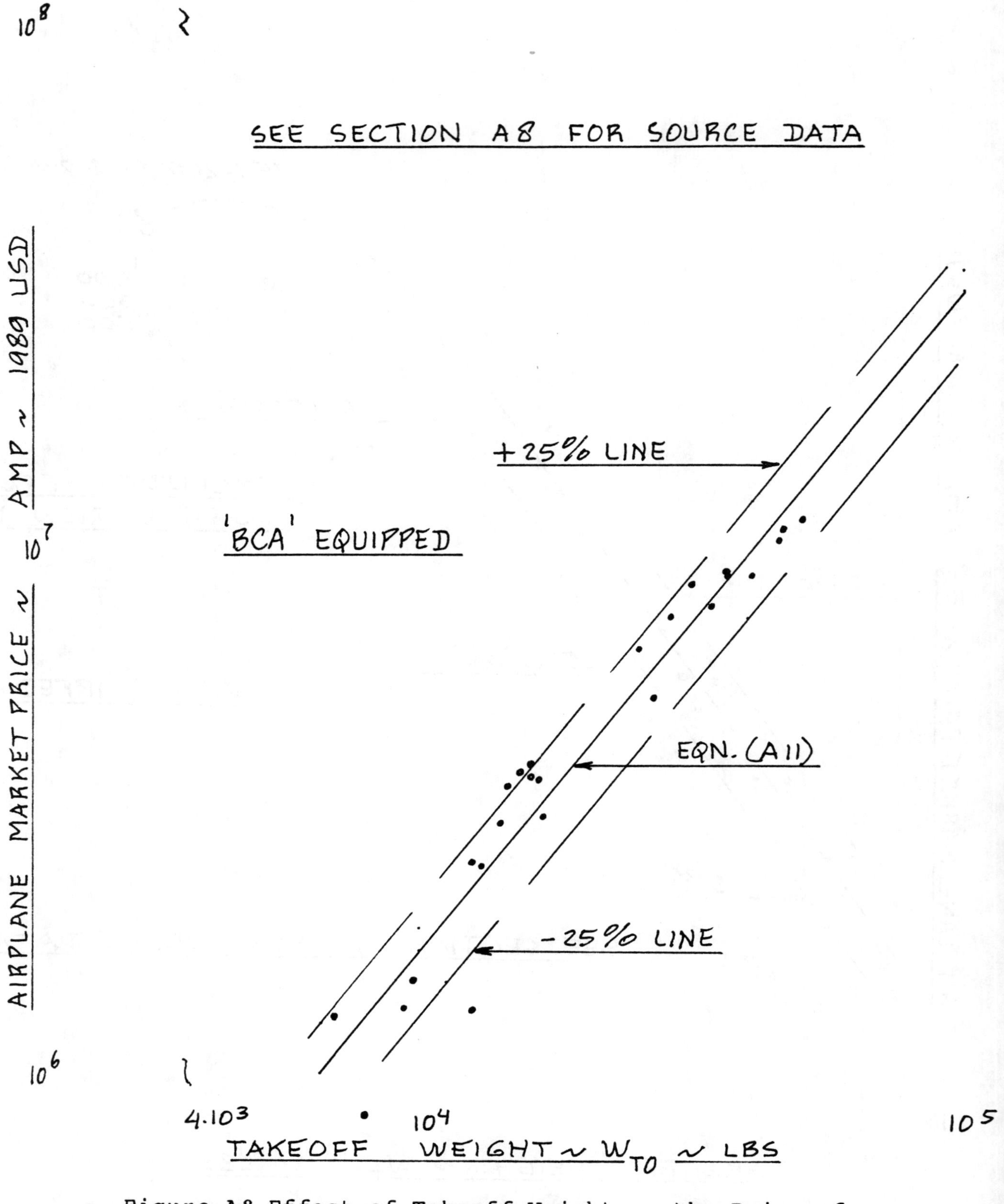

Figure A8 Effect of Takeoff Weight on the Price of Turboprop Commuter Airplanes

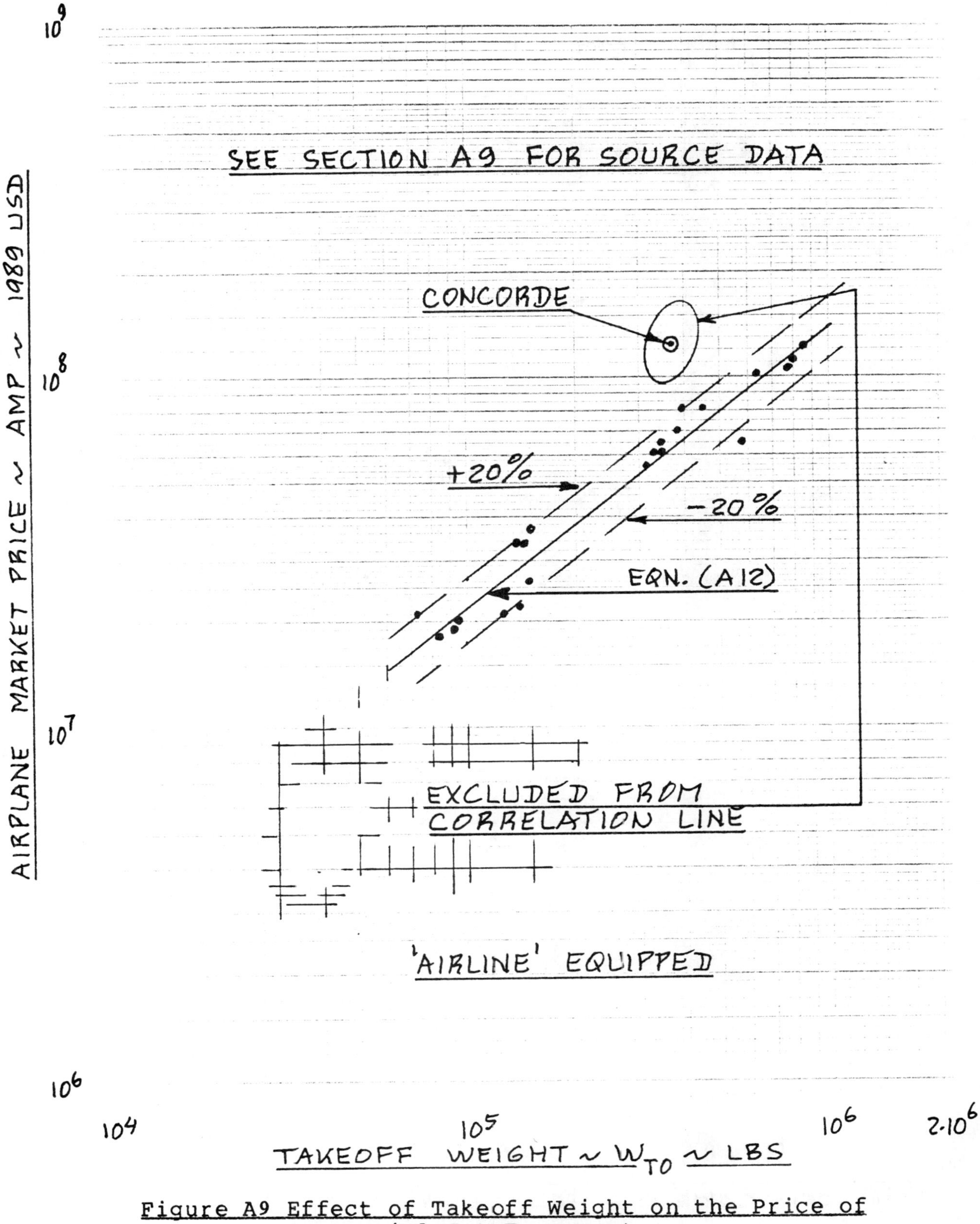

Figure A9 Effect of Takeoff Weight on the Price of Commercial Jet Transports

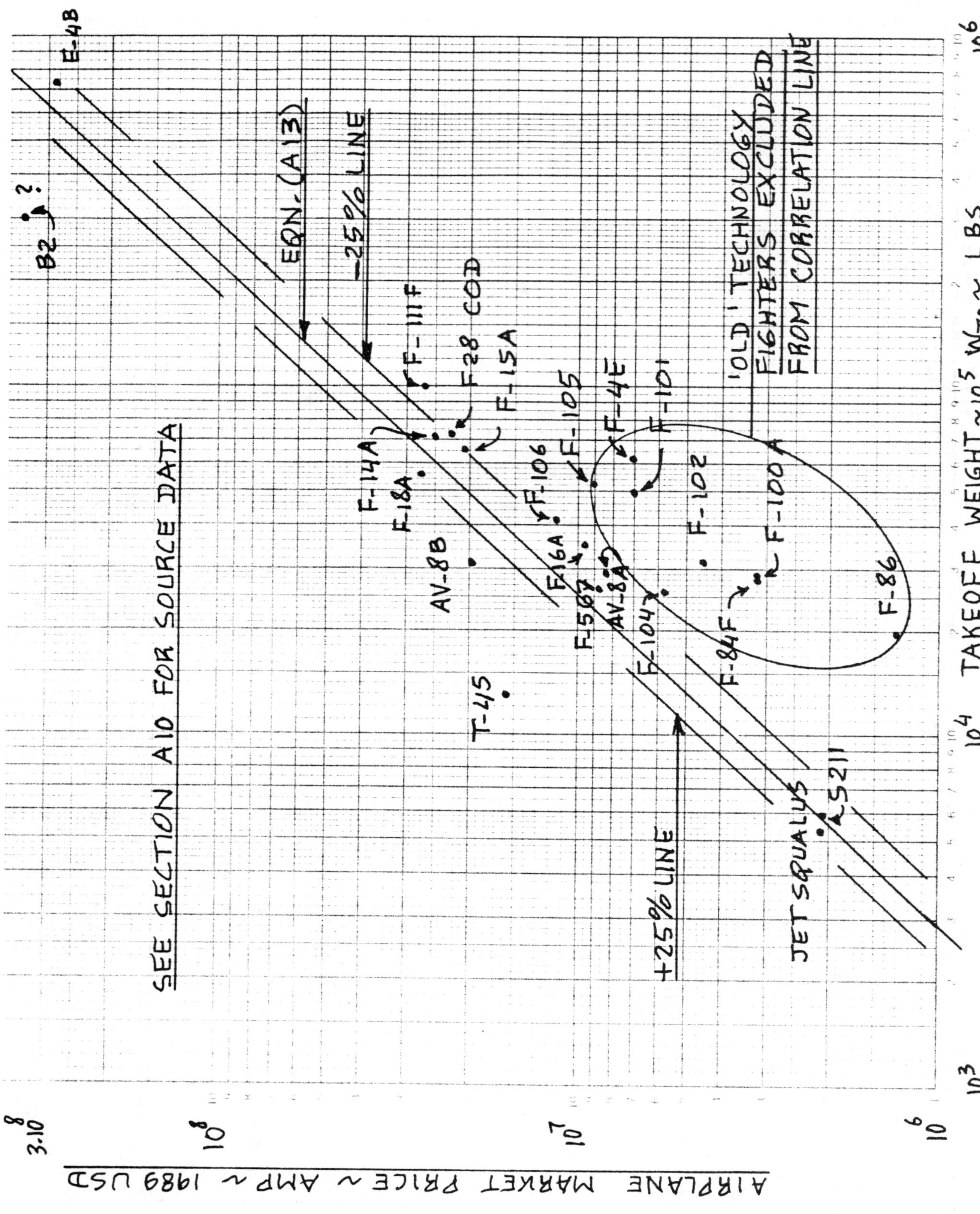

• EPSILON

Figure A10 Effect of Takeoff Weight on the Price of Military Airplanes

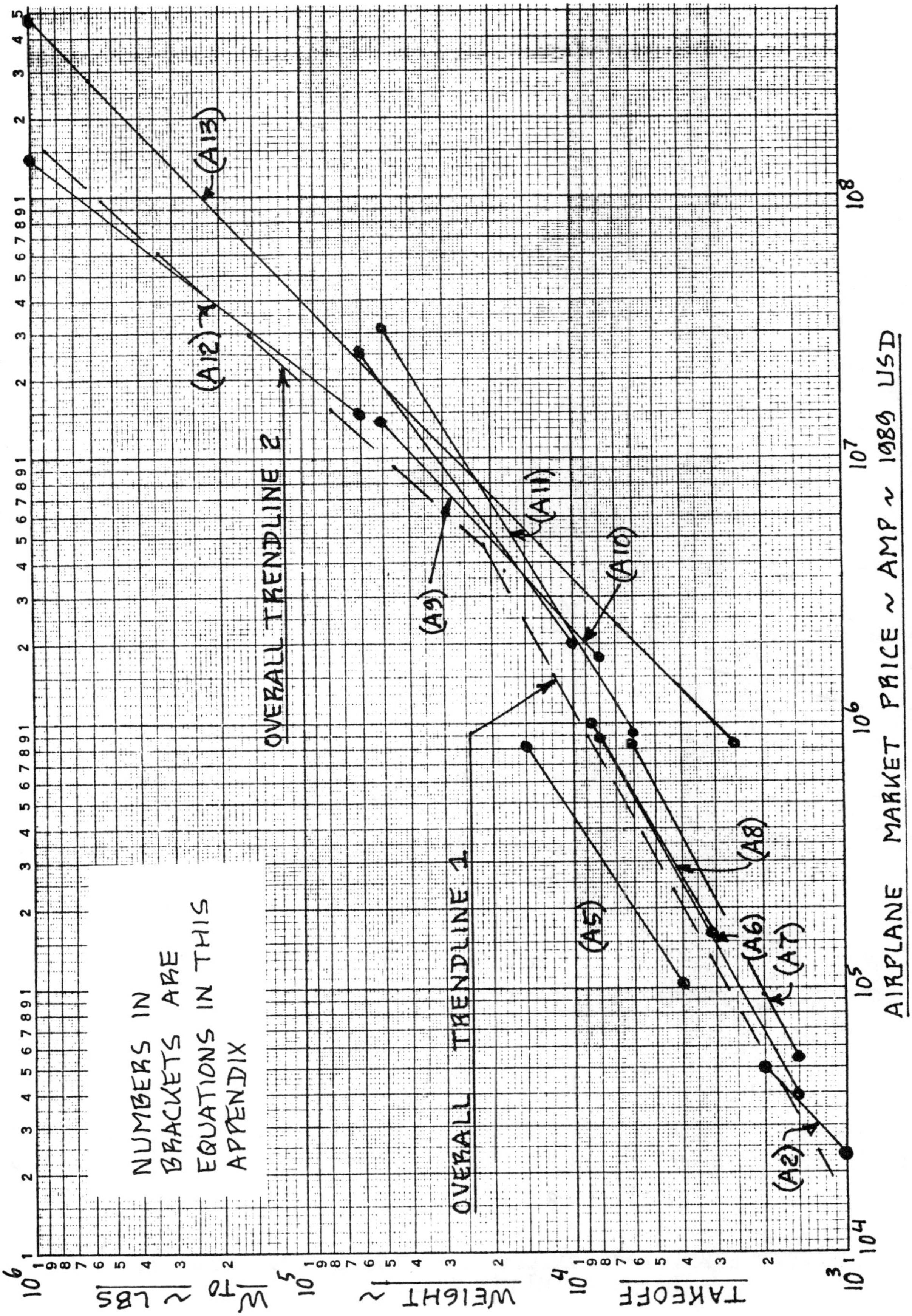

Figure A11 Composite Plot of Price Versus Takeoff Weight of all Airplane Types

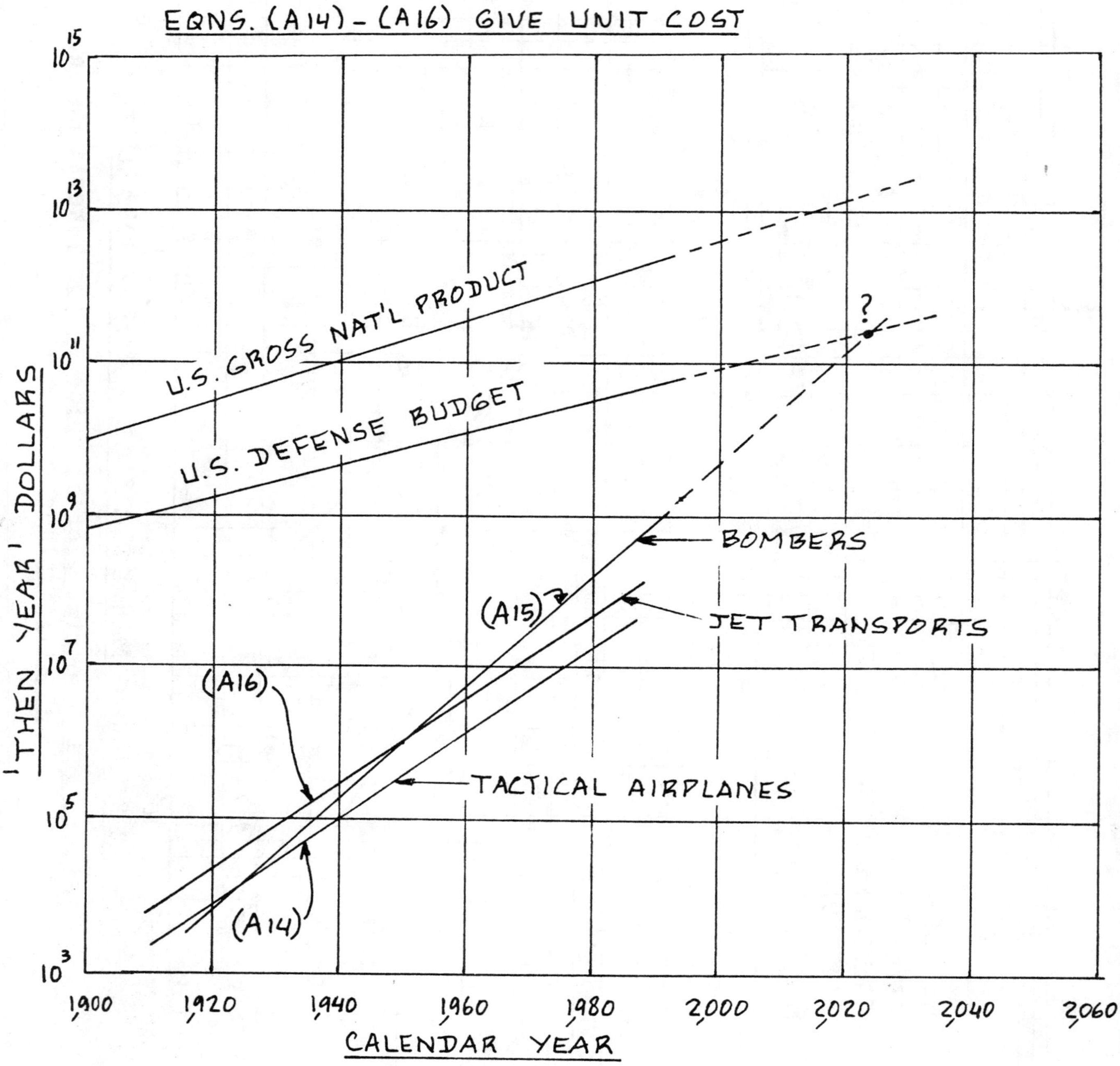

Figure A12 Effect of Calendar Year on the Price of Three Airplane Types

Table A1 Ranges of Airplane Price per Pound for 1989

Airplane Type	Range of Takeoff Weights in Lbs Range of Prices in Dollars/Lbs		
Sailplanes	600	to	2,000
Eqn. (A2)	23	to	25
Ultralight Airplane Kits	400	to	800
Eqns (A3 and A4)	3	to	10
Agricultural Airplanes	3,000	to	15,000
Eqn. (A5)	22	to	57
Single Engine Piston Airplanes			
Normally Aspirated	1,500	to	10,000
Eqn. (A6)	28	to	138
Turbocharged	1,500	to	10,000
Eqn. (A7)	37		185
Multieng. Piston Airpl.	3,000	to	8,000
Eqn. (A8)	53	to	110
Multieng. TBP Airplanes	8,000	to	50,000
(Business), Eqn. (A9)	224	to	275
Business Jets	10,000	to	60,000
Eqn. (A10)	204	to	428
Turboprop Commuter Airplanes	6,000	to	50,000
Eqn. (A11)	150		262
Commercial Transp. Jets	60,000	to	1,000,000
Eqn. (12)	242		140
Military Airplanes	2,500	to	1,000,000
Eqn. (A13)	341		485

Table A2 Ranges of Airplane Price per Seat for 1989

Airplane Type	Range of Price per Seat in Dollars/Seat		
Sailplanes	16,000	to	80,000
Single Engine Piston Airplanes	20,000	to	50,000
Multiengine Piston Airplanes	33,000	to	70,000
Multiengine Turboprop Airplanes (Business)	115,000	to	450,000
Business Jets	320,000	to	1,600,000
Turboprop Commuter Airplanes	160,000	to	360,000
Commercial Transport Jets	170,000	to	250,000

APPENDIX B ENGINE AND PROPELLER PRICE DATA

The purpose of this appendix is to provide engine and propeller price data as well as a rapid method for 'ball-parking' such prices for future designs. The sources listed in Appendix A for airplane price data (which, by the way INCLUDE the cost of the engine to the airframe manufacturer) publish engine and propeller price data only sparingly. The data presented in this appendix were primarily taken from References 55 through 59.

Propulsion system prices are affected by:

* Uninstalled engine thrust or horsepower
* Turbine and compressor technology
* Calendar time (if only because of cost escalation)
* Market conditions (demand and competition)
* Manufacturer sales strategies
* Government rules and regulations (Certification requirements, specifications and politics)

Except for the effect of cost escalation, no easy rules for predicting the effect of these factors on propulsion system prices can be given. Since all engine and propeller price data are 'dated' by calendar year, the following equation is suggested to allow for cost escalation:

$$PSP_{19XX} = PSP_{then\ year}(CEF_{19XX}/CEF_{then\ year}) \qquad (B1)$$

where: PSP_{19XX} is the desired propulsion system price in 19XX

$PSP_{then\ year}$ is the given propulsion system price in 'then year' dollars

CEF is the cost escalation factor obtained from Fig.2.7. The ratio $(CEF_{19XX}/CEF_{then\ year})$

is called the CEF-ratio

Equation (B1) may be applied to adjust prices for engines (EP) and/or propellers (PP) from one year to another. Because the price data in References 55 - 59 are in various 'then year' dollars, Equation (B1) was used frequently to adjust price data to 1989 dollars.

<u>IMPORTANT NOTE:</u> In chapters 4 - 7 the symbols used for engine and/or propeller cost to the airframe manufac-

were C_e and C_p respectively. In this appendix, the symbols used for engine and/or propeller price are EP and PP respectively. For practical purposes it is acceptable to assume:

$$EP = C_e \qquad \text{AND:} \qquad PP = C_p \qquad (B2)$$

The engine and propeller price data in this appendix are organized as follows:

B1 PISTON ENGINE PRICE DATA

B2 TURBOPROP ENGINE PRICE DATA

B3 PROPELLER PRICE DATA

B4 JET ENGINE PRICE DATA

B1 PISTON ENGINE PRICE DATA

Piston engine price data were taken from Ref.55 and adjusted to 1989 dollars with Eqn.(B1). The prices do NOT include the price of the propeller. Price data for propellers are given in Section B3.

Piston engine price data are plotted versus engine shaft horsepower at takeoff (SLS), SHP_{TO_e} in Figure B1. The solid line represents the trend line obtained from a regression analysis on all data points. The trend line does not represent the data very well. To improve this, the horsepower range was split in two ranges:

80 - 200 shp AND: 200 - 500 shp.

Separate correlation lines were determined for each range. From this, the following formulas were derived for 'ballparking' piston engine prices:

For the shaft horsepower range of 80 - 200 shp:

$$EP_{1989} = \text{invlog}\{2.9923 + 0.4536(\log SHP_{TO_e})\} \qquad (B3a)$$

For the shaft horsepower range of 200 - 500 shp:

$$EP_{1989} = \text{invlog}\{-0.7770 + 2.0917(\log SHP_{TO_e})\} \qquad (B3b)$$

where: SHP_{TO_e} is the takeoff shaft horsepower per engine at sealevel standard (SLS) conditions.

Do NOT use these equations outside the ranges of shaft horsepower indicated.

B2 TURBOPROP ENGINE PRICE DATA

Price data for turbopropeller engines (NOT including the price of the propeller) were obtained from Refs 55 - 58. These 1975 data were adjusted to 1989 dollars with the help of Eqn.(B1). The data are plotted in Figure B2. A trend line was determined with a regression analysis. The trend line is seen to represent the data very well. This leads to the following equation for 'ballparking' turbopropeller engine prices:

$$EP_{1989} = \text{invlog}\{2.5262 + 0.9465(\log SHP_{TO_e})\} \qquad (B4)$$

This equation applies to a shaft horsepower range of 400 to 5,000 shp. It should NOT be used outside that range. For larger engines, Ref.56 suggests the following equations:

$$EP_{1989} = (1.418)(2,160,000)\{0.4(SF) + 0.600\} \qquad (B5)$$

for SF > 1.0 and:

$$EP_{1989} = (1.418)(2,160,000)\{0.533(SF) + 0.467\} \qquad (B6)$$

for SF < 1.0

where: the factor 1.418 is the 1989 to 1980 CEF-ratio

2,160,000 is the 1980 base price in US dollars for a turboprop engine with 20,424 shp

SF is an engine shp scale factor determined from:

$$SF = SHP_{TO_e}/20,424 \qquad (B7)$$

NOTE: the price data in this Section INCLUDE the engine gearbox, but NOT the propeller.

B3 PROPELLER PRICE DATA

Price data for propellers were obtained from References 56 through 59 and adjusted to 1989 dollars. The data are plotted in Figure B3. It is seen that there are not enough data points to establish the effect of number of blades on propeller cost. A trend line was drawn by performing a regression analysis on all METAL propeller data. This line has the following equation which may be used to 'ballpark' METAL propeller prices:

$$PP_{1989} = \text{invlog}\{0.6119 + 1.1432(\log SHP_{TO_e})\} \tag{B8}$$

For COMPOSITE propellers the following equation may be used:

$$PP_{1989} = \text{invlog}\{0.7746 + 1.1432(\log SHP_{TO_e})\} \tag{B9}$$

These equations should NOT be used outside a range of 100 - 20,000 shp.

B4 JET ENGINE PRICE DATA

Jet engine price data were taken from Ref.55 and plotted versus maximum sealevel static thrust, T_{TO_e} in Figure B4. The data do NOT indicate a systematically higher price for turbofans than for turbojets. That is why only one regression analysis was performed, resulting in the solid line in Figure B4. The data straddle the trend line within a +/- 20 percent band. Using Eqn.(B1) to correct the jet engine price data from 1975 to 1989, the line labelled 'suggested for 1989' was obtained. The equation which represents this line is:

$$EP_{1989} = \text{invlog}\{2.3044 + 0.8858(\log T_{TO_e})\} \tag{B10}$$

This equation may be used to 'ballpark' new engine prices. It should NOT be used outside the takeoff thrust range of 1,000 to 50,000 lbs per engine.

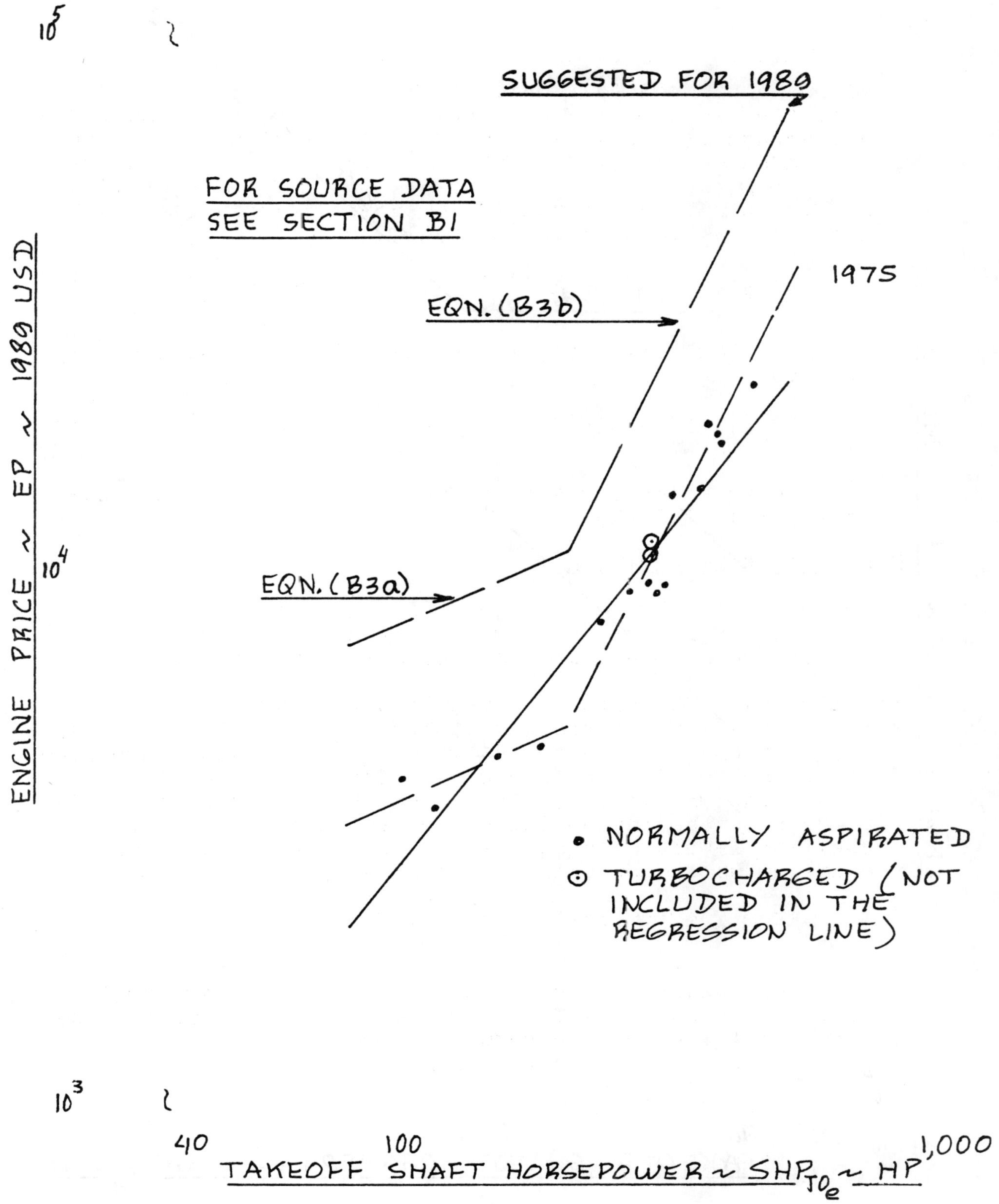

Figure B1 Effect of Shaft Horse Power on the Price of Piston Engines

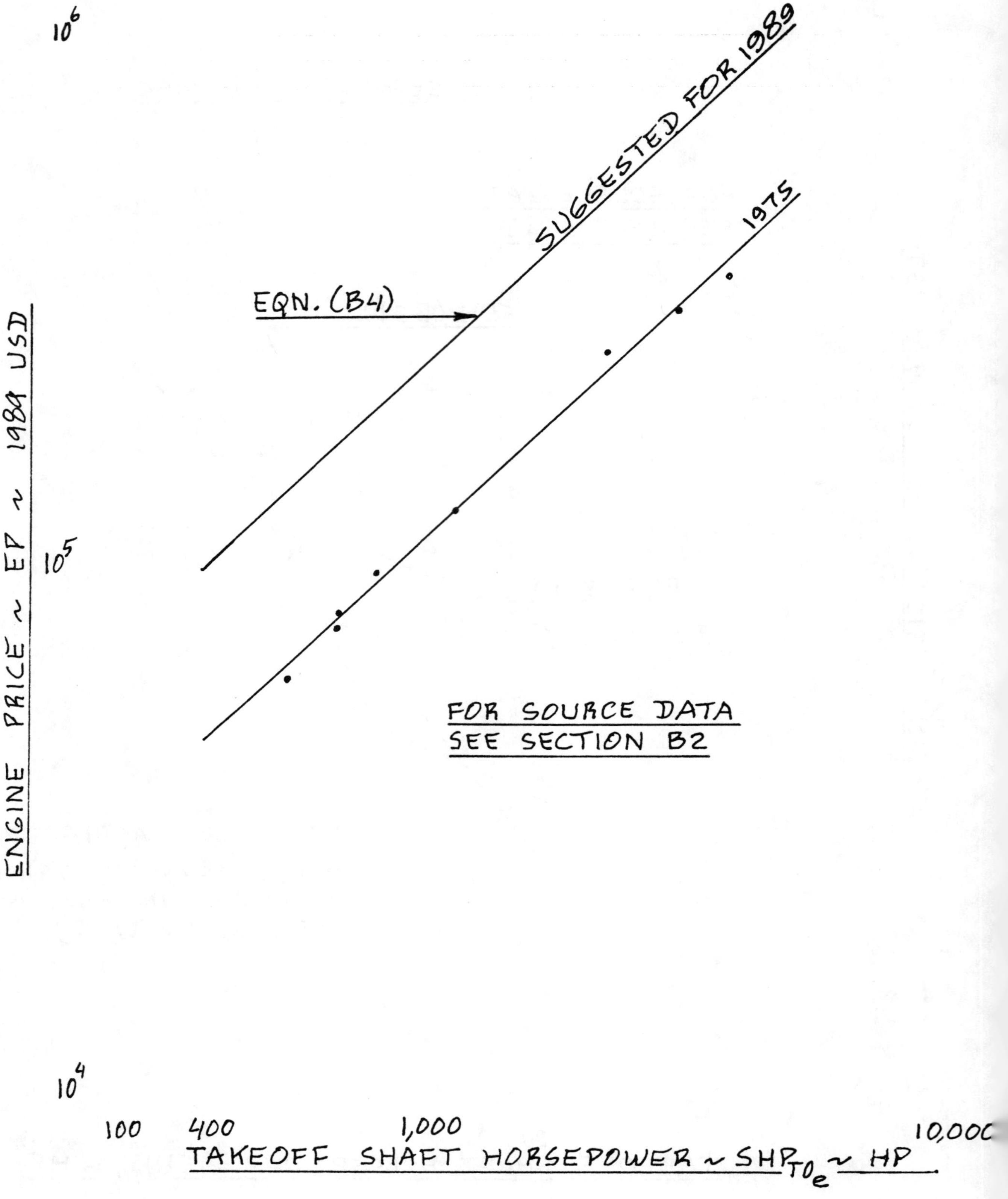

Figure B2 Effect of Shaft Horse Power on the Price of Turboprop Engines

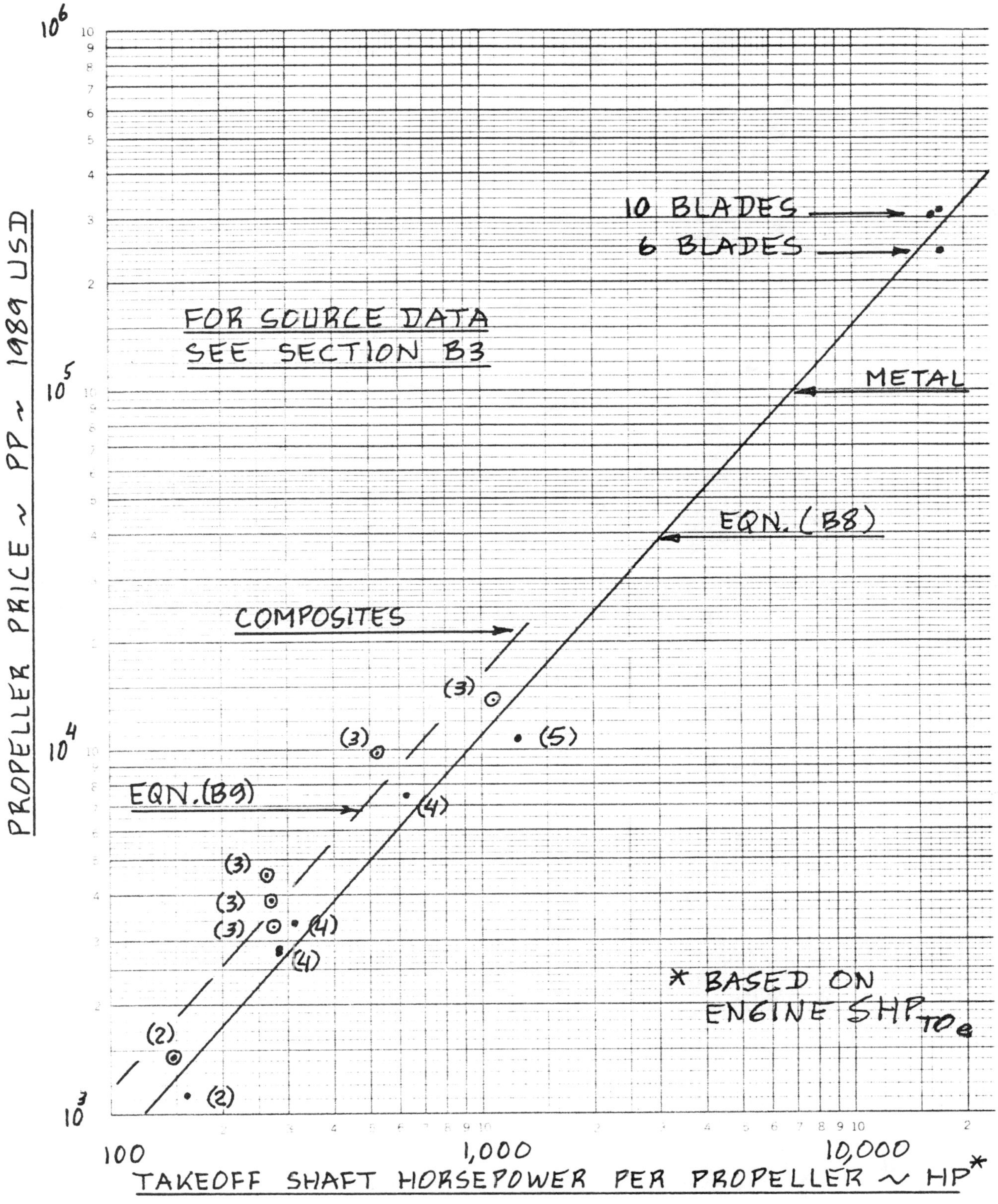

Figure B3 Effect of Shaft Horse Power on the Price of Propellers

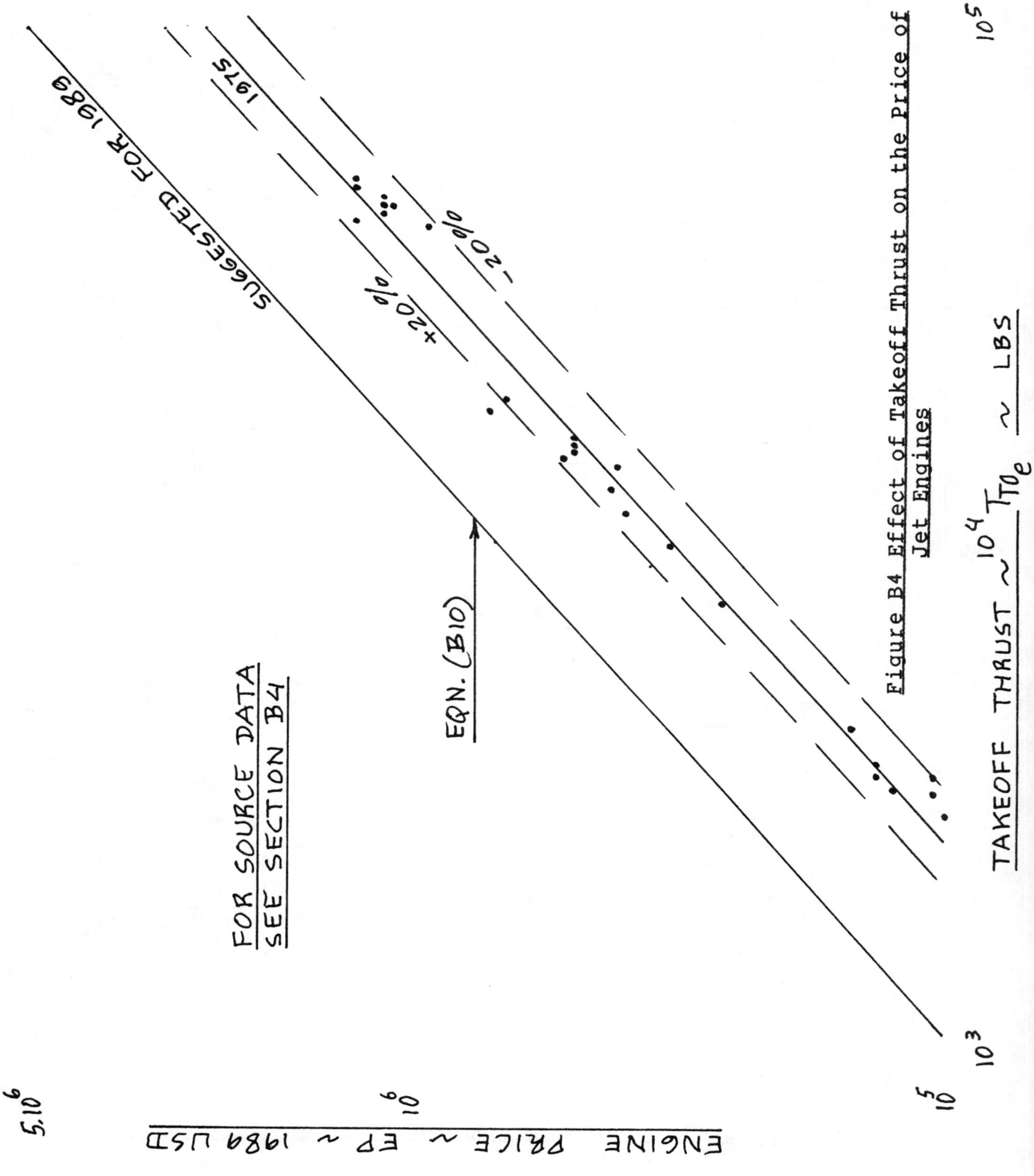

Figure B4 Effect of Takeoff Thrust on the Price of Jet Engines

APPENDIX C AVIONICS SYSTEMS PRICE DATA

The purpose of this appendix is to provide price data for airplane avionics systems.

The material is organized as follows:

C1 COMMERCIAL AVIONICS SYSTEMS PRICE DATA

C2 MILITARY AVIONICS SYSTEMS PRICE DATA

C1 COMMERCIAL AVIONICS SYSTEMS PRICE DATA

The author is grateful to Business and Commercial Aviation Magazine for their permission to copy the Avionics Section of the 1989 Planning and Purchasing Handbook. The material presented in this Section is an exact copy of the following pages of the May 1989 issue of Business and Commercial Aviation: pages 151, 152, 154, 156, 158, 160, 161, 162, 164, 166, 168, 170, 172, 174, 176, 178 and 182 through 202. These pages appear here as pages 330 through 366.

The reader should realize that any price data will vary with time and with marketing strategies of manufacturers. Whenever possible, actual and recent price data should be used. Lacking such information, the price of any avionics system listed in this appendix for 'then years' other than 1989 may be estimated from:

$$(\text{Avionics System Price})_{\text{then year}} = \qquad (C1)$$

$$= (\text{Avionics System Price})_{1989}(\text{CEF}_{\text{then year}}/\text{CEF}_{1989})$$

where: $(\text{Avionics System Price})_{1989}$ may be found from the data in pp. 330 - 366.

CEF is the cost-escalation-factor of Figure 2.7 in Chapter 2.

<u>IMPORTANT NOTE:</u> The symbol used for total avionics cost in Chapters 4 - 7 is: C_{avionics}. This cost is intended to be the SUM of the individual avionics system prices obtained from this section.

INTRODUCTION 1989 AVIONICS

Canadair's Challenger 601-3A cockpit utilizes quiet/dark design principles.

Business aircraft operators will remember the last decade of the 20th century as being the period during which digital avionics became dominant on business aircraft. Looking at the array of digital equipment in today's business aircraft, it's hard to remember that little over a decade ago, digital comm/nav/pulse radios were just being introduced. Electronic flight instrument systems (EFIS) were rarely seen in business aircraft prior to the mid-1980s. The proliferation of navigation management systems (NMS) is an even more recent phenomenon.

Integration is the single most prominent development in digital avionics today. The big three avionics houses—Bendix/King, Honeywell and Rockwell/Collins—are placing great emphasis on integrated avionics packages. For example, comprehensive Bendix/King or Honeywell installations may include EFIS, NMS, digital air data computers, comm/nav/pulse radios and weather avoidance radar—all manufactured by one company. Rockwell/Collins packages feature virtually the same broad line-up of equipment except that the firm relies on outside manufacturers for NMS such as Universal Navigation Corporation.

In late 1988, Collins announced its Pro Line 4 fully integrated avionics system that upped the ante for all competitors (see *EFIS* elsewhere in this B/CA *Planning and Purchasing Handbook*). This brand-new design features EFIS, an engine instrument and crew alerting system (EICAS) and cockpit control display units (CDUs)—all branching out from a central computer that constitutes the heart of the system. Left- and right-side navigation computer modules, radios, sensors and data concentrators are all contained in just one remote-mount computer, greatly reducing the volume, weight and number of components in the avionics bay.

Pro Line 4 accelerates the acceptance of Arinc 429 as the digital interface of choice in business aircraft. Other digital and analog interface formats will persist for many years, owing to preferences of individual avionics manufacturers, but Arinc 429 (or possibly GAMA 429) is becoming about as commonplace as the RS 232 serial digital interface used by desktop computers.

Multi-sensor NMS have all but squeezed out single-sensor navigation

The Learjet 55's flight deck boasts an impressive array of standard avionics including five-tube Collins EFIS, Bendix/King HF, Universal UNS-1 navigation management system with VLF/Omega sensor, SELCAL and a Global Wulfsberg Flitefone VI. Digital avionic designs are becoming increasingly prevalent.

systems in the more sophisticated business aircraft. Multi-sensor NMS units are able to take advantage of reliable and accurate short-range DME and VOR signals when in line-of-sight reception range of ground stations. Some systems, such as the Foster LNS 616B, are capable of using UHF TACAN bearing and distance information, thereby escaping the radial scalloping signal distortion associated with some VOR signals.

NMS provide exceptional flexibility in the choice of long-range nav sensors. Some aircraft operators may need Omega/VLF receivers for extended over-water operations. Others may opt for Loran-C sensors, a popular choice for continental U.S. operations. For frequent international operations, it's hard to beat the current generation of laser inertial reference sensors that have achieved drift rates well under the one-nautical-mile-per-hour variation design criteria.

While the launch of Navstar global positioning satellites has resumed (there were eight of a planned 21 satellites in orbit at press time), the cost of putting operational and spare satellites into orbit could be a hot topic on Capitol Hill with today's emphasis on federal budget deficit reduction. If funding constraints delay the launch of additional GPS vehicles, multi-sensor NMS designers may be forced to look at inter-operable multi-sensor technology.

An inter-operable system is quite different from an NMS that simply uses inputs from several independent long-range navigation sensors. Such a system actually combines inputs from many sources to compute the aircraft's position. For example, GPS might be used for an extremely accurate time check. One line of position might be computed from signals received from Omega or VLF stations, while a second line of position might be obtained from a pair of Loran-C transmitters. And eventually all Loran-C stations may be synchronized, permitting a Loran-C receiver to plot lines of position from any two stations in any GRI chain—without the need to receive the master station.

As for ease-of-use features, virtually all long-range and nav management systems now come equipped with comprehensive databases. Operators can expect to find airports and navaids stored in almost all of these systems, with the better units having intersections and even arrival and departure procedures.

Most manufacturers now are making database updating easy. Rather than pulling a unit out of the avionics bay or changing internal components, such as programmable read-only memory chips, many systems may be equipped with optional diskette data-loaders. One innovative approach to updating is the use of database cartridges so compact that they easily fit into a shirt pocket.

Panel-mounted Loran-C receivers will be found in the flight decks of many business aircraft for which nav management systems represent too large an investment; however, the number of active manufacturers of such systems may stabilize at about a half-dozen strong firms.

Manufacturers' top of the line systems such as ARNAV Systems' R-50, Foster Airdata Systems' F4 Phoenix, Northstar's M1 and II-Morrow's 618 SuperBandit, have airport, navaid and special-use airspace databases. Most of these units are certified for VFR operations only, except for the IFR-certified Foster F4 Phoenix. Bendix/King will enter the panel-mount Loran-C market later this year when the firm introduces its cross-chain, IFR-certifiable unit. This innovative product should prove highly capable for a panel-mount Loran-C receiver because of its reception range, its impressive database and its wide selection of features. But don't anticipate existing Loran-C manufacturers to stand still in the face of this new competitor.

By the year's end, we also anticipate that at least one manufacturer will announce a panel-mount NMS, complete with a short-range VOR/DME navigation capability as well as an internal Loran-C receiver. It is not unreasonable to conclude that a GPS sensor will find a home in compact, panel-mount nav systems in the future, right alongside the internal Loran-C sensor.

We also can look forward to the expanded use of EFIS in business aircraft because of the variety and sophistication of information that can be displayed on a single tube. In addition, reversion modes reduce the odds of being grounded by a single tube or component failure.

Growing in importance is the increasing time and cost associated with repairing aging electro-mechanical instruments. The reliability of EFIS units has increased substantially in the last several years, reducing the overall cost of ownership to a level that is becoming competitive with

David A. Kern

Dassault's Falcon 900 employs an all-DC electrical system and digital avionics throughout its spacious cockpit. Honeywell furnishes the aircraft's radar, dual laser IRSes, EFIS and dual Flight management systems. Collins Pro Line II Navcomm is standard on the aircraft.

conventional "clocks and dials." In recent years, we've witnessed a great deal of research and development activity, accompanied by subtle but effective publicity, regarding flat panel displays. However, we will need to wait beyond 1989 before seeing flat panel displays offered for business aviation.

The trend in EFIS is toward parts commonality. The number and variety of displays, instruments, dials and annunciators is declining in instrument panels. This decline is accompanied by a reduction in shop spares. Ultimately, fleet operators that have equipped all their various aircraft with the same brand of EFIS/EICAS may find their parts shelves bare of all flight and engine instruments, plus annunciators, save one common replacement display tube and maybe one spare generic symbol generator.

TCAS, the traffic alert and collision avoidance system that has been mandated for the air carriers and air taxi operators of larger aircraft, continues to be a popular topic discussed by business aircraft operators. Many industry officials believe that such operators would quickly purchase TCAS units if they were certified and in dealer inventory.

Legitimate concern exists on the part of some observers that active TCAS installed on host aircraft may overload the reply capacity of the conventional ATCRBS transponders on intruder aircraft. Such a limitation may heighten the need for development of passive collision avoidance systems, such as Avion Systems' Sentinel II (B/CA, March 1989, page 49).

Nevertheless, there is little doubt that any future derivative collision avoidance system will rely on conventional transponders as the key cooperative avionics unit needed to identify intruder aircraft.

The arrival of active TCAS is putting pressure on Mode S transponder development. Mode S transponders are used by TCAS systems in two approaching host aircraft to coordinate traffic avoidance maneuvers.

Activity associated with microwave landing systems (MLS) continues to be decidedly flat as a result of the stalemate between the FAA and its MLS ground station supplier, Hazeltine. Delivery delays, cost overruns and organizational problems have plagued the MLS program for more than three years and many business aircraft operators simply have lost interest in buying MLS avionics.

But, it should be noted that the Europeans are going ahead with their

Foster AirData Systems' F4 Phoenix Loran offers impressive performance, operating features, all-lighting readability and relatively low-cost.

MLS installation plans. On the Continent, MLS deployment is being mandated by the success of commercial FM broadcasters that are legally climbing up the frequency band into the lower end of frequencies formerly allocated to ILS localizers. Also, the rock-solid, stable signal characteristics of MLS represent a genuine improvement compared to ILS at airports subject to terrain interference or ground-plane variations caused by snow drifts or tidal changes.

In the United States, MLS deployment is likely to occur first at military and federal government airports plus a few private installations. Because of these developments, business aircraft owners that operate internationally will find a greater need for MLS avionics in the immediate future than operators that fly within the confines of North America.

The next 12 months will be exciting ones in the avionics market. While it's not likely that a third crew seat will be added to business aircraft, the intelligent functions, decision-making assistance and help in controlling the aircraft provided by today's advanced avionics may just make two pilots think that there's an extra crewmember somewhere in the cockpit.

AVIONICS GLOSSARY

ADF—Automatic direction finder, a navigation receiver providing bearing to a nondirectional beacon.
ADI—Attitude deviation indicator.
AHRS—Attitude-heading reference system, a sensor deriving aircraft attitude from a strapdown gyro or other position sensor, accelerometers, and heading from a flux valve; replaces the conventional gyro set.
Air Data—Values computed from pitot, static and temperature measurements, usually by means of a digital computer; Arinc 575 defines outputs.
AM—Amplitude modulation.
AOA—Angle of attack.
Arinc—Acronym for Aeronautical Radio, Incorporated, a nonprofit corporation owned by member airlines to define form, fit and function of avionics equipment and to provide radio communication services.
Arinc 419—A compendium of digital information transfer characteristics as applied in Arinc 500-series equipment specifications.
Arinc 429—A standard for broadcast digital information transfer systems for general applications.
Arinc 561—Inertial navigation system specifications.
Arinc 571—Inertial sensor attitude-heading reference system specifications.
Arinc 575—Digital air-data system specifications.
Arinc 600-series—Equipment specifications.
Arinc 700-series—All digital equipment specifications for new-generation transport category aircraft.
ASCB—Avionics Standard Communication Bus, a digital electronic bus developed by Sperry that provides bidirectional "broadcasting" of data between aircraft systems by means of a "ring network"; similar to, but slower than the MIL-S 1553 network.
ATI and ATR—Arinc form factors (see "Arinc Form Factor" chart).
BCD—Binary coded decimal.
Beam Width—The included angle of a weather radar interrogation signal as defined by the antenna characteristic.
BRG—Bearing.
CDI—Course deviation indicator.
CDP—Counter-drum-pointer, a type of altimeter display that uses a combination of a rotating drum with numerals and a pointer.
CDU—Control-display unit.
Coherent Detection—A technique used in certain ADF receivers to improve useful range by providing better definition of weak signals.
Composite Video—Analog VOR receiver output before processing.
Contact Digitizer—A mechanical device that converts analog information to digital codes by means of electrical contacts.
CRT—Cathode ray tube, similar to a "TV" picture tube.
CW—Continuous wave, a radio carrier broadcast that does not have modulation.
DA—Drift angle.
DG—Directional gyro.
DH—Decision height.
Digitizer—Any electronic device capable of converting information to a digital format; usually refers to encoding devices that convert sensed altitude into a transponder code.
DR—Dead reckoning.
DTK—Desired track.
Duplex—Separate channels for transmission and receiving.
E- (as a prefix)—Used with "ADI"or "HSI," to designate an electronic display using a CRT.
EAROM—Electrically alterable read-only memory, a type of digital memory device.
ECDI—Electronic course deviation indicator.
EFIS—Electronic flight instrumentation system.
EEPROM—Electrically erasable programmable read-only-memory.
Gas discharge—A type of luminescent digital display, usually seven-segment and driven by a high-voltage source; characterized by low heat generation.
FM—Frequency modulation.
GCR—Ground clutter rejection.

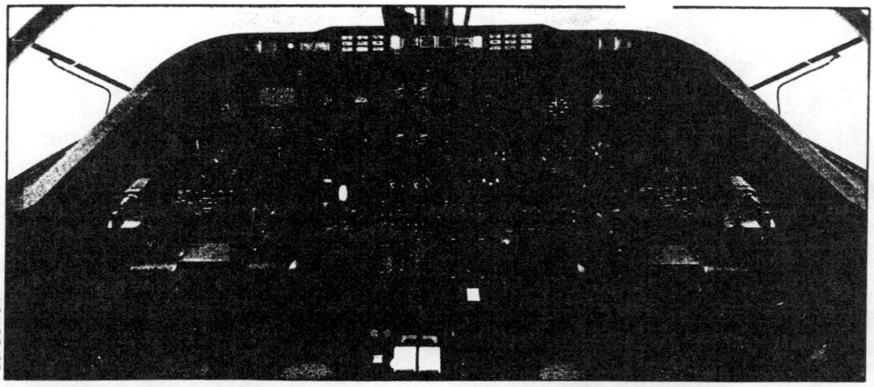

Paul Bowen

The Beechjet 400 cockpit will be the first to be equipped with Collins Pro Line 4 EFIS, an advanced integrated display system that could grow into an all-glass panel.

GMT—Greenwich Mean Time, used in navigation systems.
GS—Groundspeed.
Hot Wire—Any method for providing some continuous electrical power to a memory device from the airplane's primary battery.
HSI—Horizontal situation indicator.
Hz or Hertz—Cycles per second, used to describe radio frequencies; usually with the prefix *k* for kilo (thousands) or*M* for mega (millions).
ILS—Instrument landing system, a precision approach system using a localizer, glideslope, an outer marker beacon, middle marker beacon and approach lights.
INS/IRS—Inertial navigation system/inertial reference system.
ITU—International Telephone Union, refers to certain HF channels.
Laser IRS—An inertial reference system deriving angular rate information by measuring doppler shift between two contra-rotating light beams.
LCD or LC—Liquid crystal display.
LED—Light-emitting diode.
LOC—Localizer, the course guidance component of an ILS.
Looks per minute—Scanning or sweep rate of a weather radar antenna.
Loran-C—Hyperbolic grid navigation system that is based upon measured time differences (TD) from pulse transmissions.
Mid-Continent Gap—A lapse in Loran-C signal coverage roughly spanning the United States and Canada, east of the Rocky Mountains and west of the Mississippi River.
MFD—Multifunction display, the "third or fifth tube" in an EFIS, replaces the weather radar and displays radar data, navigation maps, checklists, and other information.
MS—Military specifications, which define the size of a standard three-inch instrument panel cutout for such mechanical displays as altimeters and airspeed indicators.
N/A—Not applicable.
NDB—Nondirectional beacon.
NMS—Navigation management system.
Nonvolatile—Describes a digital memory that retains information through system shutdown.
Omega—A very-low-frequency navigation system.
Optical Digitizer—Any device using a photosensor to read a light beam through slots or cutouts and thereby to convert analog information to digital codes.
Peak power output—The maximum transmitter power output measured over a very short time period; usually used to rate pulse transmissions.
PEP—Peak envelope power, a standard electronic rating of any AC source, including a radio transmitter.
Pulse—A transmission of very brief duration used to carry information by using time measurement or as a series of pulses representing code.
RA—Radio altitude.
RAM—Random access memory.
RMI—Radio magnetic indicator.
RMS—Root mean square.
RNAV—Area navigation.
ROM—Read-only memory.
RS-232—A serial digital format used in data transmission.
RT—Receiver-transmitter combined in a single line-replaceable unit.
SAR—Search and rescue, usually pertaining to weather radars and navigation systems.
SI—Slip indicator or inclinometer, usually a ball within a tube containing fluid.
Simplex—Single frequency for both transmitting and receiving in communications; generally implies a push-to-talk function and verbal procedures.
Slaving Rate—The rate, in degrees per minute, a compass system can adjust for gyro precession; a "fast-slave" mode is used to align a compass system when first turned on.
SSB—Single sideband, a highly efficient form of radio transmission wherein information is carried on a sideband instead of the center carrier in order to gain range performance; usually used in HF communication transceivers.
STC—(1) Sensitivity-time compensation, in weather radar a technique and circuit for calibrating a display; (2) supplemental type certificate.
Synchro—Any device capable of converting mechanical position into an (analog) electronic signal.
Synthesizer—Frequency synthesizer, an electronic circuit capable of generating multiple frequencies from a single crystal oscillator.
TAS—True airspeed.
TD—Time difference, in Loran-C navigation measurements.
3P—Three-pointer altimeter display.
TKE—Track error.
TRK—Track, or path over the ground.
TSO—Technical standard order, a performance specification and production compliance criteria applied to avionics and defined by FARs and the Radio Technical Commission for Aeronautics.
TTG, TTS or TTW—Time to go, time to station, time to waypoint.
Turbulence Mode—Flight control system mode in which a "softer" response to gust upsets is programmed.
2 X 5—Two-out-of-five, an electronic radio tuning code.
USB—Upper sideband.
VAC—Volts alternating current.
VDC—Volts direct current.
VLF—Very low frequency, refers to Navy VLF transmissions used in long-range navigation systems in combination with Omega.
VNAV—Vertical navigation, or the capability to compute pitch-axis maneuvers and altitude-related points in space.
VOR—Very high frequency omni range.
X-band—Frequency range in which most general aviation weather radars operate.
XTK—Cross track; cross track error.

ARINC FORM FACTOR

The standard air transport radio (ATR) case is shown with all optional projections included. Volumes are given, except for the forward-facing projections called "doghouses."

Not shown on this chart is the short, low-profile (SLP) case used in the Collins Pro Line. All Pro Line remote units are about 3.25 inches high and 14 inches long and vary in width.

Panel-mounted avionics are built to the "mark width," which is about 6.25 inches. Widths will vary beyond the 6.25 inches because of flanges that extend beyond the radio mounting tray.

In the avionics tables, a case size is shown as 2 ATI, 3 ATI, up through 4 ATI. Arinc 600 conversion factors: As a rule of thumb, divide Minimum Configuration Units (MCU) by eight for ATR size:

1 MCU = 1/8 ATR*
2 MCU = 1/4 ATR
3 MCU = 3/8 ATR
4 MCU = 1/2 ATR
5 MCU = 5/8 ATR*
6 MCU = 3/4 ATR
7 MCU = 7/8 ATR*
8 MCU = 1 ATR

AIR TRANSPORT INDICATOR (ATI) CASE SIZES

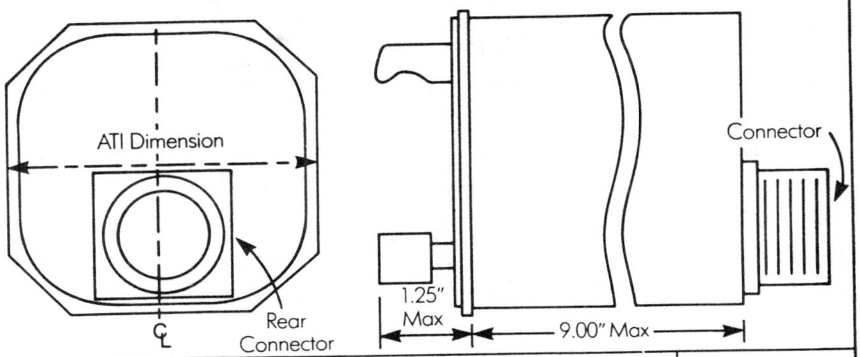

3 ATI 6.2				
Case Size Nominal	Actual	Air Transport Indicator	Length	Max Weight (lb)
2	2.175"		Expressed to nearest 10th of an inch.	3
3	3.175"			5
4	3.975"			9
5	4.975"			14
6	5.975"			20

STANDARD ATR CASE DIMENSIONS

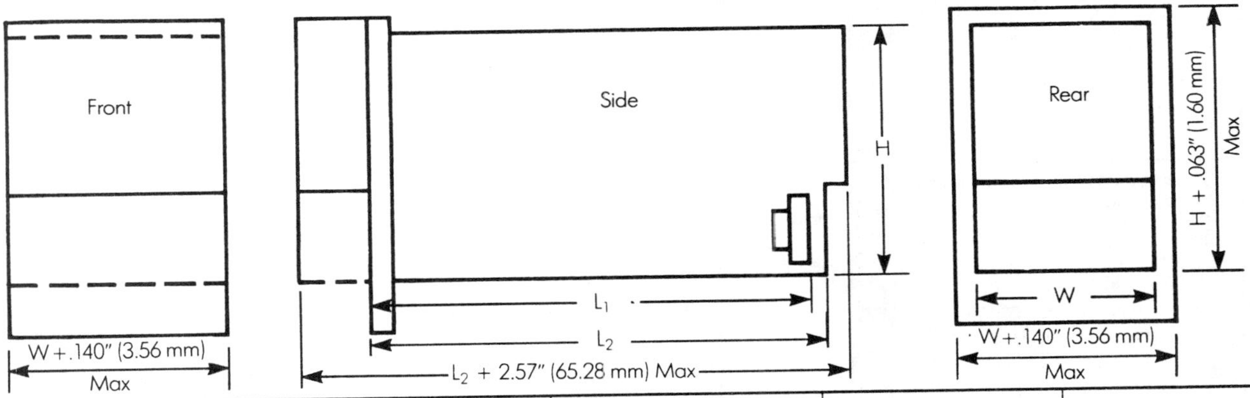

ATR Size	Approx. Vol.		W		L_1		L_2 (Max)		H (Max)	
	In³	Liter	±.03 In	±.76 mm	±.04 In	±1.0 mm	In	mm	In	mm
Dwarf	95	1.56	2.25	57.15	12.52	318.0	12.62	320.5	3.38	85.8
¼ Short	215	3.52	2.25	57.15	12.52	318.0	12.62	320.5	7.62	193.5
¼ Long	335	5.49	2.25	57.15	19.52	495.8	19.62	498.3	7.62	193.5
⅜ Short	340	5.57	3.56	90.41	12.52	318.0	12.62	320.5	7.62	193.5
⅜ Long	530	8.69	3.56	90.41	19.52	495.8	19.62	498.3	7.62	193.5
½ Short	470	7.70	4.88	123.95	12.52	318.0	12.62	320.5	7.62	193.5
½ Long	725	11.88	4.88	123.95	19.52	495.8	19.62	498.3	7.62	193.5
¾ Short	720	11.80	7.50	190.50	12.52	318.0	12.62	320.5	7.62	193.5
¾ Long	1,120	18.36	7.50	190.50	19.52	495.8	19.62	498.3	7.62	193.5
1 Short	975	15.98	10.12	257.05	12.52	318.0	12.62	320.5	7.62	193.5
1 Long	1,510	24.75	10.12	257.05	19.52	495.8	19.62	498.3	7.62	193.5
1½	2,295	37.62	15.38	390.65	19.52	318.0	19.62	498.3	7.62	193.5

B/CA Art Dept.

AUTOMATIC DIRECTION FINDERS

MANUFACTURER	MODEL TSO	MEMORY CAPACITY FREQ. (kHz) DISPLAY	POWER INPUT MEMORY TYPE	UNITS/ WEIGHT (lb) SIZE	PRICE (uninstalled)	REMARKS
Bendix/King Avionics 400 N. Rogers Rd. Olathe, KS 66062 (913) 782-0400	KR 86 none	N/A 200-1750 mechanical	14 or 28 VDC none	2/6.3 6.25 x 2.6 x 9.05	$2,760	Panel-mounted; automatic band switching; self-contained indicator; no RMI capability; KA 42B combined loop-sense antenna available at same price
	KR 87 C41c	2 200-1799 gas discharge	11 to 33 VDC EAROM nonvolatile	3/6.7 6.25 x 1.3 x 11.23	$3,855	Panel-mounted; times flight and approaches; coherent detection; includes slaved KI 227 indicator; KI 228 dual pointer indicator available, KI 229/KNI 582 RMIs available as options
	KDF 806 C41c	2/9* 190-1799 gas discharge	28 VDC EAROM non-volatile	3/7.4 5.0 x 2.0 x 10	$8,935	Digital control head displays both active/standby frequency; coherent detection; flag circuitry; strappable for 2182 kHz, Gables control optional; combined loop-sense antenna included; remote flip-flop optional; *nine-frequency pilot-programmable "A" control head optional; night vision goggle compatible "A"control head optional; two-year warranty
	DFS-43 C41c	2 190-1860* (0.5 kHz) LCD or dichroic	28 VDC non-volatile	3/8.4 4.0 x 4.0 x 12.62	$12,926	ADF left-right steering to HSI; *2182 kHz intl. distress freq std; combined antenna; 0.5 kHz spacing, analog and Arinc 429 interfaces; two-year warranty
Collins Avionics Group Rockwell International 400 Collins Rd. NE Cedar Rapids, IA 52498 (319) 395-4085	ADF-650A C41c	N/A 200-1799 mechanical	14 VDC* none	3/6.0 6.25 x 1.75 x 10.8	$4,425	Panel-mounted; includes ANT-650A combined loop-sense antenna and IND-650A indicator *28 to 14 VDC adapter — $120
	ADF-462 C41d	6 190-1799 gas discharge	28 VDC nonvolatile	3/8.0 3/8 ATR-SL	$10,105	Built-in diagnostics; compatible only with CSDB or Arinc 429 controls, digital signal processing; 2182 KHz emergency frequency; includes CTL-62 CSDB control and ANT-462A antenna; ANT-462B dual antenna optional
	ADF-60A C41c	2/6 190-1750 (½-kHz steps) gas discharge	28 VDC nonvolatile	3/8.7 3/8 ATR-SL	$10,630	Remote-mounted; coherent detection; combined loop-sense antenna; dual antenna optional; three-wire or sine/cosine RMI output available; includes CTL-60 electronic control and ANT-60A; optional CTL-62 with CAD-62 adapter provides serial digital tuning and six stored frequencies
	DF-206 C41b	N/A 190-1750 (0.5-kHz steps) none	28 VDC none	6/15.1 ¼ ATR-S	$20,816	Remote-mounted; azimuth card driven by compass system; automatic pointer stow in receive position; coherent detection; includes flush fixed loop and sense antennas, antenna cable, control head, antenna coupler; Arinc 570

AUTOMATIC DIRECTION FINDERS

MANUFACTURER	MODEL TSO	MEMORY CAPACITY FREQ. (kHz) DISPLAY	POWER INPUT MEMORY TYPE	UNITS/ WEIGHT (lb) SIZE	PRICE (uninstalled)	REMARKS
Narco Avionics Inc. P.O. Box 1807 Laguna Beach, CA 92652-1807 (714) 497-5077	ADF 841 C41c	2 200-1799 gas discharge	11-32 VDC hot-wire nonvolatile	3/5.7 6.25 x 1.5 x 11.0	$3,022	Digital ADF with dual-frequency, flip-flop capability; elapsed- and flight-timers; includes combined loop-sense antenna and indicator
Terra Corp. 3250 Pan American Freeway NE Albuquerque, NM 87107 (505) 884-2321	TDF 100 none	N/A 200-1799 mechanical	11 to 33 VDC N/A	3/6.8 3.2 x 1.6 x 11.5	$1,895	

LONG-RANGE NAVIGATION SYSTEMS

MANUFACTURER	MODEL	SYSTEM TYPE DISPLAY	NUMBER WAYPOINTS DATABASE	OUTPUTS	INPUTS	POWER INPUT MEMORY	UNITS/ WEIGHT (lb)	PRICE (uninstalled)	REMARKS
Advanced Navigation, Inc. 61 Thomas Johnson Dr. Frederick, MD 21701 (301) 695-4040	ANI-7000 Sensor	Loran-C sensor only	N/A N/A	Arinc 419/ 429 pos., GS, track, circular error probability computation	N/A	18 to 32 VDC —	2/14.6	$8,640	TSO C60a; software identical to ANI-7000; a Loran-C sensor for use with KNS 660, Tracor 7880 and other navigation management computer systems
	ANI-7000	Loran-C incandescent filament	200 nonvolatile internal battery	Lat/long and all nav data, with four analog steering signals standard; CDI / HSI; autopilot data; ARINC 429, 561, 568; EFIS, RMI, RS-232; GMT clock	HDG, TAS optional for automatic wind calculation	18 to 32 VDC (60 watts) —	3/19.6	$11,448	TSO C60a full continental US airspace IFR-approved including mid-continent area, limited non-precision approach approval AC 90-45A; AC 20-121
ARNAV Systems 16100 SW 72nd Ave. P.O. Box 23939 Portland, OR 97223 (503) 684-1600	R-25	Loran-C LED	1150 nonvolatile RAM	Waypoint alert; CDI; HSI; autopilot; RS-232	FC-10 Fuel computer; manual input of TAS & HDG for wind calculations altitude for VNAV gph, fuel remaining for fuel range	10 to 35 VDC —	2/4.0	$2,495	Plain language alphanumeric display; antenna included

LONG-RANGE NAVIGATION SYSTEMS

MANUFACTURER	MODEL	SYSTEM TYPE DISPLAY	NUMBER WAYPOINTS DATABASE	OUTPUTS	INPUTS	POWER INPUT MEMORY	UNITS/ WEIGHT (lb)	PRICE (uninstalled)	REMARKS
ARNAV Systems 16100 SW 72nd Ave. P.O. Box 23939 Portland, OR 97223 (503) 684-1600	R-50 Database	Loran-C LED	150 nonvolatile RAM	Waypoint alert warnings; PTK & APR annunciator; CDI; HSI; autopilot; RS-232	FC-10 Fuel Computer; manual input of TAS & HDG for wind calculations; altitude for VNAV advisories	10 to 35 VDC ——	2/4.0	$3,995	Plain language alphanumeric display; 20 route legs; auto extended range; CDI nav bearing; nearest 6 airports; nearest 6 VORs; TCA Alert
	R-40	Loran-C LED	200 nonvolatile RAM	Waypoint alert warnings; PTK & APR annunciator; CDI; HSI; autopilot; RS-232	None	10 to 35 VDC ——	1/4.5	$4,495	Plain language alphanumeric display; 20 route legs; Omni (pseudo VOR) extended range; CDI scaling; TSO C60a
	R-40 Database	Loran-C LED	150 nonvolatile RAM	Waypoint alert warnings; PTK & APR; annunciator; CDI; HSI; autopilot RS-232	Manual input of TAS & HDG for wind calculations altitude for VNAV	10 to 35 VDC ——	1/4.5	$4,995	IFR in compliance with AC 20-121; 10,000+ waypoint database with emergency nearest airport search; min. safe altitude; auto extended range; 20 route legs; Omni (pseudo VOR); optional aerial survey configuration (VFR certification); available with grid/nav aerial survey software — no database: $4,745
	R-60 Database	Loran-C LED	200 nonvolatile RAM	Waypoint alert warnings; PTK & APR; annunciator; CDI; HSI; autopilot; RS-232	Manual input of TAS & HDG for wind calculations altitude for VNAV	10 to 35 VDC ——	2/7.7	$9,495	Designed specifically for pedestal installations; TAS & winds aloft; VNAV; extended range; 20 route legs; Omni (pseudo VOR); available with grid/nav aerial survey software — VFR configuration without database
AZURE Technology San Jose Jet Center 1250 Aviation Ave. San Jose, CA 95110 (408) 947-2070	Long Ranger F/P	Loran-C fixed panel or portable	99 none	RS-232	None	12 VDC ——	2/4.2	$1,095	automatic chainswitching and station management; four flight plans up to 10 waypoints each. Antenna and mounting options avail.
Canadian Marconi 2442 Trenton Ave. Montreal, Quebec, Canada H3P 1Y9 (514) 341-7630	CMA-734 OSS (sensor) Mk II	Omega/VLF —	891 non-volatile RAM	Position, status, stations used, GS	synchro heading, air data (optional)	28 VDC ——	2/15.0	$23,833	Arinc 429; optional LCD CDU completes system for back-up long-range nav capability; TSO C94; three-year warranty
	CMA-764 OSS (sensor)	Omega/VLF —	891 non-volatile RAM	position, status, station used, GS	Arinc 429, heading air data	28 VDC ——	2/8.7	$23,962	Optionall LCD CDU completes system for back-up long-range nav capability; TSO C94; three-year warranty

LONG-RANGE NAVIGATION SYSTEMS

MANUFACTURER	MODEL	SYSTEM TYPE DISPLAY	NUMBER WAYPOINTS DATABASE	OUTPUTS	INPUTS	POWER INPUT MEMORY	UNITS/ WEIGHT (lb)	PRICE (uninstalled)	REMARKS
Canadian Marconi 2442 Trenton Ave. Montreal, Quebec, Canada H3P 1Y9 (514) 341-7630	CMA-734 "Arrow" Mk II	Omega/VLF liquid crystal	891 non-volatile RAM	Pos, steering, nav data, HSI, autopilot	synchro, heading, TAS	28 VDC ——	3/17.2	$27,250	Lightweight system, tracks three Omega and one VLF frequency; TSO C94; three-year warranty
	CMA-734 Alpha Mk II	Omega/VLF incan-descent filament	4,000 non-volatile RAM	Position, steering, nav data, HSI, autopilot	synchro, heading, TAS	28 VDC ——	3/20.0	$37,507	CMA-734 with alphanumeric CDU; three-year warranty
	CMA-771 Alpha Mk III	Omega/VLF incan-descent filament	4,000 non-volatile RAM	Pos, steering, nav data, HSI, autopilot, digital	synchro, heading, TAS	115 VAC, 400 Hz ——	3/33.0	$54,207	Standard CMA-771 but with alphanumeric CDU; optional HSI synchro outputs; Arinc 599 analog, Arinc 429 digital interfaces; compatible with Global-Wulfsberg NDB-2; three-year warranty
Foster Airdata Systems, Inc. 7020 Huntley Rd. Columbus, OH 43229 (614) 888-9502	LRN 500	Loran-C LCD	234 non-volatile RAM	L6R-3, 1K ohm loads; T/F-2, 200 ohm loads; Flag-3, low-level 1K ohm loads; parallel track and WPT alert annun-ciation	none	10 to 32 VDC ——	1/3.1	$1,695	TSO C60a; IFR — AC 20-121; includes panel unit, 5,400 waypoint pilot-changeable database cartridge, antenna and preamp; VNAV advisories, nearby airport locater; "Demo Mode" allows removal from aircraft; carrying case $50
	F-4 Phoenix	Loran-C LED dot matrix	234 nonvolatile RAM	see LRN 500	none	10 to 32 VDC ——	1/3.1	$2,595	See LRN 500
	LNS 616B	Loran C/ auto Tacan LED	400 non-volatile RAM	HSI, EFIS, RS-232, Arinc 561, RMI, RS-232 Tacan channeling	VOR composite, Arinc 568, Foster serial	18 to 32 VDC ——	3/12.75	$22,800	TSO C60a; provides channeling to Foster DME 670 for automatic navigation; includes NY-153 antenna; optional DME-670 $11,900
Global Systems 2144 Michelson Dr. Irvine, CA 92715 (714) 851-0119	GNS-500A Series 4A	VLF/ Omega CRT	127 NDB-2 interface	Arinc 561 and 571 to EFIS; radar; INS; including Pos; TK; DTK; DA; GS; TTW; flight plan waypoints wind; dist. mag. var.	Heading; TAS; Arinc 571; Pos from INS; EFIS; radar; Arinc 561; joystick from EFIS; AFIS performance data	28 VDC at 7.5 amp ——	3/39.0	$64,025	GPS compatible; stores 127 waypoints defined by five letters, nine flight plans up to 20 waypoints each; digital datalink with AFIS; sunlight readable, full-alphanumeric CRT CDU; Interfaces with FD, EFIS and autopilots; auto GMT and date; active flight plan capacity 50 waypoints; GNS 500A Series 1 - 3 may be converted to Series 4 configuration

LONG-RANGE NAVIGATION SYSTEMS

MANUFACTURER	MODEL	SYSTEM TYPE DISPLAY	NUMBER WAYPOINTS DATABASE	OUTPUTS	INPUTS	POWER INPUT MEMORY	UNITS/ WEIGHT (lb)	PRICE (uninstalled)	REMARKS
Global Systems 2144 Michelson Dr. Irvine, Ca. 92715 (714) 851-0119	GNS-500A Series 4M special mission navigation system	VLF/ Omega CRT	127 NDB-2 interface	Arinc 561 and 571 to EFIS; radar; INS; including Pos; TK; DTK; DA; GS; TTW; flight plan waypoints wind; dist. mag. var.; flight data recorder	Heading; TAS; Arinc 571; Pos from INS; EFIS; radar; Arinc 561; joystick from EFIS; AFIS performance data; forward looking infra-red sensor	28 VDC at 7.5 amp —	3/39.0	$66,525	same capabilities as Series 4A + quick-initialization for scramble launch; special search / surveillance flight patterns; law enforcement, military configuration
Honeywell Inc. P. O. Box 29000 Phoenix, AZ 85039 (602) 863-8000	LASEREF	Laser IRS —	none none	Arinc 429, Arinc 419, Arinc 561 Incl: attitude, heading (mag & true), GS, wind, track, present position	TAS Alt. V. Spd.	115 VAC, 400 Hz or 28 VDC backup —	2/48	$146,742	Price includes installation kit; primarily analog interfaces to flight director (EFIS or EM), autopilot, radar, NMS:GNS-1000, UNS-1, KNS-660; fast align; no pressurization required
	LASEREF II	Laser IRS —	none none	ASCB, Arinc 429, Arinc 419 Incl: attitude, heading (mag & true); GS, wind, track, present, position	TAS Alt V. Spd.	115 VAC, 400 Hz or 28 VDC back-up —	2/48.0	$149,563	Primarily digital interfaces to Honeywell FMS, EFIS, autopilot; fast align; no pressurization required
Internav Box 1261 Sydney, Nova Scotia B1P 6J9 Canada (902) 564-2043	LC1210 (sensor)	Loran-C none	— —	Position	—	12/28 VDC, 20W —	2/10.0	$8,000	Loran-C sensor for navigation computer systems; continuous dual-chain operation; Output/ Control via Arinc 429 (dual input, triple output)
	LC1200	Loran-C vacuum fluorescent	200 nonvolatile RAM and EPROM	Pos; track error; TTW; GS; HSI; wind; digital interface	Heading; TAS, VOR-DME	12/28 VDC, 40W —	3/15.0	$12,000	Continuous dual chain operation, Arinc format CDU; 1,200 programmed waypoint database; multiroute storage, optional VOR/DME back-up mode; five character alphanumeric waypoint identifiers position output for air-ground datalink; D/R mode usable outside Loran signal areas; optional SAR program

LONG-RANGE NAVIGATION SYSTEMS

MANUFACTURER	MODEL	SYSTEM TYPE / DISPLAY	NUMBER WAYPOINTS / DATABASE	OUTPUTS	INPUTS	POWER INPUTS / MEMORY	UNITS/ WEIGHT (lb)	PRICE (uninstalled)	REMARKS
Litton Aero Products 6101 Condor Road Moorpark, CA 93021 (805) 378-2000	LTN-311	VLF/Omega 8-line LED	99 none	Digital pos; track; GS; TTG, HSI; Wind; XTD; TAE; DA; TH	Heading; TAS; ADC valid	115 VAC, 400 Hz —	3/41.0	$44,190	Direct ARINC replacement for LTN-211; integrated GPS optional; ACARS route entry with OAT option; AC/DC noninterruptible P.S.; XTK discrete; Digital Air Data (575/429); std ARINC 599; BITE/STATUS in English language; pre-heated oscillator for rapid start; 1 year warranty
	LTN-211	Omega —	9 none	Pos; track error; TTW; GS; HSI; wind	Heading; TAS	115 VAC, 400 Hz —	3/34.0	$44,580	HSI synchro and digital outputs standard; Arinc 599; optional VLF receiver available; 1 year warranty
	LTN-710	GPS LED matrix	— —	Digital ARINC 429; pos; track; vert track; velocity; acceleration Time (> 100 ns); course steering	Altimeter precise clock	115 VAC, 60-400 Hz —	3/20.0	$50,000	Designed for aerial survey; present position display only; operates over entire range of signal levels; fixed freq. signals; single correlator; two IF freqs.; phase sampler; time accuracy to 900 nanoseconds; operational worldwide; 24 hr. all-weather coverage; pos. accuracy 40 SEP 3-D; 1 year warranty
	LTN-72	INS —	9 none	Pos; track error; TTW; GS; HSI; wind	TAS	115 VAC, 400 Hz 28 VDC back-up —	3/62.0	$172,285	Expandable memory; may be transverse-mounted; standby battery optional; Arinc 561; 1 year warranty
	LTN-92	Laser INS 5-line LED matrix	99 none	Arinc 561 digital & analog Arinc 704 digital	VOR/DME; Tacan; Arinc 429; autopilot; steering; EFIS, PMS; Radar; pos; trk error; TTW; GS; HSI; wind	115 VAC / 400 Hz 28 VDC back-up —	3/63.6	$174,360	Pin for pin replacement for gymbaled-gyro inertial nav systems — Arinc 561 in 1 ATR configuration; accommodates Omega/VLF acceleration advisory or GPS; interfaces with bulk storage units; solid-state design; 3 year warranty
	LTN-72R	INS/ RNAV —	9 none none	Pos; track error; TTW; GS; HSI; wind	TAS; VOR-DME	115 VAC, 400 Hz 28 VDC back-up —	3/62.0	$175,090	Combined INS andvortac-based RNAV; compatible with Global-Wulfsberg NDB-2; meets requirements for enroute, terminal and approach; may be transverse-mounted; standby battery optional; Arinc 561; 1 year warranty

LONG-RANGE NAVIGATION SYSTEMS

MANUFACTURER	MODEL	SYSTEM TYPE / DISPLAY	NUMBER WAYPOINTS / DATABASE	OUTPUTS	INPUTS	POWER INPUTS / MEMORY	UNITS/ WEIGHT (lb)	PRICE (uninstalled)	REMARKS
Litton Aero Products 6101 Condor Road Moorpark, Ca. 93021 (805) 378-2000	LTN-90 -100	Laser IRS sensor	N/A N/A	Arinc 429; Pos; attitude; heading; angular rates; acceler-ations; drift angle; VS; GS; flight path angle, wind track	Arinc 429 air data	115 VAC, 400 Hz 28 VDC back-up —	2/43.0	$190,030	Optional control display unit; primary inertial sensor for navigation management computer systems; complies with Arinc 704 spec for IRS; growth capability to ADIRS; 10 MCU size; for digital applications; 3 year warranty
	LTN-72RL	INS/ RNAV —	99 none	Pos; track error; TTW; GS; HSI; wind	TAS; VOR-DME	115 VAC, 400 Hz 28 VDC back-up —	3/62.0	$192,070	Combined INS and VORTAC-based RNAV; compatible with Global-Wulfsberg NDB-2; meets requirements for enroute, terminal and approach; transverse-mounted; standby battery optional; bulk storage with five-line alphanumeric LED display; Arinc 561; 1 year warranty
	LTN-91	Laser IRS sensor	N/A N/A	Arinc 429; Pos; attitude; heading; angular rates; acceler-ations; drift angle; VS; GS; flight path angle, wind track	Arinc 429 air data	115 VAC, 400 Hz 28 VDC back-up —	2/43.0	$196,030	Optional control display unit; primary inertial sensor for navigation management computer systems; complies with Arinc 704 spec for IRS; growth capability to ADIRS; 10 MCU size; additional analog outputs; 3 year warranty
Northstar Avionics 30 Sudbury Rd. Acton, MA 01720 (617) 897-6600	M1	Loran-C LED	250 non-volatile RAM	analog	rotary knob	10 to 30 VDC EPROM	2/4.5	$3,995	Great ergonomics; VFR-only; no TSO; 20,000+ waypoint database; update with 8 EPROM chips — $75, dealer replaceable; first update included in purchase price; all 6,500 US public airports, 1,150 VORs, 1,600 NDBs and 4,800 low-altitude intersections; 20 waypoint flight plan; 3 year warranty
Narco Avionics Inc. P.O. Box 1807 Laguna Beach, CA 92652-1807 (714) 497-5077	LRN 840	Loran-C LCD	200 non-volatile RAM	External CDI autopilot	RS-232	11 to 32 VDC RAM with five yr. battery	2/4.0	$1,050	Automatic selection of GRI; secondaries MAG VAR; database available with airports, VORs, other fixes; database update cartridge — $75, plugs in back of unit
STS Avionic Products, Inc. 11600 Lilburn Park Road St. Louis, MO 63146 (314) 567-0304	120MC	Loran-C LCD	120 non-volatile RAM	analog, CDI	encoded altitude, heading, air data	10 to 32 VDC —	2/5.0	$1,995	Cross-chain, master-independent operation; concentric knob controls; 15,000 waypoint database; database updates with unit installed; includes antenna, coupler and mounting hardware; new super-twist LCD; 3 year warranty

LONG-RANGE NAVIGATION SYSTEMS

MANUFACTURER	MODEL	SYSTEM TYPE DISPLAY	NUMBER WAYPOINTS DATABASE	OUTPUTS	INPUTS	POWER INPUTS MEMORY	UNITS/ WEIGHT (lb)	PRICE (uninstalled)	REMARKS
Terra Corp. 3520 Pan American Freeway NE Albuquerque, NM 87107 (505) 884-2321	TLC 120 none	Loran-C gas discharge	120 none	CDI autopilot	rotary knob	10 to 40 VDC —	2/2.5	$1,595	2.0-in high by 6.25-in wide by 11.0-in long; panel-mounted; includes whip antenna with CT6 preamp
Tracor, Inc. Avionics Products 6500 Tracor Lane MD 27-23 Austin, TX 78725 (512) 929-2233 Exclusive Distributor: DAC International Inc. 6702 McNeal Dr. Austin, TX 78729 (512) 331-5323	TA-7880	Omega/ VLF color CRT	2,480 30,000 waypoints optional	Pos; track error; TTW; GS; HSI; winds; CDI; autopilot roll steering; radar; cabin display	heading; TAS	21.5 VDC to 36 VDC —	3/17.0	$29,583*	*base price; 2,480 pilot-defined waypoints, 170 flight plans with up to 96 waypoints each, alphanumeric identifiers; optional outputs: analog — HSI, autopilot, synchro and roll steering; digital — Arinc 429, 561; EFIS compatible; Jeppesen world-wide database optional — $11,275; TA-2000 3½" disk database loader — $3,360; TA-7900 large CDU ¢6,797, exchange; includes E-field antenna and installation kit; optional DME/DME nav, GPS
	TA-7900	Omega/VLF color CRT	2,480 30,000 waypoints optional	Pos; track error TTW; GS; HSI; winds; CDI; autopilot roll steering; radar; cabin display	heading; TAS	21.5 VDC to 36 VDC or 115 AC, 400 Hz —	3/23.0	$44,973*	*base price; Arinc 599 spec'n Omega/VLF; price includes ADF coupler; 2,480 pilot-defined waypoints; 170 flight plans with up to 96 waypoints each, alphanumeric identifiers; standard outputs: analog — HSI, autopilot, synchro and roll steering; digital — Arinc 419, 429, 561; EFIS compatible; 60,000 waypoint Jeppesen database optional — $11,275; TA-2000 3½" disk database loader — $3,360
II Morrow, Inc. P.O. Box 13549 Salem, OR 97309 (503) 581-8101	604	Loran-C LED	200 —	CDI, HSI, autopilot, RS-232	concentric knobs, push buttons	6.5-48 VDC RAM with five-year battery	1/3.68	$1,795	VFR only; TSO C60a; panel-mounted; 6.25W x 2.0H; 350 KTAS A-16 antenna with preamp included; manual magnetic variation input; world-wide coverage — 14 station chains; manual secondary selection; upgradable to 604TCA configuration; includes installation kit; 26-month warranty

LONG-RANGE NAVIGATION SYSTEMS

MANUFACTURER	MODEL	SYSTEM TYPE / DISPLAY	NUMBER WAYPOINTS / DATABASE	OUTPUTS	INPUTS	POWER INPUTS / MEMORY	UNITS/ WEIGHT (lb)	PRICE (uninstalled)	REMARKS
II Morrow, Inc. P.O. Box 13549 Salem, OR 97309 (503) 581-8101	612B	Loran-C LEC	100 EPROM	CDI, HSI, autopilot RS-232	concentric knobs, push buttons	6.5-48 VDC RAM with five-year battery	1/3.75	$2,695	VFR only; 6.25in. wide by 2.0-in high; panel-mounted; includes A-16 antenna with preamp; TSO C60a; standard database for 48 United States and Alaska; optional database includes 620 Canadian airports; dealer-installed EPROM database updates — $75; alphamatic sort; offset waypoint mode; non-volatile two-leg flight plan; couples to Shadin fuel management system, Glatzer ELT; VNAV option; Model 612C also available in 3 ATI package — $2,995; includes installation kit; 26-month warranty
	604TCAep	Loran-C LED	100 cartridge	CDI, HSI, autopilot, external annunicators, RS-232	concentric knobs, push buttons	6.5-48 VDC RAM with five-year battery	1/3.75	$2,995	VFR only; 6.25-in wide by 2.0-in high; includes A-16 antenna with preamp; all 19 Loran-C chains; database cartridge user-installed in rear panel — includes 7,000 airports, VORs, VORTACs and NDBs in US, Canada and Alaska; update cartridges available every 56 days, each — $75, annual subscription — $200; couples to Shadin fuel management system, Glatzer ELT; 26-month warranty
	618	Loran-C LED	100 EPROM	CDI, HSI autopilot, external annunciators RS-232	concentric knobs, push buttons	6.5-48 VDC RAM with five-year battery	2/3.75	$3,295	VFR only; 6.25W x 2.0H; panel-mounted; A-16 antenna with preamp included; 20,000+ WP database for lower-48 US; updates — $75; auto secondary search; emergency airport search; couples to Shadin fuel management system, Glatzer ELT; comprehensive special use airspace intrusion feature; 26-month warranty
	612	Loran-C LED	100 EPROM	CDI, HSI autopilot, external annunciators RS-232	concentric knobs, push buttons	6.5-48 VDC RAM with five-year battery	2/3.75	$3,295	TSO C60a; 6.25W x 2.0H; panel-mounted; A-16 antenna with preamp included; IFR — AC 20-121; 2,600 WP database for lower-48 US; updates — $75; auto secondary search; emergency airport search; couples to Shadin fuel management system, Glatzer ELT; NPA software available; 26-month warranty

LONG-RANGE NAVIGATION SYSTEMS

MANUFACTURER	MODEL	SYSTEM TYPE DISPLAY	NUMBER WAYPOINTS DATABASE	OUTPUTS	INPUTS	POWER INPUTS MEMORY	UNITS/ WEIGHT (lb)	PRICE (uninstalled)	REMARKS
II Morrow, Inc. P.O. Box 13549 Salem, Or. 97309 (503) 581-8101	614R	Loran-C LED	100 EPROM	CDI, HSI, autopilot RS-232	N/A	6.5-48 VDC RAM with five-year battery	2/4.75	$6,995	IFR — AC 20-121; IFR demo approach certification; remote, standard Dzus rail mount for CDU; includes 521 KTAS-rated, A-17 antenna with preamp; TSO C60a; 0 — 50,000' MSL; VNAV standard; all other specifications same as 612IFR

VHF NAVCOMS

MANUFACTURER	MODEL TSO	DISPLAY MEMORY CHANNELS STORABLE	VOR OUTPUT	POWER OUTPUT (watts)	POWER INPUT	UNITS/ WEIGHT (lb) SIZE	PRICE (uninstalled)	REMARKS
Aire-Sciences, Inc. 216 Passaic Ave. Fairfield, NJ 07006 (201) 228-1880	RT-553A none	mechanica none none	CDI/HSI	6	14 VDC*	1/5.5 6.5 x 3.25 x 10.9	$2,175	Built-in converter-indicator; output for remote converter-indicator; built-in intercom; *28 to 14 VDC adapter, $145
	RT-563A none	mechanical none none	CDI/HSI	5	14 VDC*	1/7.8 6.5 x 3.25 x 10.9	$2,995	Automatic omni; built-in converter-indicator; output for remote converter-indicator optional *28 to 14 VDC adapter, $145
Bendix/King Avionics 400 N. Rogers Rd. Olathe, KS 66062 (913) 782-0400	KX 155 C34c C36c C37b C38b C40a	gas discharge EAROM non-volatile 2	CDI/RMI	10	14 or 28 VDC	2/7.0 6.25 x 2.5 x 10.25	$3,240	Auto squelch; includes KI 209 VOR/ LOC/ GS converter-indicator; price includes built-in 40-channel GS receiver; optional audio amplifier; 760 channel;
	KX 165 C34c C36c C37b C38b C40a	gas discharge EAROM non-volatile 2	CDI/RMI HSI	10	14 or 28 VDC	2/7.0 6.25 x 2.0 x 10.25	$7,225	Auto squelch; includes KI 206 VOR/LOC/ GS indicator; price includes built-in VOR converter and GS receiver; price less GS with indicator $6,350; digital radial readout
Narco Avionics Inc. P.O. Box 1807 Laguna Beach, CA 92652-1807 (714) 497-5077	Escort IIA none	gas discharge none none	none	5	14 VDC,	1/3.0 3.25 x 3.25 x 10.75	$1,149	Auto omni, digital OBS, ECDI, and digital bearing; 28 VDC configuration — $1,229

Footnote: All prices and weights, unless noted in Remarks, include a nav indicator.

VHF NAVCOMS

MANUFACTURER	MODEL TSO	DISPLAY MEMORY CHANNELS STORABLE	VOR OUTPUT	POWER OUTPUT (watts)	POWER INPUT	UNITS/ WEIGHT (lb) SIZE	PRICE (uninstalled)	REMARKS
Narco Avionics Inc. P.O. Box 1807 Laguna Beach, CA 92652-1807 (714) 497-5077	MK-12D C34c C37b C38b C40a	gas discharge hot-wire non-volatile 2/2	CDI/HSI	8	14 or 28 VDC	2/4.4 6.25 x 2.5 x 11.0	$3,569	Includes GS receiver and separate ID-825 VOR/LOC/GS indicator; $2,955 without GS and with ID-824 VOR/LOC indicator; includes 2x5 DME channeling; 8-watt transmitter; 720 channels; Cessna ARC form factor,direct-replacement design — $1,795
	MK-12E C34c C37b C38b C40a	gas discharge hot-wire non-volatile 2/2	CDI/HSI	8	14 or 28 VDC	2/5.4 6.25 x 2.5 x 11.0	$3,955	Includes GS receiver and separate ID-825 VOR/LOC/GS indicator; $3,343 without GS and with ID-824 VOR/LOC indicator; includes 2x5 DME channeling; 8-watt transmitter; 760 channels
	NCS-812 C34c C37b C38b C40a	gas discharge hot-wire non-volatile 2/2	CDI/HSI	8	14 or 28 VDC	2/5.4 6.25 x 2.5 x 11.0	$5,510	Includes separate 25-watt IDME-895 DME/ VOR/LOC/GS indicator and GS receiver; 8-watt VHF comm transmitter; 760 channels
Terra Corp. 3520 Pan American Freeway NE Albuquerque, NM 87107 (505) 884-2321	TXN 920 none	mechanical/ ECDI none none	CDI	6	14 VDC*	2/2.5 6.25 x 1.6 x 10.0	$1,795	Pushbutton tuning; manual or auto squelch; with GS — $2,195; 2x5 DME channeling; *28 to 14 VDC adapter — $85
	TXN 960 none	mechanical/ ECDI none none	built-in	6	14 VDC*	1/3.7 6.25 x 3.3 x 10.0	$1,795	Pushbutton tuning; auto squelch, manual override; with GS — $2,195; 2x5 DME channeling: *28 to 14 VDC adapter — $85

Footnote: All prices and weights, unless noted in Remarks, include a nav indicator.

VHF PANEL-MOUNTED NAV RECEIVERS

MANUFACTURER	MODEL TSO	DISPLAY MEMORY CHANNELS STORABLE	INCLUDES GS MB	POWER INPUT VOR OUTPUT	UNITS/ WEIGHT (lb) SIZE	PRICE (uninstalled)	REMARKS
Aire-Sciences, Inc. 216 Passaic Ave. Fairfield, NJ 07006 (201) 228-1880	R-662 C36c C40a	mechanical none none	no no	14 or 28 VDC DC to CDI	2/4.9 3.25 x 2.62 x 11.4	$945	interfaces with ECDI-541 converter-indicator — $995; glideslope and 2x5 DME channeling optional
	R-552 none	mechanical none none	no no	14 or 28 VDC DC to CDI	2/5.0 3.2 x 2.62 x 11.44	$1,745	Includes converter-indicator; glideslope and 2x5 DME channeling optional; includes ECDI-554A display

VHF PANEL-MOUNTED NAV RECEIVERS

MANUFACTURER	MODEL TSO	DISPLAY MEMORY CHANNELS STORABLE	INCLUDES GS MB	POWER INPUT VOR OUTPUT	UNITS/ WEIGHT (lb) SIZE	PRICE (uninstalled)	REMARKS
Aire-Sciences, Inc. 216 Passaic Ave. Fairfield, N.J. 07006 (201) 228-1880	R-554 none	mechanical none none	yes no	14 or 28 VDC DC to CDI	2/5.4 3.25 x 2.62 x 11.44	$2,620	Includes indicator; 2x5 DME channeling
	R-664 C34c C36b C40a	mechanical none none	yes no	14 or 28 VDC DC to CDI	2/5.4 3.25 x 2.62 x 11.44	$2,885	Interfaces with OBS indicator; 2x5 DME channeling
Bendix/King Avionics 400 N. Rogers Rd. Olathe, KS 66062 (913) 782-0400	KN 53 C34c C36c C40a	gas discharge EAROM non-volatile 2	no no	11 to 33 VDC CDI*	2/4.7 6.25 x 1.3 x 10.5	$4,250	Displays both active and preset frequency; price includes *KI 203 indicator; $5,100 with GS and *KI 204 indicator
Collins Avionics Div. Rockwell International 400 Collins Rd. NE Cedar Rapids, IA 52498 (319) 395-4085	VIR-351 C36c C40a	incandescent filament none none	no no	14 VDC* OBS resolver ORZ'd at 300° 0.05 Vrms to RNAV	2/4.2 3.12 x 2.6 x 12.45	$4,925	Displays either frequency or to/from bearing; electronic frequency readout; includes IND-350A indicator; includes glideslope and DME channeling; Arinc 547 *28 to 14 VDC adapter — $120
Narco Avionics Inc. P.O. Box 1807 Laguna Beach, CA 92652-1807 (714) 497-5077	NAV-824 C36c C40a	gas discharge hot-wire non-volatile 2	no no	11 to 32 VDC Arinc: .5 Vms at 0°	1/3.3 6.25 x 1.5 x 11.0	$2,132	RMI function; ID 824 indicator included in price; 2x5 DME channeling
	NAV-825 C34c C36c C40a	gas discharge hot-wire non-volatile 2	yes no	11 to 32 VDC N/A	1/3.3 6.25 x 1.5 x 11.0	$2,625	RMI function; ID 825 indicator included in price; 2x5 DME channeling
Terra Corp. 3520 Pan American Freeway, NE Albuquerque, NM 87107 (505) 884-2321	TN 200 none	mechanical none none	optional no	14 VDC* N/A	1/1.5 3.2 x 1.6 x 9.0	$695	2x5 DME channeling; with GS option $995; *28 to 14 VDC adapter — $85

VHF REMOTE-MOUNTED NAV RECEIVERS

MANUFACTURER	MODEL TSO	DISPLAY MEMORY CHANNELS STORABLE	INCLUDES GS MB	POWER INPUT VOR OUTPUT	UNITS/ WEIGHT (lb) SIZE	PRICE (uninstalled)	REMARKS
Bendix/King Avionics 400 N. Rogers Rd. Olathe, KS 66062 (913) 782-0400	VNS-41 C40b C36c C34c C35d	LCD or dichroic non-volatile 2	yes yes	28 VDC analog, Arinc 429	2/5.5 4.0 x 4.0 x 12.62	$12,695	Digital RMI readout, analog or Arinc 429 interface; includes LCD control head; two-year warranty
	KNR 634 C40a C36c C34c C35d	gas discharge EAROM non-volatile 2*	yes yes	28 VDC CDI, RMI, HSI	2/5.8 2.0 x 5.0 x 10.0	$13,310	Includes GS and MB receivers; self-contained converter; price includes KFS 564 with simultaneous digital display of active and standby frequency; Gables control option; King Serial Data tuning; also available without GS or MB; requires 26 VAC for sine/cosine output; night vision compatible "A" control head available; *optional control head with 9 programmable frequencies
	KNR 634A C40a C36c C34c C35d	gas discharge EAROM non-volatile 2*	yes yes	28 VDC Arinc 429 CDI, RMI, HSI	2/6.5 3.0 x 5.0 x 10.0	$17,485	synchro-interface RMI optional; price includes GS and MB receivers, control head; night vision compatible "A" control head available; *optional control head with 9 programmable channels; two-year warranty
	KTU 709 C66a	252 gas discharge EAROM non-volatile none	no no	28 VDC CDI, RMI	2/6.6 3.0 x 5.0 x 10.0	$18,425	Tacan: includes DME; bearing information derived from DME signal; price includes KDI 572 DME indicator; King serial data tuning; control head optional; night vision compatible "A" control head available; two-year warranty
Collins Avionics Div. Rockwell International 400 Collins Rd. NE Cedar Rapids, IA 52498 (319) 395-4085	VIR-32 C34c C35d C36d C40b	gas discharge non-volatile 6	yes yes	28 VDC, 26 VAC, 400 Hz sine/cosine serial digital	2/5.7 ⅜ ATR-SL	$13,390	RMI compatible; replaces VIR-30A directly; includes CTL-32 serial digital control; digital signal processing; built-in diagnostics; rotor modulation filter optional (strap)
	VIR-432 C34d C35d C36d C40b	gas discharge non-volatile 6	yes yes	28 VDC, 26 VAC, 400 Hz serial digital	2/5.7 ⅜ ATR-SL	$15,055	Built-in diagnostics; compatible only with CSDB or Arinc 429 controls; includes CTL-32 CSDB control; digital signal processing; auto instrumentation for use with RMI; rotor modulation filter optional (strap)
	51RV-4 C34c C36c C40a	mech. synchro none none	yes no	28 VDC, 26 VAC, 400 Hz electronic synchro driver	1/9.9 ½ ATR-S	$17,184	Self-contained converter; RMI capability; RNAV compatibility; options for glideslope receiver and monitoring of any or all functions; digital circuitry, output and tuning; coherent detector in receiver; Arinc 547; 2x5 DME channeling

VHF REMOTE-MOUNTED NAV RECEIVERS

MANUFACTURER	MODEL / TSO	DISPLAY / MEMORY / CHANNELS STORABLE	INCLUDES GS / MB	POWER INPUT / VOR OUTPUT	UNITS/ WEIGHT (lb) / SIZE	PRICE (uninstalled)	REMARKS
	VIR-31A **C34c** **C35d** **C36c** **C40a**	**gas discharge** **non-volatile** **2**	**yes** **yes**	**28 VDC; 26 VAC, 400 Hz** **three-wire synchro**	**2/8.0** **½ ATR-SL**	**$18,425**	**RMI compatibility; includes CTL-30 electronic control; 2x5 DME channeling; three-wire synchro VOR output; rotor modulation filtering optional**
Collins Avionics Div. Rockwell International 400 Collins Rd. NE Cedar Rapids, IA 52498 (319) 395-4085	VIR-30A C34c C35d C36c C40a	gas discharge nonvolatile 2	yes yes	28 VDC; 26 VAC, 400 Hz sine/cosine	2/6.8 ⅜ ATR-SL	$18,840	RMI compatibility; superflag output; includes CTL-30 electronic control; 2x5 DME channeling; four-wire sine-cosine VOR output
Honeywell Inc. P. O. Box 29000 Phoenix, AZ 85038 (602) 863-8000	RNZ-850 C34d C35d C36d C40b C41d C66b	Color CRT non-volatile RAM 1 to 12	yes yes	28 VDC, 26 VAC 400 Hz serial digital, analog	2/—	—	Nav subsystem of SRZ-850 Primus II integrated, digital radio system; full dual system price — $128,427; includes RM-850 color CRT radio management unit, digital audio system, NV-850 VHF nav receiver, DF-850 ADF with AT-860 combined loop sense antenna, DM-850 precision format compatible DME (+/−100-ft) with station ID display and 6 channel scanning; built-in diagnostics, maintenance log; 1 to 12 stored channels depending on function; self-contained converter; digital signal processing; RTCA DO-160B environmental, DO-178A software

VHF PANEL-MOUNTED TRANSCEIVERS

MANUFACTURER	MODEL / TSO	DISPLAY	MEMORY / CHANNELS STORABLE	NOMINAL POWER OUTPUT (watts)	POWER INPUT	UNITS/ WEIGHT (lb) / SIZE	PRICE (uninstalled)	REMARKS
Aire-Sciences, Inc. 216 Passaic Ave. Fairfield, NJ 07006 (201) 228-1880	RT-551A none	mechanical	none none	6	14 VDC*	1/2.7 3.25 x 2.62 x 10.66	$2,050	Manual squelch; portable version available *28 to 14 VDC adapter, $145
	RT-661A C37b C38b	mechanical	N/A N/A	6	14 or 28 VDC*	1/2.7 3.25 x 2.62 x 11.44	$2,250	Manual squelch; portable version available *Adapter, $145

VHF PANEL-MOUNTED TRANSCEIVERS

MANUFACTURER	MODEL TSO	DISPLAY	MEMORY CHANNELS STORABLE	NOMINAL POWER OUTPUT (watts)	POWER INPUT	UNITS/ WEIGHT (lb) SIZE	PRICE (uninstalled)	REMARKS
Bendix/King Avionics 400 N. Rogers Rd. Olathe, KS 66062 (913) 782-0400	KY 96A C37c C38c	LCD	EEPROM 9	7	28 VDC	1/2.9 6.25 x 1.3 x 10.5	$1,140	Auto leveling; auto squelch; stuck mike protection; remote mike keying and channel call-up; 760 channel
	KY 97A C37c C38c	LCD	EEPROM 9	7	14 VDC	1/2.9 6.25 x 1.3 x 10.5	$1,140	Auto leveling; auto squelch; stuck mike protection; remote mike keying and channel call-up; 760 channel
Bendix/King Avionics 400 N. Rogers Rd. Olathe, KS 66062 (913) 782-0400	KY 92 C37b C38b	mechanical	none none	7	14 VDC*	1/3.2 6.25 x 1.5 x 10.5	$2,160	Auto squelch; self-test *28 to 14 VDC adapter, $300; 760 channel
	KY 196A C37b C38b	gas discharge	EAROM non-volatile 9	16	28 VDC	1/3.2 6.25 x 1.3 x 10.5	$3,060	Auto squelch; displays active and standby frequency; remote flip-flop; optional remote call-up; 760 channel; optional extended frequency range available
	KY 197A C37b C38b	gas discharge	EAROM non-volatile 9	7	14 VDC	1/3.2 6.25 x 1.3 x 10.5	$3,060	Same as KY 196A, but 14 VDC power input
Collins Avionics Div. Rockwell International 400 Collins Rd. NE Cedar Rapids, IA 52498 (319) 395-4085	VHF-253 C37b Class II; C38b	LCD	EAROM non-volatile 6	10	10 to 32 VDC	1/2.7 6.3 x 1.3 x 12.1	$3,910	Remote frequency management; visible active and standby frequencies plus four stored presets; liquid-crystal displays; optional broadband version (VHF-253S) $4,440
General Aviation Electronics, Inc. 8284 E. 250th Hampton, MN 55031 (507) 263-4002	Alpha 720 none	incandescent filament	none none	5	14 VDC	1/4.0 6.5 x 2.5 x 10.0	$1,000	Vehicle- or ground-station versions available; 25kHz spacing; 10-watt carrier option — $200
Narco Avionics Inc. P.O. Box 1807 Laguna Beach, CA 92652-1807 (714) 497-5077	COM-810/811 C38b Class C	gas discharge	volatile (.2mA hotwire) 2	8	14 or 28 VDC	1/3.6 6.25 x 1.5 x 11.0	$1,295	Auto squelch; audio leveling: speaker, mike; 10-watt amp included
Terra Corp. 3520 Pan Am Freeway NE Albuquerque, NM 87107 (505) 884-2321	TX 720 none	mechanical	none none	5	14 VDC*	1/1.2 3.2 x 1.6 x 9.0	$945	Auto squelch; lighted; intercom; portable, vehicle or ground station version available *28 to 14 VDC adapter — $85

VHF REMOTE-MOUNTED TRANSCEIVERS

MANUFACTURER	MODEL TSO	DISPLAY	MEMORY CHANNELS STORABLE	POWER OUTPUT (watts)	POWER INPUT	UNITS/ WEIGHT (lb) SIZE	PRICE (uninstalled)	REMARKS
Bendix/King Avionics 400 N. Rogers Rd. Olathe, KS 66062 (913) 782-0400	KTR 908 C37b C38b	gas discharge	EAROM 2*	20	28 VDC	2/4.3 1.8 x 5.0 x 10.0	$8,310	Includes KFS 598 control with digital display of active and standby frequencies; Gables control option; remote flip-flop *optional control head with 9 programmable frequencies and remote frequency call-up function; 760 channel; 118.0 - 151.975 MHz option; night vision compatible "A" control head available; 2-year warranty
	VCS-40 C37c C38c	LCD or dichroic	non-volatile 2	20	28 VDC	2/5.5 4.0 x 4.0 x 12.62	$10,775	Arinc 429 or analog interface; continuous transmit capable; SELCAL and ACARS compatible; 760 channel; 118.0 - 151.975MHz option; 2-year warranty
Collins Avionics Div. Rockwell International 400 Collins Rd. NE Cedar Rapids, IA 52498 (319) 395-4085	VHF-21A C37c C38c	gas discharge	non-volatile 8	20	28 VDC	2/5.8 ⅜ ATR-SL	$8,900	Directly replaces VHF-20A; built-in diagnostics; 2x5 serial tuning; includes CTL-22 serial tuning control; VHF-21B, 118.0 - 151.975 MHz — $11,635
	VHF-22A C37c C38c	gas discharge	non-volatile 8	20	28 VDC	2/5.8 ⅜ ATR-SL	$8,900	Built-in diagnostics; 2x5 serial digital tuning — includes CTL-22 serial tuning control; VHF-22B, 118.0-151.975MHz — $11,635
	618M-3 C37b C38b	none	none none	25	28 VDC	1/10.1 ½ ATR-S	$9,748	Arinc 566A
	VHF-422A C37c C38c	gas discharge	non-volatile 8	20	28 VDC	2/5.8 ⅜ ATR-SL	$9,895	Built-in diagnostics; compatible only with CSDB or Arinc 429 controls; includes CTL-22 CSDB control; VHF-422B, 118-151.975 MHz — $12,795
	VHF-20A C37b C38b	gas discharge	non-volatile 2	20	28 VDC	2/6.8 ⅜ ATR-SL	$10,705	CTL-20 electronic control included; VHF-20B, 118 - 151.975 MHz — $13,635
Global-Wulfsberg Systems 2144 Michelson Dr. Irvine, CA 92715 (714) 851-0119	WT-200B C37b C38b	mechanical	none none	20	28 VDC	1/5.6 2.9 x 5.2 x 13.28	$7,982	Arinc and VHF 20A conversion mount available; 1,360 channels at 25 KHz spacing from 118.000 to 151.975 MHz; narrow or wide IF bandwidth models available; all solid state

Footnote: Arinc form factor is shown in the Units/Weight/Size columns. (See Form Factor explanation at the front of the Avionics section.)

VHF REMOTE-MOUNTED TRANSCEIVERS

MANUFACTURER	MODEL TSO	DISPLAY	MEMORY STORABLE CHANNELS	POWER OUTPUT (watts)	VOLTS INPUT	UNITS/ WEIGHT (lb) SIZE	PRICE (uninstalled)	REMARKS
Honeywell Inc. P. O. Box 29000 Phoenix, AZ 85038 (602) 863-8000	RCZ-850 C37c C38c C112	color CRT	non-volatile RAM 2 to 12	20, 500*	28 VDC	2/— 14.0 x 6.8 x 3.36	—	Comm subsystem of SRZ-850 Primus II integrated, digital radio system; full dual system price — $128,427; includes RM-850 color CRT radio management unit, 20-channel AV-850 or -851 digital audio system, TR-850 VHF transceiver with 118.0 - 151.975MHz range, XS-850 500-Watt Mode S transponder; built-in diagnostics, maintenance log; 2 to 12 stored channels depending on function; digital signal processing; RTCA DO-160B environmental, DO-178A software

Footnote: Arinc form factor is shown in the Units/Weight/Size columns. (See Form Factor explanation at the front of the Avionics section.)

HF TRANSCEIVERS

MANUFACTURER	MODEL TSO	CHANNELS FREQ CONTROL	FREQ. RANGE (MHz)	POWER OUTPUT (watts)	POWER INPUT	UNITS/ WEIGHT (lb)	PRICE COUPLER	REMARKS
Aircom Electronics, Ltd 9501 Ryan Ave. Dorval, Quebec H9P 1A2 Canada (514) 636-3874	SSB 10/20 none	10 crystal	2-29.999 SSB/AM	100 PEP	28 VDC	1/7.3	$10,080* $1,344*	*$12,000 Canadian, $1,600 Canadian; remote-, panel- and direct- (side) mount configurations available; 14 - 28 VDC remote coupler, fully compatible with other HF radios;
Bendix/King Avionics 400 N. Rogers Rd. Olathe, KS 66062 (913) 782-0400	KHF 950 C31c C32c	280,000 synthesizer	2-29.9999 SSB/AM	150 PEP (SSB) 35 (AM)	28 VDC	3/20.2	$16,860 —	Automatic antenna coupler included; 99 operator programmable, 176 ITU radiotelephone channels with standard, KCU 951 control head; optional, space saving KFS 594 control head provides 19 operator programmable, 176 ITU radiotelephone channels; optional dual HF $hunt antenna; optional SELCAL compatibility; optional speech processing; explosion proof; two-year warranty
	KHF 990 C31c C32c	280,000 synthesizer	2-29.9999 SSB/AM	150 PEP (SSB) 35 (AM)	28 VDC	3/22.5	$19,270 —	Designed for helicopter use; includes digital antenna coupler/ probe antenna in single unit; 19 preset, 176 ITU radiotelephone channels with standard KFS 594 control head; explosion proof, salt-spray proof; two-year warranty

HF TRANSCEIVERS

MANUFACTURER	MODEL TSO	CHANNELS FREQ. CONTROL	FREQ. RANGE (MHz)	POWER OUTPUT (watts)	POWER INPUT	UNITS/ WEIGHT (lb)	PRICE COUPLER	REMARKS
Collins Avionics Div. Rockwell International 400 Collins Rd. NE Cedar Rapids, IA 52498 (319) 395-4085	HF-230 C31c C32c	280,000 synthesizer	2-29.9999 SSB/AM AM	100 PEP (SSB)	28 VDC	4/24.4	$16,640 —	Half-duplex; compatible with maritime radiotelephone network; additional channels: 40 operator programmable, 176 ITU radiotelephone preprogrammed channels; optional SELCAL; PAC-230 antenna coupler for helicopters, $17,710 system price; weight 23.2 lbs.
	HF-9000 C31d C32d	280,000 synthesizer	2-29.9999 SSB/AM AM data	175 PEP,	28 VDC	3/27.5	$28,200 —	Fiber-optic interface; 99 operator programmable, 176 ITU radiotelephone programmed channels; selectable power output — 10, 50, 175-Watt PEP; rapid-tune antenna coupler (40 msec computer tuning); BITE
	628T-3A C31c C32c	280,000 synthesizer	2-29.9999 SSB/ AM/CW	400 PEP (SSB) 125 (AM) 25 (CW)	28 VDC; 115 VAC, 400 Hz	3/26.0	$30,336 —	Panel-mounted control; remote-mounted ½ ATR-S transceiver and antenna coupler; power amplifier coupler; shock mounts included; frequencies automatically tuned in 0.1-kHz increments; Arinc 533, 404; Varies with installation
Sunair Electronics 3101 S.W. 3rd Ave. Fort Lauderdale, FL 33315 (305) 525-1505	ASB-500 C31c C32c	32,000 synthesizer	2-17.999 SSB/AM	100 PEP (SSB)	28 VDC	2/36.3	$9,495* —	*Factory dealer price; panel-mounted LED control unit; remote-mounted transceiver; SELCAL compatibility; includes automatic, full HF range ACU-150D microprocessor controlled antenna coupler
	ASB-850 C31c C32c	284,000 synthesizer	2-29.9999 SSB/AM	100 PEP (SSB)	28 VDC	3/30.0	$11,850 —	Integrated package including SCU-80 control-display unit, RE-800 remote-mounted HF transceiver and ACU-810 microprocessor-controlled antenna coupler; SELCAL compatibility

DISTANCE MEASURING EQUIPMENT

MANUFACTURER	MODEL TSO	CHANNELS OUTPUT	DISPLAY	QUANTITIES DISPLAYED	PEAK POWER OUTPUT (watts)	VOLTS INPUT	UNITS/ WEIGHT (lb) SIZE	PRICE (uninstalled)	REMARKS
Bendix/King Avionics 400 N. Rogers Rd. Olathe, KS 66062 (913) 782-0400	KN 64 none	200 King serial data	gas discharge	dist.: 0-389 nm gnd. spd.: 0-999 kt TTS: 0-99 m.i.n.	35	11 to 33 VDC	1/2.4 6.25 x 1.3 x 12.2	$2,085	Distance accuracy: ±0.1 nm nominal to 99 nm, ±1.0 nm 100 to 200 nm; panel-mounted; includes antenna and installation kit; accepts remote nav channeling

Footnote: Arinc form factor is shown in the Units/Weight/Size columns. (See Form Factor explanation at the front of the Avionics section.)

DISTANCE MEASURING EQUIPMENT

MANUFACTURER	MODEL TSO	CHANNELS OUTPUT	DISPLAY	QUANTITIES DISPLAYED	PEAK POWER OUTPUT (watts)	VOLTS INPUT	UNITS/WEIGHT (lb) SIZE	PRICE (uninstalled)	REMARKS
Bendix/King Avionics 400 N. Rogers Rd. Olathe, KS 66062 (913) 782-0400	KN 62A C66a	200 King serial data	gas discharge	dist.: 0-389 nm gnd. spd.: 0-999 kt TTS: 0-99 min	100	11 to 33 VDC	1/2.4 6.25 x 1.3 x 12.2	$4,515	Distance accuracy: ±0.1 nm nominal to 99 nm, ±1.0 nm 100 to 200 nm; panel-mounted; includes antenna and installation kit; accepts remote channeling
	KN 63 C66a	200 King serial data	gas discharge	dist.: 0-389 nm gnd. spd.: 0-999 kt TTS: 0-99 min	100	11 to 30 VDC	2/3.6 6.5 x 1.9 x 11.55	$6,850	Distance accuracy: ±0.1 nm nominal to 99 nm, ±1.0 nm 100 to 200 nm; includes KDI 572 indicator; optional slaved indicator
	KDM 706 C66a	252 King serial data	gas discharge	dist.: 0-389 nm gnd. spd.: 0-999 kt TTS: 0-99 min	250	28 VDC	2/5.5 2.5 x 5.25 x 11.8	$11,295	Distance accuracy: ±0.1 nm to 99 nm, ± 100 to 200 nm; includes KDI 572 indicator; optional slaved indicators; remotely mounted; 2-year warranty
	KDM 706A C66b	252 Arinc 429, 568	gas discharge	dist.: 0-389 nm gnd. spd.: 0-999 kt TTS: 0-99 min	250	28 VDC	2/6.3 5.0 x 3.0 x 10.0	$13,385	KDM 706 with Arinc 429, 568 interface; KDI 572 included; optional slaved indicators; remotely-mounted; night vision compatible "A" control head available; 2-year warranty
	DMS-44 C66	252 2 x Arinc 568 2 x 40mv/nm Arinc 429 40-bit serial	LCD or dichroic	dist.: 0-300 nm gnd. spd.: 0-999 kt TTS: 0-19:59 min	325	28 VDC	2/6.6 5.0 x 4.0 x 12.62	$15,690	Triple scanning DME, displays Nav 1 and Nav 2 distance simultaneously; Arinc 429 or analog interface; displays RNAV distance; remotely-mounted; 2-year warranty
	KTU 709 C66a	252 King serial	gas discharge	dist.: 0-389 nm gnd. spd.: 0-999 kt TTS: 0-99 min	250	11 to 33 VDC	6.6 5.0 x 3.0 x 10.0	$17,260	Tacan (DME with Tacan-based bearing independent of VOR); optional KFS-579A control head — $4,180; KDI 572 indicator included; night vision compatible "A" control head available; 2-year warranty
Collins Avionics Div. Rockwell International 400 Collins Rd. NE Cedar Rapids, IA 52498 (319) 395-4085	DME-450 C66a	200 40 mv/nm	incandescent filament	dist.: 0-199.9 nm gnd. spd.: 0-399 kt TTS: 0-120 min	100	14 or 28 VDC	3/6.1 3.5 x 5.0 x 11.0	$7,605	Distance accuracy: ±0.1 nm or 1%; includes ANT-451 antenna and IND-450 indicator
	DME-451 C66a	200 40 mv/nm	incandescent filament	dist.: 0-199.9 nm gnd. spd.: 0-399 kt TTS: 0-120 min	100	14 or 28 VDC	3/6.4 3.5 x 5.0 x 11.0	$9,215	Distance accuracy: ±0.1 nm or 1%; built-in timer based on GMT clock; includes ANT-451 antenna and IND-451 indicator

Footnote: Arinc form factor is shown in the Units/Size/Weight column. (See Form Factor explanation in the front of the Avionics section.)

DISTANCE MEASURING EQUIPMENT

MANUFACTURER	MODEL TSO	CHANNELS OUTPUT	DISPLAY	QUANTITIES DISPLAYED	PEAK POWER OUTPUT (watts)	VOLTS INPUT	UNITS/WEIGHT (lb) SIZE	PRICE (uninstalled)	REMARKS
Collins Avionics Div. Rockwell International 400 Collins Rd. NE Cedar Rapids, IA 52498 (319) 395-4085	DME-42 C66b	252 40 mv/nm Arinc 568B serial digital	gas discharge	dist.: 0-300 nm gnd. spd.: 0-999 kt TTS: 0-120 min ID: 2,3,4 letters	300	28 VDC	2/6.2 ½ ATR-SL	$12,935	Includes two-digit-code diag nostics; directly replaces DME-40; tracks three channels simultaneously when used with CTL-32 and IND-42A (CTL-32 controls nav receiv er); serial digital, 2 x 5 tuning; digital signal processing; echo monitor; includes IND-42A indicator
	DME-442 C66b	252 40 mv/nm Arinc 568B serial digital	gas discharge	dist.: 0-300 nm gnd. spd.: 0-999 kt TTS: 0-120 min ID: 2,3,4 letters	300	28 VDC	2/6.2 ½ ATR-SL	$14,510	Tracks three channels simultaneously when linked to CTL-32, IND-42A; decodes / displays station ident; compatible interfaces: Collins Serial Data Bus (CSDB) or ARINC 429, only; digital signal processing; echo monitor; built-in diagnostics; includes IND-42A
	DME-40 C66a	252 40 mv/nm Arinc 568 pulse pair	gas discharge	dist.: 0-250 nm gnd. spd.: 0-999 kt TTS: 0-99 min	300	28 VDC	2/8.1 ½ ATR-SL	$15,950	Distance accuracy: ±0.1 nm or 0.01%; includes IND-40 indicator
	860E-5 C66a	252 Arinc 568B	N/A	dist.: 0-390 nm	700	115 VAC, 400 Hz	1/16.0 ½ ATR-SL	$26,016	Distance accuracy: ±0.1 nm nominal; indicator not included
	860E-4 C66a	252 Arinc 521D	N/A	dist.: 0-390 nm	700	115 VAC, 400 Hz	1/20.0 ½ ATR	$30,080	Distance accuracy: ±0.2 nm; channeling and lock-on less than 2.5 seconds; indicator not included
Foster Airdata Systems, Inc. 7020 Huntley Rd. Columbus, OH 43229 (614) 888-9502	DME 670 C66b	252 RS-422, Arinc 568, DME pulse pair, 40 mv/nm, 400 HzAC	LED dot matrix	dist.: 0-400 nm gnd. spd.: 0-720 kt TTS: 0-99 min. bearing: (Tacan) 0-359° ID: 3, 4 letter Morse	250	28 VDC	2/6.8 10.25 x 3.25 x 5.0 1/2-3ATI x 6.5	$11,900	Tracks up to four stations; displays distance and ident or Tacan radial from two stations; distance, ident, Tacan radial and speed from single station; also available with ARINC 429 output for interface with long-range navigation systems
Narco Avionics Inc. P.O. Box 1807 Laguna Beach, CA 92652-1807 (714) 497-5077	DME 890 none	200 Narco digital	gas discharge	dist.: 0-160 nm gnd. spd.: 44-600 kt TTS: 0-99 min	25	11 to 32 VDC	1/3.3 6.25 x 1.5 x 11.0	$1,949	Microprocessor controlled; meets all minimum operating characteristics; Narco RNAV 860, NS-801 compatible
	IDME 891 none	200 Narco digital	LED	dist: 0-160 nm gnd. spd.: 44-600 kt	25	11 to 32 VDC	1/26 three-inch cutout	$2,639	Requires MK 12D; package with GS, $3,650; includes VOR/LOC indicator and marker lights; compatible with Mark 12D, NAV 825, RNAV 860, NS-801

Footnote: Arinc form factor is shown in the Units/Weight/Size column. (See Form Factor explanation in the front of the Avionics section.)

TRANSPONDERS

MANUFACTURER	MODEL TSO	INTERRO-GATION MODES	POWER OUTPUT (watts)	VOLTS INPUT	UNITS/ WEIGHT (lb) SIZE	PRICE (uninstalled)	REMARKS
Aire-Sciences, Inc. 216 Passaic Ave. Fairfield, NJ 07006 (201) 228-1880	RT-787 C74c Class 1A	A,C	500	14 VDC*	2/3.4 6.5 x 1.8 x 9.9	$1,495	Manual reply-light dimmer *28 to 14 VDC adapter, $95
	RT-887 C74c Class 1A	A,C	500	14 VDC*	2/4.0 3.25 x 3.25 x 10.12	$1,795	Manual reply-light dimmer; electronic digit display; mounts into 3.25-inch instrument cutout *28 to 14 VDC adapter, $95
Bendix/King Avionics 400 N. Rogers Rd. Olathe, KS 66062 (913) 782-0400	KT 76A C74c Class 1A	A,C	250	14 or 28 VDC*	1/3.1 6.25 x 1.6 x 10.0	$1,225	Automatic reply-light dimmer; system test; remote ident capability *Adapter, $30
	KT 79 C74c Class 1A	A,C	250	11 to 33 VDC	1/3.4 6.25 x 1.3 x 10.5	$3,315	Auto 1200 select; displays encoded altitude; all solid-state, no cavity tube; remote ident capability
	KXP 756 C74c Class 1A	A,C	250	28 VDC	2/4.7 5.0 x 1.75 x 10.0	$6,975	Includes KFS 576 digital control head with auto 1200 select; test displays encoded altitude; solid-state; no cavity tube; Gables control optional; remote ident capability; night vision compatible "A"control head available; 2-year warranty
	TRS-42 C74c	A,B,C	325	28 VDC	2/5.8 4.0 x 4.0 x 12.62	$10,125	Analog or Arinc 429; controls two RTs; auto 1200 select; digital Mode C altitude readout; 2-year warranty
Collins Avionics Div. Rockwell International 400 Collins Rd. NE Cedar Rapids, IA 52498 (319) 395-4085	TDR-950 C74c Class 1A	A,C	250	14 VDC*	2/2.0 6.3 x 1.6 x 8.2	$2,010	Manual reply-light dimmer; optional remote ident *28 to 14 VDC adapter, $125
	TDR-90 C74b C74c	A,C	500	28 VDC	2/4.9 ¼ ATR-SL	$8,045	Remote-mounted; manual reply-light dimmer; includes CTL-90 electronic control; optional CTL-92 with CAD-62 adapter provides serial digital tuning, two stored codes
	621A-6A C74a	A,B,C (D opt.)	700	115 VAC, 400 Hz	3/15.0 ⅜ ATR-S	$17,808	Remote-mounted; must add control head with reply-light; Arinc 572
	TPR-720 C74c C112	A,C,S	250-631	115 VAC, 400 Hz	1/12.0 ½ ATR	$19,348	Mode S transponder; 4 MCU package (approx. ½ ATR); LCD CDU — 5.0 lbs

ENCODING ALTIMETERS/DIGITIZERS

MANUFACTURER	MODEL TSO	INTERRO-GATION MODES	POWER OUTPUT (watts)	VOLTS INPUT	UNITS/ WEIGHT (lb) SIZE	PRICE (uninstalled)	REMARKS
Narco Avionics Inc. P.O. Box 1807 Laguna Beach, CA 92652-1807 (714) 497-5077	AT-150 C74C Class 1A	A, C	250	14 or 28 VDC*	1/2.3 6.25 x 1.75 x 11.3	$1,249	Manual reply-light dimmer; remote ident available *Adapter, $27
Terra Corp. 3520 Pan American Freeway, NE Albuquerque, NM 87107 (505) 884-2321	TRT 250 C74C Class 2a	A,C	200	10 to 30 VDC	1/1.7 3.2 x 1.6 x 10.0	$1,195	All solid-state; no cavity tube; remote ident capability; push-button code selection; 3" ATI configuration available

ENCODING ALTIMETERS/DIGITIZERS

MANUFACTURER	MODEL TYPE	RANGE (x 100 ft)	DISPLAY CASE	LENGTH WEIGHT	DIGITIZER DESIGN	POWER INPUT	PRICE (uninstalled)	REMARKS
ACK Technologies 424 W. Julian St. San Jose, CA 95110 (408) 287-8021	A-30 altitude digitizer	−10+307	— —	6.0 .45	solid state	10 to 32 VDC	$325	TSO C88a; remotely mounted
Aero Mechanism 20327 Nordhoff St. Chatsworth, CA 91311-6161 (818) 709-2851	8141B-35LW —	−10+350	CDP 3" ATI	6.26 .75	optical	14 or 28 VDC	$4,743	Millibar baro setting optional; mating connector included; internal lighting: 5, 14, 28 VDC available; TSO C10b, C88
Aerosonic Corp. P.O. Box 4627 Clearwater, FL 33515 (813) 461-3000	102200 -01812 —	−10+200	3P 3 ATI	5.3 2.0	optical	14 or 28 VDC	$1,391	Millibar baro setting, dual scale, internal lighting and altitude alerter optional; TSO C10b, C88
	102200 -01818 —	−10+350	3P 3 ATI	5.1 2.0	optical	14 or 28 VDC	$1,470	Millibar baro setting, dual scale, internal lighting and altitude alerter optional; TSO C10b, C88
	102200 -01831 —	−10+350	CDP 3 ATI	6.4 2.4	optical	14 or 28 VDC	$2,531	Millibar baro setting, dual scale, internal lighting and altitude alerter optional; TSO C10b, C88
	101400-01235 —	−10+500	CDP 3 ATI	5.2 7.7	optical	14 or 28 VDC	$2,575	Millibar baro setting, dual scale, internal lighting and altitude alerter optional; other configurations available; TSO C10b, C88

ENCODING ALTIMETERS/DIGITIZERS

MANUFACTURER	MODEL TYPE	RANGE (x 100 ft)	DISPLAY CASE	LENGTH WEIGHT	DIGITIZER DESIGN	POWER INPUT	PRICE (uninstalled)	REMARKS
Bendix/King Avionics 400 N. Rogers Rd. Olathe, KS 66062 (913) 782-0400	KE 127 altitude digitizer	−10+200	— —	3.25 1.00	optical	14 or 28 VDC	$1,075	Digitizer only; remotely-mounted;
	KEA 129 encoding altimeter	−10+2000	3P MS	5.57 1.90	optical	14 or 28 VDC	$2,585	Millibar / inches baro setting; kit and rack included; internally lighted
	KEA 130 encoding altimeter	−10+350	3P MS	5.57 1.90	optical	14 or 28 VDC	$2,850	Millibar / inches baro setting; internally lighted; kit and rack included; KEA 130A with altitude pre-select / alerter compatibility $3,035
	KEA 346 servoed encoding altimeter	−10+500	CDP 3 ATI	8.94 2.90	optical	28 VDC	$8,800	Internal lighting; altitude-alert outputs; DH annunciation; dual baro set in millibars or inches
Kollsman Avionics Division 220 Daniel Webster Hwy. Merrimack, NH 03054 (603) 889-2500	B4515 Alti-coder	−10+200	CDP MS	4.75 2.50	optical	28 VDC + 115 VAC	N/A	Price dependent on specific aircraft; internal white or red lighting; three-digit display; 35,000-ft model available
	B4450 Alti-coder II	−10+350	3P MS	5.30 2.50	optical	14 or 28 VDC	N/A	Price dependent on specific aircraft; optional internal lighting and failure warning flag
	24929 encoding altimeter	−10+200	CDP 3 ATI	6.9 2.9	optical	14 or 28 VDC	$14,400*	* base price; red, white or blue lighting; 35,000-ft / 50,000-ft models available
	28007 encoding altimeter	−10+500	CDP 3 ATI	8.3 3.5	contact	26 VAC 400 Hz	$15,504*	* base price;,above options, features; includes altitude alert, altitude rate outputs
	28702 encoding altimeter	−10+350	CDP 3 ATI	7.0 3.5	contact	28 VDC	$16,044*	* base price; above options, features
	29007 encoding altimeter	−10+50	CDP 3 ATI	8.0 3.8	contact	26 VAC 400 Hz	$17,536*	* base price; above options, features; includes radio altimeter (RA) display; links to GPWS; requires separate RA receiver/transmitter
	29702	−10+350	CDP 3 ATI	8.0 3.8	contact	28 VDC	$18,512*	* base price; optional altitude alert light; extended baro range; above options, features
Narco Avionics Inc. P.O. Box 1807 Laguna Beach, CA 92652-1807 (714) 497-5077	AR-850 altitude digitizer	−10+250 −10+307 altitude digitizers	— —	6.15 0.7	solid state	11 to 32 VDC	$260 $300	−10;250 model TSO C88a −10;307 model not TSOed

ENCODING ALTIMETERS/DIGITIZERS

MANUFACTURER	MODEL TYPE	RANGE (x 100 ft)	DISPLAY CASE	LENGTH WEIGHT	DIGITIZER DESIGN	POWER INPUT	PRICE (uninstalled)	REMARKS
Pointer, Inc. 1027 N. Stadem Dr. Tempe, AZ 85281 (602) 966-1674	L115 altitude digitizer	−10+200	— —	7.3 1.1	solid state	14 or 28 VDC	$575	tentative price: remotely mounted; installation kit included
Terra Corp 3520 Pan American Freeway, N.E. Albuquerque, NM 87107 (505) 884-2321	AT 3000 blind encoder	−10+300	— —	6.2 0.5	solid state	14 or 28 VDC	$495	Digitizer only; remote-mounted
Trans-Cal Industries, Inc. 16141 Cohasset Street Van Nuys, CA 91406 (818) 787-1221	D120P2-T D125P2-T D130P2-T D135P2-T D150P2-T altitude digitizers	−10+200 −10+250 −10+300 −10+350 −10+500	— —	3.75 1.0	optical	14 or 28 VDC	$765 $850 $850 $925 $1,700	replaceable electronic module for ease of repair; uses generic aneroid designed for sensitive baro altimeters
United Instruments 3625 Comotara Ave. Wichita, KS 67226 (316) 685-9203	5035 encoding altimeter	−10+200 −10+350	3P MS	5.50 1.80	optical	14 or 28 VDC	$1,617 $1,760	Optional white or blue-white lighting

HORIZONTAL SITUATION INDICATORS/COMPASS SYSTEMS

MANUFACTURER	MODEL TSO	DISPLAY CASE GYRO	SLAVING RATE (norm. fast)	VOLTS INPUT AUTOPILOT OUTPUT	UNITS/ WEIGHT (lb)	PRICE (uninstalled)	REMARKS
Aeronetics 455 Keho Blvd. Carol Stream, IL 60188 (312) 668-3040	8000 C6c, C34c C36c, C40b	3 ATI remote	3 degrees/min 150 degrees/min	28 VDC Arinc analog	3/6.7	$7,370	Arinc format; automatic emergency mode function (flux valve input); autopilot outputs; with bearing pointer — $8,460; 3 wire synchro boot strap output to RMI; two-year warranty
	421 Series C6c, C34c C36c, C40b	4 ATI remote	3 degrees/min 300 degrees/min (nominal)	28 VDC Arinc analog	3/7.5	$7,800	Arinc format; with bearing needle — $8,480; with bearing needle and digital distance and course readout — $10,080; individual components available; Arinc 568 or King Serial Digital; 40 mv/nm optional; two-year warranty
Bendix/King Avionics 400 N. Rogers Rd. Olathe, KS 66062 (913) 782-0400	KCS 305 C6c	4 ATI remote	3 degrees/min 240 degrees/min	115 VAC 400 Hz Hdg. 393 mv AC/degree Course: 393 mv AC/degree	4/10.4	$5,755*	*base price; includes autopilot outputs; many options available; includes boot strap heading in HSI; night vision goggle compatible instruments available; 2-year warranty

HORIZONTAL SITUATION INDICATORS/COMPASS SYSTEMS

MANUFACTURER	MODEL	DISPLAY CASE TYPE GYRO	SLAVING RATE (norm. fast)	VOLTS INPUT AUTOPILOT OUTPUT	UNITS/ WEIGHT (lb)	PRICE (uninstalled)	REMARKS
Bendix/King Avionics 400 N. Rogers Rd. Olathe, KS 66062 (913) 782-0400	KCS 55A C6c	3 ATI remote	3 degrees/min 180 degrees/ min	14 or 28 VDC Hdg.±550 mv DC/ deg.: max. ±12.5v Course: ±220 mv DC/deg.: max ±12.5v	4/9.4	$7,750	Includes autopilot outputs; RMI output $815 additional; night vision goggle compatible instruments available
Century Flight Systems Mineral Wells, TX 76067 (817) 325-2517	NSD-360A-15 C5c C9c C52a	3 ATI panel mounted	non-slaved	14 or 28 VDC potentiometer	1/4.5	$3,805	Includes autopilot outputs; requires 5-in vacuum
	NSD-360A (S) -26 C6c C9c C52a	3 ATI panel mounted	2 degrees to 3 degrees/min N/A	14 or 28 VDC potentiometer	4/5.1	$5,525	Includes autopilot outputs; requires 5-in vacuum; NSD-360A-35(S) provides RMI output, requires 26 VAC, 400 Hz power — $6,135
Collins Avionics Div. Rockwell International 400 Colli28,375ns Rd. NE Cedar Rapids, IA 52498 (319) 395-4085	MCS-65 C6c	— remote	2 degrees/min 7 degrees/sec	28 VDC three-wire synchro	3/6.7	$9,055	Includes autopilot outputs; digital interface with EHSI-74
	PN-101 C6c	3 ATI remote,	3 degrees/min 300 degrees/ min	14 or 28 VDC three-wire synchro	4/12.7	$24,520	Includes autopilot outputs, slaving accessory and 331A-3G course indicator

Footnote: Arinc form factor is shown in the Units/Weight/Size column. (See Form Factor explanation in the front of the Avionics section.)

RADIO ALTIMETERS

MANUFACTURER	MODEL TSO	ALTITUDE RANGE PITCH/ROLL LIMITS	ACCURACY	UNITS SIZE/WEIGHT	DISPLAY	PRICE (uninstalled)	REMARKS
Bendix/King Avionics 400 N. Rogers Rd. Olathe, KS 66062 (913) 782-0400	KRA 10A FM/CW none	-20-2,000 ft ±20 degrees/20 degrees	40-500 ft ±5 ft or 5%; 500-2,000 ft ±7%	rcvr./xmtr. 3.1 x 3.5 x 8 in; ind. 3 ATI; 1 ant. 3.8 lb	pointer and dial type; linear scale expanded from 0-500 ft	$4,635*	*28 VDC system; continuous DH set; "skewed" antenna optional for mounting on angled aircraft surface; 14 VDC configuration — $4,807
	KRA 405 FM/CW C87	-20-2,000 ft ±20 degrees/ ±25 degrees	-20-500 ft ±3 ft or 3%; 500-2,000 ft ±5%	rcvr./xmtr. 11.5 x 5 x 3 in; ind. 3 ATI; 2 ant.; 10.7 lb	pointer and dial type; linear scale expanded from 0-500 ft	$11,490	Continuous DH set; tracking to 2,500 ft; expanded helicopter version available; night vision goggle compatible version available; 2-year warranty
Collins Avionics Div. Rockwell International 400 Collins Rd. NE Cedar Rapids, IA 52498 (319) 395-4085	ALT-50A CW C87	-0+2,000 ft ±40 degrees/ ±50 degrees	±2 ft or 2%;	rcvr./xmtr. ⅜ ATR-SL; indicator, .83 x 3.38 in two antennas; 6.7 lb	digital	$11,935	DH light and four trips settable over entire range; includes DRI-55 indicator and two ANT-52 antennas
	ALT-55B CW C87	-0+2,500 ft ±40 degrees/ ±50 degrees	±2 ft or 2%;	rcvr./xmtr. ⅜ ATR-SL; indicator, .83 x 3.38 in two antennas; 6.7 lb	digital	$14,010	DH light and four trips settable over entire range; includes DRI-55 indicator and two ANT-52 antennas
	AL101 FM/CW C87	-20+2,500 ft ±40 degrees/ ±50 degrees	-20+500 ft ±2 ft or 2%; 500-2,500 ft ±5%	rcvr./xmtr. ½ ATR-S; ind. 3 ATI; 2 ant. 7-in dia. x 2.75 in; 22.5 lb	pointer and dial type; linear -20+500 ft; logarithmic 500-2,500 ft	$39,236	ARINC 552, 552A specification design, includes vertical tape instrument
Honeywell Inc. P.O. Box 29000 Phoenix, AZ 85038 (602) 863-8000	AA-300 pulse C87	0-2,500 ft ±45 degrees/ ±45 degrees	0-100 ft ±5ft; 100-500 ft ±5%; 500-2,500 ft ±7%	rcvr./xmtr. 10.7 x 4 x 7.5 in; ind. 3.3-in dia. x 4 in; 2 ant. 8.25 lb	pointer and dial type; linear scale expanded from 0-500 ft or 0-200 ft	$14,307	Altitude rate optional; continuous DH set; tracking to 3,000 ft; includes RT-300 receiver/transmitter, RA-315 indicator and AT-300 antenna
Terra Corp. 3520 Pan American Freeway NE Albuquerque, NM 87107 (505) 884-2321	TRA 3000 FM/CW none	40-2,500 ft ±20 degrees/ ±30 degrees	40-100 ft ±5 ft; 100-500 ft ±5%; 500-2,500 ft ±7%	rcvr./xmtr. 1.0 x 5.0 x 7.6 in; indicator 3.5 x 3.25 x 4.0 in. antenna built into R/T unit; 2.5 lb	—	see remarks	*Prices: with TRI 20 LED indicator — $2,395; with TRI 30 analog indicator — $2,995; with TRI 40 LED indicator, with 100' AGL gear warning — $3,295; continuous DH set; aux analog output and STWT
	TRA 3500 FM/CW none	0-2,500 ft ± 20 degrees/ ± 30 degress	0-10 ft ± 1 ft; 10-100 ft ± 5 ft; 100-2,500 ft ± 5-7%	rcvr./xmtr. 3.0 x 3.0 x 6.9 in; indicator: 3.5 x 3.25 x 4.0 in; —	pointer and dial type; linear 40-2,500 ft; (enlarged linear 40-500 ft)	$5,495	Price includes TRI 40 LED indicator, including 100 ft gear warning; TSO C87 pending; anti-hover circuit design

Footnote: Arinc form factor is shown in the Units/Size/Weight column. (See Form Factor explanation in the front of the Avionics section.)

THUNDERSTORM DETECTION SYSTEMS

MANUFACTURER	MODEL	RANGE (nm)	DISPLAY SIZE VIEWING ANGLE	DISPLAY TYPE	UNITS/ WEIGHT (lb)	PRICE (uninstalled)	REMARKS
Honeywell Inc. P.O. Box 29000 Phoenix, AZ 85038 (602)863-8000	LSZ-850	Same as radar display to 100 nm	Interfaces on radar display 120 or 360 degrees	Quantitative electrical dis-charge activity displayed on CRT	3 11.0	$19,703	Frequency of lightning discharge causes one of three unique symbols to be displayed at the relative location of such activity; integrated radar & lightning system sensor controls available
3M Stormscope Systems 6530 Singletree Dr. Columbus, OH 43229 (614) 885-3310	WX-8	100	3-in dia. 135 degrees	liquid-crystal display	1 + ant. 4.0	$3,975	Panel-mounted unit/display; three-color liquid-crystal display range resolution; storm intensity enhancing; 10 - 30 VDC
	WX-1000 Stormscope Series II	25 50 100 200	3-in ATI 120 or 360 degrees	vertical electrical discharge activity on CRT	2 + ant. 10.9	$8,995	Displays only vertical electrical discharge activity associated with thunderstorms; up to six checklists; clock/calendar mode with elapsed and flight time; built-in test and error code display; future expansion capability; 10.5 - 32 VDC; 28 Watts; TSO C110
	WX-1000+ Stormscope Series II	25 50 100 200	3 in. ATI 120 or 360 degrees	vertical electrical discharge activity on CRT	2 + ant. 10.9	$9,995	Heading stabilized — 5 to 50 VAC, 400 Hz; displays only vertical electrical discharge activity associated with thunderstorms; up to six checklists; clock/calendar mode with elapsed and flight time; built-in test and error code display; future expansion capability; 10.5 - 32 VDC; 28 Watts; TSO C110

MLS RECEIVERS

MANUFACTURER	MODEL TSO	CHANNELS CONTROL/ DISPLAY	SIZE/WEIGHT	POWER	PRICE (uninstalled)	REMARKS
Bendix/King Avionics 400 N. Rogers Rd. Olathe, KS 66062 (913) 782-0400	MLS-21 C104	200 ICAO standard rotary knob LCD	receiver: 11.75 x 5 x 3.9 in. control: 2.5 x 3.125 x 2.5 in. (2) antenna: flush-mount system weight: 8.7 lbs	28 VDC/ .75 amps	$20,162	Arinc 429 compatible; ICAO magnetic course reference; price includes installation tray, kit and two antennas; PDME compatible & EFIS compatible; night vision goggle compatible control head available; two-year warranty
Canadian Marconi 2442 Trenton Ave. Montreal, Quebec, Canada H3P 1Y9 (514) 341-7630	CMA-2000 NA	200 ICAO standard rotary knob LCD	receiver: 3 MCU control: 3.75 x 5.75 x 2.375 in. (3) antenna system weight: 11.5 lbs	115 VAC 400 Hz, 75 VA max	$31,040	In full compliance with Arinc 727 and ICAO specifications; analog and Arinc 429 interfaces; manual and fully automatic self-test; auto/manual angle modes; three-year warranty
Honeywell Inc. P.O. Box 29000 Phoenix, AZ 95038 (602) 863-8000	MLZ-850 C104	200 ICAO standard rotary knob LED	receiver: 13.9 x 3.3 x 3.9 in. control: 2.375 x 2.625 x 5.5 in. (2) antenna: 1.2 x 2.5 in. flush aperture system weight: 6.6 lbs	28 VDC	$23,033	Price includes tray and installation kits; analog output to ADI, HSI, FCS; CSDB digital interface; 2x5 DME channeling; updatable to future ICAO magnetic course reference; Primus II MLS sensor configuration, 5.5 lbs — $19,520

NAVIGATION MANAGEMENT SYSTEMS

MANUFACTURER	MODEL TSO	TOTAL WAYPOINTS FLIGHT PLANS / WAYPOINTS DATA ENTRY	PRIMARY SENSOR(S) ADDITIONAL SENSORS	INPUTS OUTPUTS	POWER INPUT COOLING	ATR SIZE / WEIGHT / UNITS	PRICE	REMARKS
Bendix / King Avionics 400 N. Rogers Rd. Olathe, KS 66062 (913) 782-0400	KNS 660 C40b C90a	3,000 400 / 25 keyboard (non-volatile)	DME/DME VOR/DME Omega/VLF TACAN GPS Loran C IRS	analog digital analog digital	28 VDC 26 VAC / 400 Hz forced air required	2/20.2 —	$28,570*	*47,995 with optional Omega/VLF sensor; world-wide database: VORs, TACANs, airports, intersections; airport data, SIDs, STARs, with updating every 28 days with Bendix / King supplied diskette; monochrome CRT; VNAV advisory; optional full-alpha keyboard; coupled VNAV with KFC 400; compatible with Bendix / King GPS sensor; direct interface with Bendix EFIS; radar mapping available through Bendix / King weather radar with optional GC 360A radar graphics unit; optional rubidium frequency standard; night vision compatible CDU available; special packages may be available; two-year warranty

NAVIGATION MANAGEMENT SYSTEMS

MANUFACTURER	MODEL TSO	TOTAL WAYPOINTS FLIGHT PLANS / WAYPOINTS DATA ENTRY	PRIMARY SENSOR(S) ADDITIONAL SENSORS	INPUTS OUTPUTS	POWER INPUT COOLING	ATR SIZE / WEIGHT / UNITS	PRICE	REMARKS
Canadian Marconi 2442 Trenton Ave. Montreal, Quebec, Canada H3P 1Y9 (514) 341-7630	CMA-900 pending	35,000 400 / 99 keyboard disk data loader	DME / DME Omega/VLF IRS MLS GPS Loran-C DADC	analog Arinc 429 analog Arinc 429 Arinc 561	28 VDC / 60 Watts or 115 VAC / 80 VA no external cooling required	Color CDU: 5.75 x 6.8 x 7.8; 8 lbs. NMU: 2 MCU; 10 lbs.	$50,000	price varies with number and type of interfaces and optional long-range sensors; internal Omega/VLF included, may be replacedwith optional, internal GPS; external Omega/VLF sensor available; 10 Arinc 429 input ports; fuel computer optional; frequency management optional; 3.5" diskette data loader optional; ColorCDU capable of controlling other sub-systems
Global-Wulfsberg Systems 2144 Michelson Dr. Irvine, CA 92715 (714) 851-0119	GNS X C115	256 49 / 30* keyboard	DME/DME VOR/DME Loran C VLF/Omega IRS	Arinc 429 Arinc 571 CSDB analog Arinc 429 Arinc 561 Arinc 571 CSDB analog	28 VDC / 1.8 Amps 26 VAC / 400 Hz** no external cooling required	Computer: ¼ATR-S / 8.4 lbs CDU: 4.5 x 5.75 x 6.5 / 5.5 lbs	$41,852	*up to 50 route waypoints on active flight plans; **AC power for analog interfaces only; price includes installation kit, Loran C sensor board, 50,000 waypoint, worldwide internal database, fuel flow interface, full alpha keyboard CDU; no VNAV; frequency management system standard; AFIS compatible, optional Data Management Unit — $26,700, first year subscription included; 3.5" diskette Data Transfer Unit — $2,800; with AFIS option, flight plans and messages can be interchanged via datalink; optional VLF/Omega RPU — $22,500
	GNS 1000 C115 C109 C94a	256 49 / 30* keyboard	DME/DME VLF/Omega VOR/DME IRS	Arinc 429 Arinc 571 CSDB analog Arinc 429 Arinc 561 Arinc 571 CSDB analog	28 VDC / 5.2 Amps 26 VAC / 400 Hz** no external cooling required	Computer: ½ATR-S / 15.0 lbs Receiver / Processor: ½ATR-S / 16.0 lbs NDB-2: ¼ATR-S / 9.1 lbs CDU: 4.5 x 5.75 x 6.5 / 5.5 lbs	$88,143	*up to 50 route waypoints on active flight plans; ** AC power for analog interfaces only; price includes installation kit, H-field antenna; dual systems — $159,177; monochrome CRT; optional full alpha keyboard — $1,220; no VNAV; frequency management system standard; AFIS compatible, optional Data Management Unit — $26,700, includes first year subscription; 3.5" diskette Data Transfer Unit — $2,800; with AFIS option, flight plans and messages can be interchanged via datalink; compatible with almost all fuel flow sensors or indicators; 25,000 waypoint NDB-2 database includes North American / European or world-wide data; remove database from aircraft every 28 days to update; VLF/Omega Receiver Processor Unit included in price, but may be deleted for credit

NAVIGATION MANAGEMENT SYSTEMS

MANUFACTURER	MODEL TSO	TOTAL WAYPOINTS FLIGHT PLANS / WAYPOINTS DATA ENTRY	PRIMARY SENSOR(S) ADDITIONAL SENSORS	INPUTS OUTPUTS	POWER INPUT COOLING	ATR SIZE / WEIGHT / UNITS	PRICE	REMARKS
Honeywell Inc. P.O. Box 29000 Phoenix, AZ 85038 (602) 863-8000	LASERNAV II —	255 20 / 20 keyboard	Laser IRS VOR DME VLF/Omega	Arinc 429 Arinc 419 Arinc 429 Arinc 561 Arinc 571 CSDB	28 VDC / 5 VDC or 115 VAC / 400 Hz no external cooling required	INU: modified 10 MCU / 48.5 lbs CDU: Arinc 704-4 / 7.0 lbs	$163,214	certificated as a primary sole-source, long-range navigation system; supplies attitude and heading reference to flight instruments; ring-laser-gyros used for inertial reference; monochrome display; NDB-2 database, updated with Honeywell-supplied tape; accommodates 2 x VORs, 2 x DMEs, 2 x VLF/Omega sensors; compliance with RTCA DO-160A
	FMZ-800 C115	1,400 100 / 50 3.5" disk data loader	DME/DME VOR Omega/VLF IRS AHRS (MLS) (GPS) air data	2 x ASCB 3 x Arinc 429 6 x CSDB ASCB Arinc 429 CSDB	28 VDC / 2.7 Amps no external cooling required	Color CDU: 5.4 x 7.5 x 10.0; 9.4 lbs Computer: ½ATR-S; 14.3 lbs	$92,326	system "learns" operating characteristics of aircraft from previous flights; 200 pilot-defined waypoints in addition to those listed; non-volatile EEPROM memory; flight plans computed by Lockheed JetPlan and down loaded via 3.5" diskette; mono display also available; 3-D navigation with full VNAV; linked J-route / V-airway mode; radial intercept mode; full-autothrottle growth mode; ACARS / datalink growth mode; Loran-C capability — TBA; no TACAN; USIFR-RNAV, NAT MNPS, RNAV Approach, sole-source over-water nav approvals; full Jeppesen database, updated with 3.5" diskette data loader; standard equipment on Canadair 601-3A, Gulfstream IV and Falcon 900 aircraft; 30,000 waypoint database standard; optional 1.25-Mbyte NZ-910 international nav database including 4,000+ paved airports, airways, SIDs/ STARs; 28-day updates take less than 8 minutes; 1-year warranty
Universal Navigation Corporation 3260 East Lerdo Rd. Tucson, AZ 85706 (602) 741-2300	UNS-1 NMS C115 C66b C94a C60a	3,000 200 / 98 keyboard alpha or line-item selection (3.5" disk data loader)	DME/DME VOR IRS Omega/VLF Loran-C GPS TACAN DADC	analog Arinc 429 Arinc 575 CSDB analog Arinc 429 Arinc 561 Arinc 571 Arinc 575 ASCB	28 VDC / 1.8 Amps 26 VAC / 400 Hz no external cooling required	CDU: 5.75 x 4.5 x 6.75, 4.5 lbs NCL: 2 MCU, 7.0 lbs 5.75W x	$30,000	price varies with up to three optional long-range sensors; 40,000 waypoint worldwide database; auto-scanning DME / DME / DME; diskette data loader — $3,600; advisory VNAV; direct-to key; two engine fuel management — DC-analog inputs; optional Super CDU to upgrade to UNS-1A configuration — mono CRT, adds 3.0 lbs and $7,000, color CRT, adds 3.0 lbs and $12,000; direct-to key; 3-D approach mode; direct-to mode; optional frequency management system; multi-chain Loran-C sensor — $12,000; UNS-RRS DME/VOR/ TACAN sensor — $16,000; optional UNS-764 VLF/Omega sensor — $22,000; three-year warranty includes sensors

NAVIGATION MANAGEMENT SYSTEMS

MANUFACTURER	MODEL TSO	TOTAL WAYPOINTS FLIGHT PLANS / WAYPOINTS DATA ENTRY	PRIMARY SENSOR(S) ADDITIONAL SENSORS	INPUTS OUTPUTS	POWER INPUT COOLING	ATR SIZE / WEIGHT / UNITS	PRICE	REMARKS
Universal Navigation Corporation 3260 East Lerdo Rd. Tucson, Az. 85706 (602) 741-2300	UNS-1A Compact FMS C115 C66b C94a C60a	3,000 200 / 98 keyboard alpha or line item select (3.5" disk data loader)	DME / DME VOR IRS Omega/VLF Loran-C TACAN GPS DADC	analog Arinc 429 Arinc 575 CSDB analog Arinc 429 Arinc 561 Arinc 571 Arinc 575 ASCB	28 VDC / 2.1 Amps 26 VAC / 400 Hz no external cooling required	SuperCDU: 5.75 x 6.4 x 7.88 7.6 lbs NCU: 2 MCU, 7.8 lbs	$55,000	price varies with CDU and up to five long-range sensors; 40,000 waypoint worldwide database; auto-scanning DME/DME/DME and VOR/DME; monochrome super CDU, deduct $5,000; frequency management unit — $10,000; dual system data cross-talk mode; off-line flight planning; diskette data loader — $4,500; plus all features and options of UNS-1 NMS above; three-year warranty includes sensors
	UNS-1A FMS C115 C66b C94a C60a	3,000 200 / 98 keyboard alpha or line item select (3.5" disk data loader)	DME / DME VOR IRS Omega/VLF Loran-C TACAN GPS DADC	analog Arinc 429 Arinc 575 CSDB analog Arinc 429 Arinc 561 Arinc 571 Arinc 575 ASCB	28 VDC / 2.3 Amps 26 VAC / 400 Hz no external cooling required	SuperCDU: 5.75 x 6.4 x 7.88 7.6 lbs NCU: 2 MCU, 11.0	$72,500	price varies with CDU and up to five optional long-range sensors; includes color Super CDU; optional SID/STAR/Approach database — $10,000; fuel management with up to four inputs; all features and options of UNS 1A Compact FMS listed above; three-year warranty includes sensors

AIRBORNE TELEPHONE SYSTEMS

MANUFACTURER	MODEL COLORS	MODE OF OPERATION	POWER OUTPUT (watts) MAX ALT	POWER INPUT	UNITS/ WEIGHT (lb)	PRICE (uninstalled)	REMARKS
Global-Wulfsberg Systems 2144 Michelson Dr. Irvine, CA 92715 (714) 851-0119	Flitefone VI 7	full duplex, HF optional	10 51,000 ft	28 VDC	2/9.5	$7,664	Remote handsets and controls available; direct-dial capable
Terra Corp. 3520 Pan American Freeway, N.E. Albuquerque, NM 87107 (505) 884-2321	Jetfone II 4	full duplex	8 40,000 ft	28 VDC	1/4.25	$3,000	Self-contained; panel-mounted unit; includes handset
	TD-3000 N/A	full duplex, AGRAS	10 55,000 ft	28 VDC	2/6.8	$7,500	Direct dial compatible; automatic channel selection, operator alert; additional handsets, $535; digital channel readout; includes cockpit control unit

C2 MILITARY AVIONICS SYSTEMS PRICE DATA

The prices of military avionics systems depend strongly on the following factors:

* Requested performance
* Operational environment (particularly high 'g' levels and extreme temperatures)
* Frequency of modifications requested by various military services
* Number of systems procured

In some cases the military will purchase 'off-the-shelf' commercial avionics equipment. In that case the price data of Section C1 might be applicable.

In most cases military avionics systems are developed for specific military purposes. Since prices of such equipment are not published regularly in the open literature, another approach is recommended. That approach is to use fractional cost data based on published information.

Figures C1 - C3 are examples of such fractional data. It appears from these data that military avionics costs typically range from:

$$C_{avionics/military} = 0.05(AEP) \qquad (C1)$$

to:

$$C_{avionics/military} = 0.40(AEP) \qquad (C2)$$

where: AEP is the airplane estimated price as estimated from Equation (4.3)

The reader has to apply 'judgment' to determine the magnitude of the fraction (0.05 to 0.40) which applies to a new design.

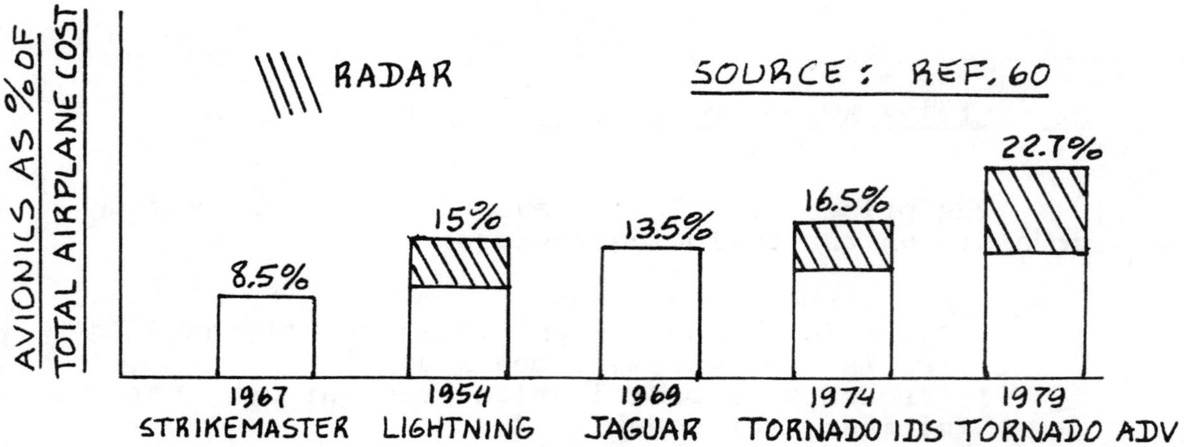

Figure C1 Avionics as a Percentage of Total Airplane Cost

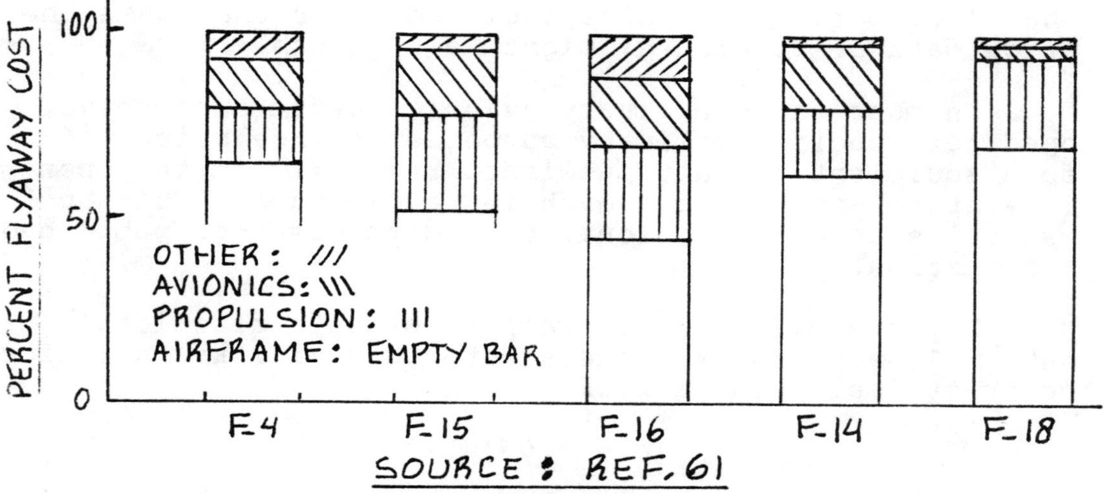

Figure C2 Avionics and Other Equipment as a Percentage of Flyaway Cost

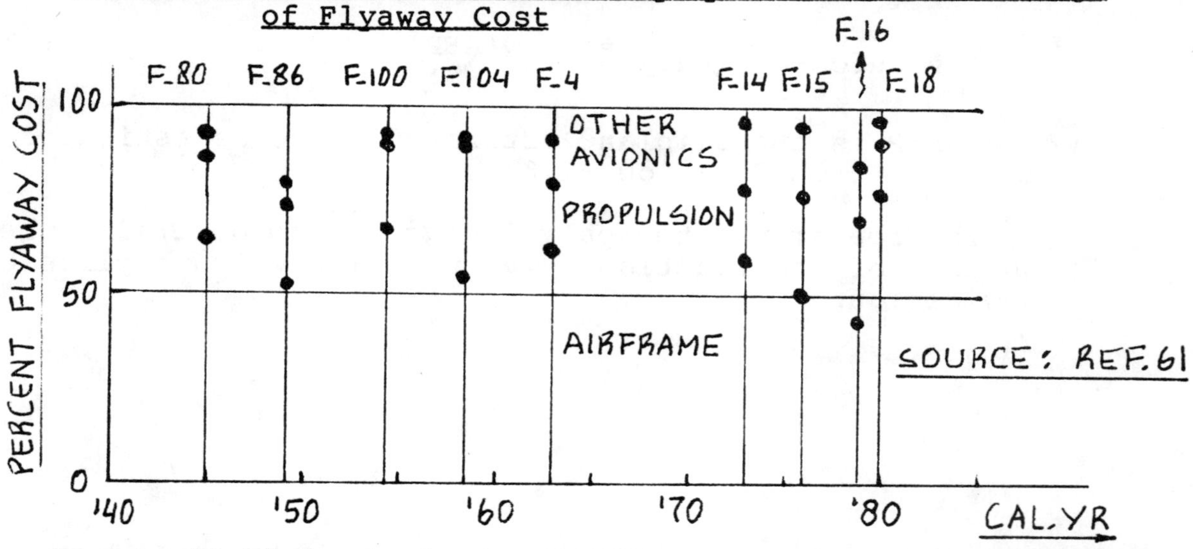

Figure C3 Historical Distribution of Avionics and Other Equipment as a Percentage of Flyaway Cost

Notes

Notes

DARcorporation
Design • Analysis • Research

Information Request Form

FOR FASTEST SERVICE, CALL 1-800-327-7144 (U.S and Canada only)
(785) 832-0434
FAX (785) 832-0524
e-mail info@darcorp.com
www.darcorp.com

Date ____________

Phone ________________________________

Fax ________________________________

Name ________________________________

Title ________________________________

Company/University ____________________

____

Address ________________________________

____

____

E-mail ________________________________

- ☐ I would like to receive your *free* Newsletter
- ☐ Please add my name to your mailing list
- ☐ Please send me a book order form

Consulting/Contracting Services

Please have a representative contact me. I am interested in the following areas:

- ☐ New Design
- ☐ Modification Analysis
- ☐ Verification Analysis

Advanced Aircraft Analysis

- ☐ Please send me a detailed product description of AAA
- ☐ I would like to arrange for a free 30-day evaluation copy of AAA
- ☐ Please have a representative contact me

Please check all that apply:

☐ Professional ☐ Professor ☐ Homebuilder ☐ Student

no. employees: ________ no. students in class: ________

AIRPLANE DESIGN AND ANALYSIS

--FIRST FOLD--

--SECOND FOLD THEN TAPE--

NO POSTAGE
NECESSARY
IF MAILED
IN THE
UNITED STATES

BUSINESS REPLY MAIL

FIRST CLASS MAIL PERMIT NO.697 LAWRENCE, KS

POSTAGE WILL BE PAID BY ADDRESSEE

DESIGN, ANALYSIS AND RESEARCH CORPORATION
120 E 9TH ST STE 2
LAWRENCE KS 66044-9799